D1355787

Illustrated Dictionary of Botany

Illustrated Dictionary of Botany

This dictionary has been compiled
by Kenneth A. Beckett

PICTURE CREDITS

We should like to credit the following people and agencies who supplied us with photographs for this book, with particular thanks to Gillian Beckett and the John Topham Picture Library who, between them, supplied the greatest number:

Ardea Photographics; The Arnold Arboretum of Harvard University; A–Z Botanical Collection; J. K. Burras; Bruce Coleman (Jane Burton); Camera Press; Crown Copyright, Ministry of Agriculture, Fisheries and Food; Crown Copyright, pictures reproduced with the permission of the Controller of Her Majesty's Stationery Office, and of the Director Royal Botanic Gardens, Kew; Dr P. Echlin; Brian Furner; Geological Collections, Oxford University Museum; F. N. Hepper; Anthony Huxley; Dr B. E. Juniper; Koninklijk Instituut voor de Tropen; G. Lewis; The Linnean Society; Malasian Rubber Research and Development Board; Mary Evans Picture Library; National Portrait Gallery; Paul Popper; John Procter; Dr A. W. Robards; Royal Horticultural Society (Jane Brown/*Observer*); The Sports Turf Research Institute; Harry Smith Collection; John Topham Picture Library (Fasmer, Foord, Markham, and Scowen); ZEFA.

Line drawings are by Thea Nockels.

First published in Great Britain in 1977 by Triune Books, London, England.
Created, designed and produced by Trewin Copplestone Publishing Ltd, London

Printed by New Interlitho, Italy

ISBN: 0 85674 029 2

How to use this book

More than 3000 plants or plant groups, terms and leading world botanists are given space in this dictionary. They can be reached directly from their alphabetical ordering which is by the English common name where the plant has one. The Latin scientific name appears immediately afterwards where this can be done with no ambiguity: thus **OAK** (*Quercus*) announces a description of the genus, all members of which are *Quercus*. However, with, for example, Pitcher Plants, several different genera are involved and their scientific names are for that reason taken one by one.

For ease of treatment and to save space, certain sub-topics worthy of appearance in the alphabetical list are entered in it but described elsewhere: thus, **GEAN** *see* CHERRY. Topics of parallel or related interest are shown in small capitals both within and at the end of many entries. Such cross-references will be particularly helpful with the more complex technical terms.

A comprehensive list of Latin names is given at the end of the book and this identifies some 1200 genera.

Standard works of reference exist for every part of botany, and if only for this reason the present work has had to be extremely selective. What the ILLUSTRATED DICTIONARY OF BOTANY aims at is to answer first-level enquiries on as wide a range, internationally, as possible. The choice of which species and varieties (even of which families) to discuss has been made with frequency and likelihood of encounter in mind, just as the technical terms are those most often met (though less often explained) in books, catalogues and articles intended for a general audience.

Throughout the book, each illustration or diagram has been carefully tied to its text, avoiding the need for separate captions. They are located by symbols: thus ▲ next to an illustration, relates to ▲ in the text nearby.

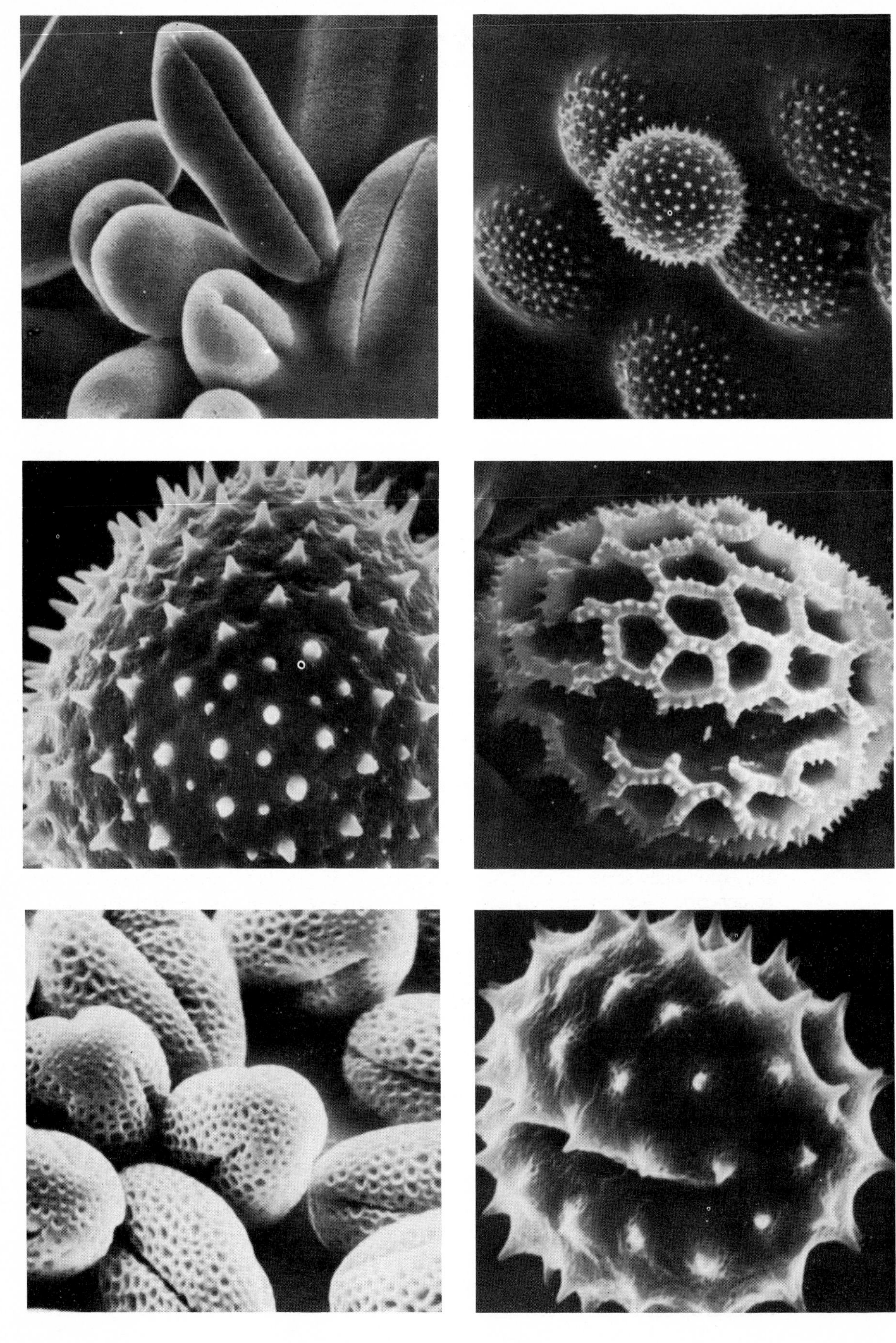

What is Botany?

Apart from agreeing that botany is the study of plants, it would be difficult to get a group of botanists to define the word more precisely. To the taxonomist, it is the study of naming and classifying plants; to the physiologist, that of examining their internal structure and finding how they function; while the bio-chemist investigates their chemical processes and tries to explain life itself. For the ecologist, plants must be studied where they grow in their natural surroundings, while the plant geographer is interested in their distribution and why some plants are abundant all over the world and others so rare.

All these are specialists, the real scientists. But they are certainly outnumbered by those who are simply interested in plants: fascinated by their shapes and patterns, the way they grow, the flowers and fruits they bear, and curiosities of the plant world like the insect-eating or carnivorous species. The amateur wants to know something about all the plants he or she sees or reads about, and how to understand the technical jargon used in describing their features.

These technical terms do seem daunting to many people, but upon consideration they obviously serve an important purpose. At first glance it may seem unnecessary to describe a leaf by such a mouthful of a word as *imparipinnate*, but how much more cumbersome to keep to 'simple English' and say of it 'a leaf divided into several small leaflets which grow opposite each other in pairs and have an extra one at the end of the leaf stalk'. Using language like that, every book on plants would be three times as bulky and expensive. So the vocabulary is made to be used. It is much easier to remember than one would think, and most important of all it is precise. All the most useful of these terms, with their meanings, are to be found in this dictionary, as are the most frequently encountered Latin names. These last are almost international in usage and have meaning to botanists all over the world.

This book does not set out to help with the identification of plants (there are many available that will do so). Rather, it provides information about those which are 'only names', as are many of the more exotic fruits and vegetables now seen in shops; about plants we read of but cannot visualise; and also about those seen in the wild and in cultivation. To give an example, everyone is familiar with the brazil nut at Christmas time, but how much more interesting to know that the nuts grow tightly packed inside a large, rounded shell like the sections of an orange – whose shape they resemble – and that they are still gathered where they fall on the ground in the tropical forest of South America.

Now that most of our lives are so remote from the land, it is difficult to appreciate how important plant lore was in the lives of ancestors. To be able to recognise every possible source of food and distinguish the many plants which are partly or wholly poisonous was vital. To help fix them in the memory, tales and legends grew up, some of which are dimly remembered in superstitions and rhyme. Once man became a cultivator, the herb patch must soon have come into existence, and useful medicinal plants were grown close to the home where they could be gathered when needed. This interest in plants which have curative properties led some men to specialise in their study and uses. In primitive tribes these were the witch doctors; as man became more civilised, so the paths of formal medicine and botany developed together, and the first real botany book was written in Greece before 300BC by Theophrastes. In all he wrote fifteen volumes on *The History of Plants* and *The Causes of Plants*. These largely covered plants of use to man. No other similar work has been left to us from these early times, and after the decline of the Greek and Roman civilisations botany slipped back to its witch doctor or wise woman domain.

It was not until the later Middle Ages that any plant book was again produced, and it is not surprising that those that did appear (mainly in England and Germany) were about herbs. These early herbals are remarkable rather for their strange accounts and even more strange illustrations then for any scientific accuracy. They are on a par with early carvings and drawings of tropical animals, which were part fact and part fancy. It was not until the period about 1500 that descriptions and drawings made from living plants began to appear and plants started to be studied for themselves rather than for their value in medicine. The names of Culpepper and Gerard are familiar from a century later, though the works of both retain a quirky personal note to their observations.

The stimulus needed to make botany truly scientific came in the 17th and 18th centuries, when travellers returned from all over the world to Western Europe with exotic fruits and sometimes plants and pressed flowers. With this wealth of new discoveries, some real organisation of the plant kingdom was necessary if it were ever to be properly studied, and the moment produced the man. This was Carl Linnaeus, a Swedish botanist. He brought into use the system of naming that we use today – the binominal system – and which is applied in a modified way to all forms of life, not only to plants. He also arranged plants into groupings, using the numbers of stamens and stigmas as his basis for classification. Although this system has now been superseded, its very existence meant that later botanists had a foundation upon which to work and build up the knowledge we have inherited.

Gardening, too, has its origins in the cultivation of medicinal herbs, and was likewise brought into new prominence with the introduction of plants first from Europe, then from the Americas and further afield. At first such introductions were either of economic plants or just the whims of travellers; but from the 18th and especially during the 19th and early 20th centuries, plant collecting became a systematic pursuit, culminating in the great gathering of seeds and living material from Eastern Asia. It is quite remarkable how many of our familiar garden plants came from China for the first time during that period.

So today, with botanical studies so actively pursued and with the enjoyment if not the practice of gardening on hand for so many, we have a better chance than at any earlier time of enjoying the marvels and beauties of the plant kingdom. By understanding more of what we see, we may appreciate many things which would otherwise pass us by.

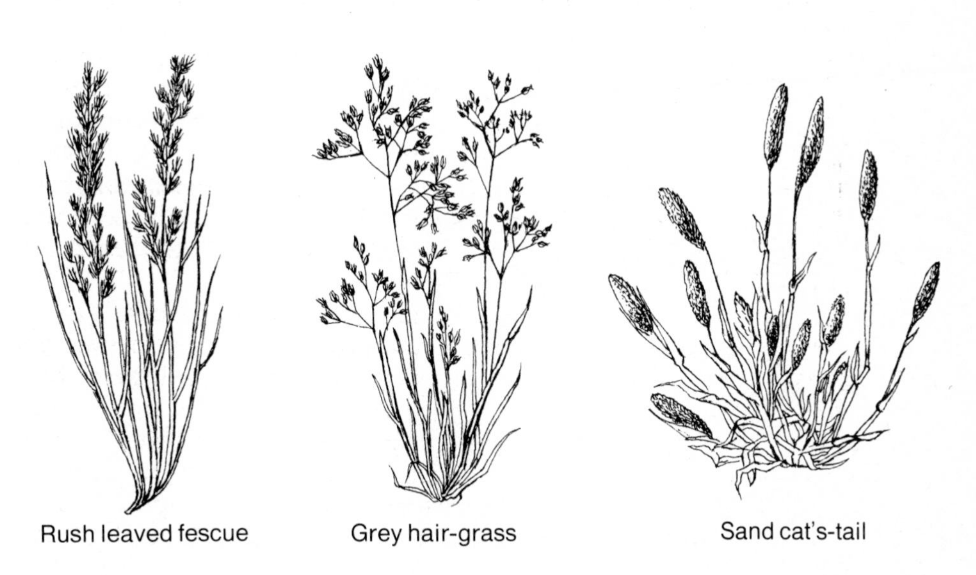

Rush leaved fescue   Grey hair-grass   Sand cat's-tail

ABBREVIATIONS AND SYMBOLS

Fam	family, families	**lflet**	leaflet	**·|·**	zygomorphic flower
Gen	genus, genera	**P**	perianth	**§**	section of genus etc.
Sp	species – singular	**K**	calyx	**!**	plant seen by the author
Spp	species – plural	**C**	corolla	**2n**	chromosome number (diploid)
Ssp	sub-species	**A**	androecium	**✱**	sign of ambiguous use
var	variety	**G**	gynoecium	**⊥**	at right angles to
cv	cultivar – cultivated variety	$\underline{G}$	superior ovary	**||**	parallel to
f	form	$\bar{G}$	inferior ovary	**μ**	micron = $\frac{1}{1000}$ m.
fl	flower	**X**	hybrid	**=**	equal
infl	inflorescence	**♂**	male	**±**	more or less
fr	fruit	**♀**	female	**>**	more than
lf	leaf	**⚥**	hermaphrodite	**<**	less than
lvs	leaves	**⊕**	actinomorphic flower		

ABAXIAL The side of an organ away from the axis or stem, e.g. the underside of a leaf.

● **ABBREVIATIONS** When describing plants, botanists often abbreviate the long scientific names and terms used. For example, when the name of a genus followed by the species is repeated, all but the first letter of the generic name is omitted, e.g. *B. perennis* is used instead of *Bellis perennis*. The most frequent abbreviation of this kind is that of the person who first described the plant. '*Bellis perennis* Linnaeus' is the full name of the common daisy, Carl Linnaeus being the first person to have described the plant in this way. In floras and botanical papers, '*Bellis* (or *B.*) *perennis* L.' is standard usage. In popular books the abbreviation is often omitted completely. There are also generally accepted abbreviations that are represented by symbols.

ABELE
see POPLAR

ABERRANT A plant, or part of a plant, which differs markedly from the normal, e.g. having double flowers.
see MUTANT.

ABORT Failure to develop properly; usually referring to stamens, petals or fruits.

▲ **ABSCISS LAYER** Also known as the separation layer, this is a diaphragm of shrunken, rounded cells that forms across the base of a leaf prior to leaf fall. Just below this zone a plate of corky cells forms, so that when the leaf falls the wound is protected.

ABSINTH
see ARTEMISIA

■ **ACACIA** A genus of about eight hundred mainly evergreen trees and shrubs in the *Leguminosae*, coming from the tropics and sub-tropics, particularly Australia. They have basically BIPINNATE leaves, but these often give way to leaf-like entire PHYLLODES after the seedling stage. The yellow flowers are very small, but are aggregated into pompon or catkin-like heads which in their turn may be carried in PANICLES. Several species are cultivated for their attractive flowers: *A. armata* (kangaroo thorn) has prickly pointed, LINEAR, 13–25mm long phyllodes, and solitary or paired globose flower heads 6mm wide; *A. dealbata* (silver wattle), has silver-grey bipinnate leaves and panicles of small, globose flower heads.

ACANTHUS A genus of 50 species of erect perennials in the *Acanthaceae* mainly from the Mediterranean region, also Africa and Asia. They have LANCEOLATE to OBOVATE leaves, usually PINNATIFID, and spike-like heads of tubular flowers. The latter have protruding lower lips and rise in the AXILS of broad, overlapping bracts. The seed CAPSULES explode on ripening. Several species are cultivated for their ornamental qualities: *A. spinosus* has spiny leaves which are thought to be the pattern for the decoration on Corinthian capitals. The flowers are purple flushed.

ACAUL|IS, -E Stemless, or appearing to be so.

ACCRESCENT Enlarging after flowering, usually referring to the CALYX as in BLADDER CHERRY.

○

■ ▲

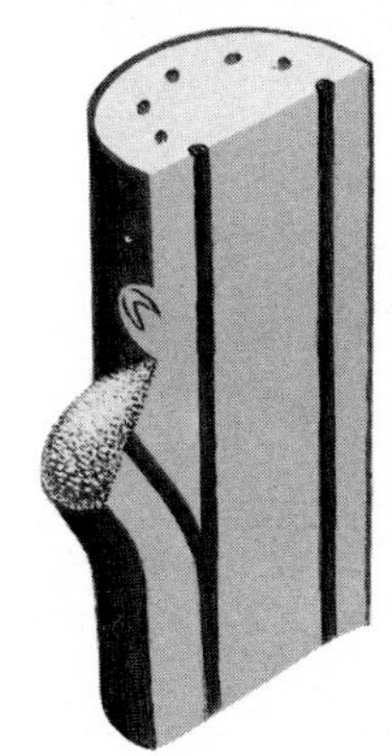

AC|ER, -RIS, -RE Acrid to the taste.

ACEROSE [ACEROSUS] Needle-like, e.g. pine leaves.

ACETABULIFORMIS Saucer-shaped, e.g. flowers of nightshade.

ACETOS|US, -A, -UM [ACETOSELLA] With an acid taste, e.g. sorrel leaves.

ACHENE A one-seeded, hard, dry fruit that does not open when ripe, e.g. cinquefoil and the 'seeds' of strawberry.

ACICULAR|IS, -E Pointed like a needle.

ACINACIFORMIS Shaped like a curved sword in outline.

ACONITE Members of the genera *Aconitum* and *Eranthis*, both in the *Ranunculaceae*. *Aconitum* has 300 species from the northern temperate zone, comprising climbing and erect herbaceous perennials with tuberous roots and PALMATE leaves. The flowers are strongly ZYGOMORPHIC with petal-like sepals, the upper forming a helmet-like hood. Two of the petals are modified to NECTARIES. All species are very poisonous: *A. napellus* (monk's hood) has deeply lobed leaves and RACEMES of blue-purple flowers; *A. lycoctonum* (wolf's bane) has dull yellow flowers. *Eranthis* has 7 species of dwarf, tuberous rooted perennials from Europe and Asia. The flowers are white or yellow, cup-shaped, appearing to sit on a ruff of leaves.

○ **ACTINOMORPHIC** Of regular shape, referring to flowers such as buttercup which can be cut into identical halves at any two points which pass through the middle.

ACULEAT|US, -A, -UM Bearing prickles.

ACUMINAT|US, -A, -UM Tapering to a slender point.

ACUTIFOLI|US, -A, -UM With pointed leaves.

ACUT|US, -A, -UM Narrowing to a short point.

ADAM'S NEEDLE Several species of yucca, evergreen somewhat palm-like plants in the *Agavaceae* from southern USA, Mexico and the West Indies. They are either clump-forming or tree-like with short, thick trunks. The leaves are LINEAR to narrowly LANCEOLATE, often spine-tipped. The large cream or white bell-shaped flowers are pendulous in large terminal PYRAMIDAL PANICLES. They are pollinated by the pronuba moth which deliberately carries a ball of pollen to the stigma, thus assuring pollination and a good crop of seeds, some of which become the food of its larvae (caterpillars). Commonly seen in gardens are: *Y. recurvifolia*, with arching leaves and *Y. gloriosa* with stiff erect ones.

ADAXIAL The side of an organ facing the axis or stem, e.g. the upperside of a leaf.

ADDER'S TONGUE *(Ophioglossum vulgatum)*. A curious fern of damp grassland in Asia, Europe, N. Africa and N. America. It has usually solitary, stalked OVATE fronds 100–200mm tall from underground RHIZOMES. The SPORANGIA are borne in spikes united to the leaf stalk.

●

▲

ADNATE Closely attached, e.g. when the stamen FILAMENT is united with the petal.

● **ADONIS** A genus of 20 species of annual and perennial herbaceous plants in the *Ranunculaceae* from Europe and Asia. They have erect stems bearing leaves cut into many LINEAR segments, and terminal red or yellow cup-shaped flowers. Several species are cultivated as ornamentals: *A. autumnalis* (pheasant's eye) is a 300mm tall plant with blood red, 13mm wide flowers; *A. amurensis* is perennial to 450mm with yellow or white 38–50mm flowers; *A. vernalis* is similar with bright yellow flowers.

ADUNCAT|US, -A, -UM, ADUNC|US, -A, -UM Hooked, e.g. spines and prickles.

ADVENTITIOUS A plant organ that arises from an unexpected position, e.g. aerial roots that spring from a stem as in sweet corn, or shoots that arise directly from true roots as in raspberry.

AECIDIAL CUP Cup-shaped structures of HYPHAE, formed by rust fungi *(Puccinia)*, and in which are formed the AECIDIOSPORES.

AECIDIOSPORES The spores of a RUST FUNGUS that will only grow on the alternative host plant.

AEQUAL|IS, -E Of equal size.

AERENCHYMA Plant tissue with large intercellular spaces or air chambers, usually found in water plants such as mare's tail and water lily.

AERIAL ROOTS Roots that grow from stems, a common feature of some tropical trees, climbers and orchids.

AEROBES Referring to BACTERIA that need oxygen to grow.

AERUGINOS|US, -A, -UM Verdigris-coloured.

AESTIVAL|IS, -E Of the summer.

AESTIVATION The way that petals and sepals are folded in the flower bud.

AFFIN|IS, -E Similar or related to.

AFRICAN LILY *(Agapanthus)*. About five species of evergreen and deciduous perennials in the *Alliaceae* from S. Africa. They have tufts of strap-shaped leaves and umbels of narrowly bell-shaped flowers in shades of blue or purple.

AFRICAN MARIGOLD *(Tagetes erecta)*. A 600–900mm annual in the *Compositae* from Mexico, with deeply PINNATIFID leaves and daisy-like flower heads having bell-shaped INVOLUCRES and broad yellow RAY FLORETS.

AFRICAN VIOLET *(Saintpaulia ionantha)*. An evergreen perennial in the *Gesneriaceae*, native to E. Africa. It has fleshy, broadly OVATE, long stalked leaves in rosettes. The ZYGOMORPHIC, 5-petalled flowers are violet-shaped.

AGAVE A genus of 300 plants in the *Agavaceae* from the southern USA to northern S. America. They have rosettes

■ 

of narrow, often spine-tipped leaves and tall, branched INFLORESCENCES. Several species are grown as ornamentals, others have economic value: *A. sisalana* yields sisal fibre; *A. americana* (century plant) produces an abundant sap which when fermented becomes pulque, the national drink of Mexico.

AGGLOMERAT|US, -A, -UM Borne together in a tight cluster.

AGGLUTINAT|US, -A, -UM Seeming glued together.

AGGREGATE FRUIT A fruit such as a blackberry or raspberry composed of many small fruitlets or carpels.

AGRESTR|IS, -E Growing in fields or cultivated land.

AGRIMONY *(Agrimonia)*. About fifteen species of herbaceous perennials in the *Rosaceae* from Europe, Asia, N. and S. America, and Africa. They have unequally PINNATE leaves and spikes of 5-petalled, yellow flowers followed by top-shaped fruits bearing hooks and containing two achenes. *A. eupatoria* is the common agrimony of Europe, Asia and N. Africa.

AIR PLANTS
see EPIPHYTES

▲ **AIR SPACES** The tiny spaces between the cells in plant tissue.

AKEE TREE *(Blighia sapida)*. Named for William Bligh of 'H.M.S. Bounty', this 9m evergreen tree is a native of Guinea and much grown in the West Indies. It has pinnate leaves and PANICLES of small, white flowers. The pear-shaped fruit contains a seed surrounded by a red, fleshy ARIL which is edible, especially when cooked.

ALBESCENS Becoming white.

ALBICANS Whitish or off-white.

ALBID|US, -A, -UM White or whitish.

ALBINO [ALBINISM] Lacking in CHLOROPHYLL or colouring pigments as in white flowered MUTANTS of coloured plants.

ALBUMEN The store of food around the embryo in some seeds.

ALBUMINOUS
see ENDOSPERM

ALB|US, -A, -UM White.

■ **ALDER** *(Alnus)*. About thirty-five species of deciduous trees in the *Betulaceae* from the northern temperate zone and the Andes. They have alternate, toothed, OVATE leaves and tiny flowers in catkins. The male catkins are pendent and flexible; the female short and rigid, on ripening looking like small cones. *A. glutinosa* (common alder) from Europe and N. Africa has reddish male catkins before the leaves in early spring.

ALECOST
see COSTMARY

ALEURITES
see CANDLE NUT

● **ALEXANDERS** *(Smyrnium olusatrum)*. A biennial plant in the *Umbelliferae* from S.W. Europe and the Mediterranean region. Once cultivated as a pot herb in Britain and now naturalised there, especially by the sea. It has 3-TERNATE, COMPOUND leaves and 0.9–1.5m stems bearing UMBELS of yellow-green flowers. The broadly OVOID 6mm long ribbed seeds are black.

ALFALFA
see LUCERNE

ALGA [ALGAE] Primitive plants, either aquatic or living in damp places. They are of diverse form, ranging from the single celled *Pleurococcus* which forms green 'powder' on old wood and tree trunks, to the giant marine seaweeds which can grow to 30m in length. Most algae are green owing to the presence of CHLOROPHYLL, but in the red and brown seaweeds this is masked by colouring matter. Apart from the seaweeds, the most common alga is blanket weed, which forms dense masses of green threads in ponds and tanks. *See also* LICHENS.

ALGAL BLOOM A condition in ponds, lakes and the sea when UNICELLULAR algae multiply in such abundance as to colour the water red or green.

ALGAROBA
see CAROB

ALIEN A plant believed or known to have been introduced by man, and now naturalised.

ALKANET Two plants are known by this vernacular name: *Pentaglottis sempervirens* and *Alkanna tinctoria*, both in the *Boraginaceae*. The first comes from S.W. Europe and is a tufted perennial with OVATE pointed leaves to 300mm long, and CYMES of bright blue, tubular flowers 3mm long with 5 rounded petal lobes. *Alkanna* is a spreading perennial about 150mm high with oblong leaves and the flowers having deep blue petal lobes and a red tube. It is native to S. Europe.

ALLANTOIDES Sausage-shaped, e.g. leaves or fruit.

ALLIUM A genus of 450 mainly bulbous rooted perennials in the *Alliaceae*, well distributed around the northern hemisphere. They have either grassy, tubular, strap-shaped or elliptic leaves, and UMBELS of starry or bell-shaped flowers. All but a few have the familiar onion smell. *See also* GARLIC, LEEK, ONION, CHIVES, WELSH ONION. Several species are grown as ornamentals: *A. moly* has yellow, starry flowers and grey-green elliptic leaves; *A. narcissiflora* has LINEAR green leaves and nodding rose bell flowers. Both are 150–300mm tall.

ALLSEED *(Radiola linoides)*. A 25–75mm, slender stemmed annual from Europe, Asia and the mountains of tropical Africa. It has tiny, OVATE, elliptic leaves, numerous minute, whitish flowers. Favours damp, sandy or peaty soils. Many seeded goosefoot is also known as allseed.

ALLSPICE *(Pimenta officinalis)*. A small evergreen tree to 9m in the *Myrtaceae* from tropical America and the West Indies. It has 75–150mm, elliptic leaves and 9mm wide cream and white flowers in CYMES. The pea-sized fruits are purple. Allspice is the dried unripe fruits, its flavour and aroma blending cinnamon, nutmegs and cloves.

ALMOND *(Prunus dulcis)*. A deciduous tree of 6m in the *Rosaceae* from the eastern Mediterranean to Asia. It has toothed, LANCEOLATE leaves and 38–51mm wide pink flowers opening before the leaves expand. The fruits are ovoid like green peaches, the stone being the familiar almond.

ALOE A genus of 275 mainly shrubby, succulent plants in the *Liliaceae* from Africa and Arabia. They have triangular to sword-shaped, fleshy leaves clustered at the ends of the branches or at ground level. The red, yellow or orange flowers are tubular, carried in simple or branched RACEMES. Several species are cultivated as ornamentals: *A. ferox* has erect stems to 3.6m and glaucous leaves, spiny-toothed on the margins and the back. The flowers are orange; *A. variegata* (partridge-breasted aloe) is clump-forming to 300mm or so with triangular, dark green leaves banded silvery white. The flowers are red.

ALOIDES Like ALOE.

ALPESTR|IS, -E Growing on the mountains.

ALPICOL|US, -A, -UM From the high mountains.

ALPINE ROSE *(Rhododendrum ferrugineum* and *R. hirsutum)*. Dwarf, evergreen shrubs in the *Ericaceae* from the mountains of Europe. They have 38mm oval to oblong leaves and TERMINAL clusters of funnel-shaped, pink flowers: *R. hirsutum* is distinguished by hairy marginal leaves and pointed CALYX lobes.

ALPIN|US, -A, -UM Growing in the mountains above the timber-line.

ALSIKE
see CLOVER

ALTERNATION OF GENERATIONS Generally applied to ferns where each species has two totally unlike generations. Fern plants do not produce sexual organs and seeds like flowering plants. They bear minute SPORES, each one of which grows into a small, leafy body known as a prothallus. Upon the prothallus are borne the sex cells or gametes which on fusion give rise to a new fern plant.

▲ **ALTERNAT|US, -A, -UM [ALTERN|US, -A, -UM]** One leaf at each NODE at opposite sides of the stem.

ALTISSIM|US, -A, -UM Very tall.

ALVEOLATE Honeycombed or pitted.

AMABIL|IS, -E Lovely; very pleasing.

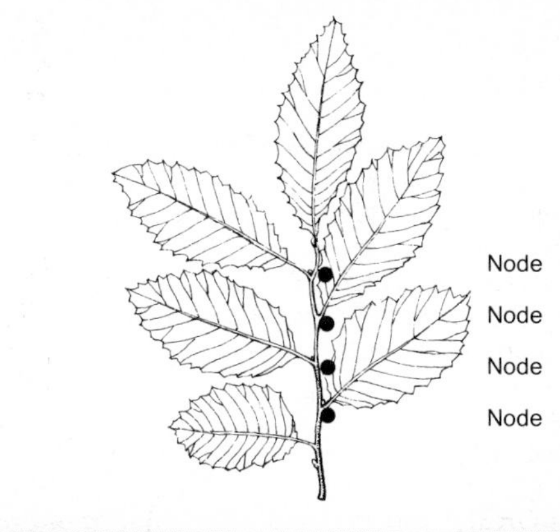

AMANITA A genus of fungi belonging to the *Basidiomycetes*, most of which are highly poisonous. They are widespread in woods, particularly among pine, birch, beech and oak throughout the northern hemisphere. Among the most poisonous species are: *A. muscaria* (fly agaric) a red capped fungus with prominent white warts or patches and a white STIPE bearing a skirt-like membranous ring at its junction with the cap. This fungus has been used by various peoples as an hallucinagen and though rarely fatal is a dangerous drug; *A. phalloides* (death cap) is deadly poisonous, having a pale yellowish cap with a slight green tinge and a somewhat darker stipe bearing a prominent skirt-like ring; *A. virosa* (destroying angel) is deadly poisonous, wholly white with a smooth, sticky cap and a rough stipe bearing a small frayed ring. *See also* BLUSHER.

AMARANTH Several species of *Amaranthus*, annual and perennial plants in the *Amarantaceae* and almost cosmopolitan in distribution. They have simple leaves and small green, red or purple tinted flowers in dense PANICLES. Several species are weeds of wasteland and a few are cultivated for their ornamental flowers and variegated leaves: *A. caudatus* (love-lies-bleeding) a 600–900mm tall, annual species with OVATE, pale green leaves and spectacular pendent spikes of red flowers; *A. gangeticus* is an annual to 900mm tall with ovate, pale green leaves which in a number of CULTIVARS are splashed with shades of red and yellow, the flowers are insignificant; *A. hypochondriacus* (Prince's feather) is a 1.2–1.5m tall annual with oblong LANCEOLATE leaves and dense panicles of deep crimson flowers.

AMAR|US, -A, -UM Bitter-tasting.

AMARYLLIS A genus of one species of bulbous plant in the *Amaryllidaceae* from S. Africa. This is *A. belladonna* whose strap-shaped leaves die down in summer and are followed in autumn by heads of lily-like, rose-red flowers on leafless, stout stems 450–760mm tall.

AMBIGU|US, -A, -UM Of uncertain origin or name.

AMBROSIA A genus of about forty species of annual and shrubby plants in the *Compositae*, mainly from South America, but also cosmopolitan. They have lobed or deeply dissected leaves and erect RACEMES of separate sexed flower heads. The male flowers usually occupy the upper part of the raceme, the fewer female ones at the base. The air-borne pollen causes hay fever. Several species are weeds of wasteland in USA and occasionally in Europe.

AMELANCHIER A genus of 25 species of deciduous shrubs and trees in the *Rosaceae*, mainly from N. America but also Europe and Asia. They have entire, toothed, ovate leaves and five-petalled white flowers in short racemes. The berry-like, black-purple fruits are edible and the leaves colour brightly in autumn. Several species are grown as ornamentals: *A. canadensis* (shadbush, Juneberry) is a tree to 7.5m with CORDATE, ovate leaves, woolly beneath, and strap-shaped petals, 16mm long; *A. laevis* (shadbush) similar to *A. canadensis* and confused with it, but has the leaves smooth and hairless, even when young.

AMENTACEOUS Bearing catkins.

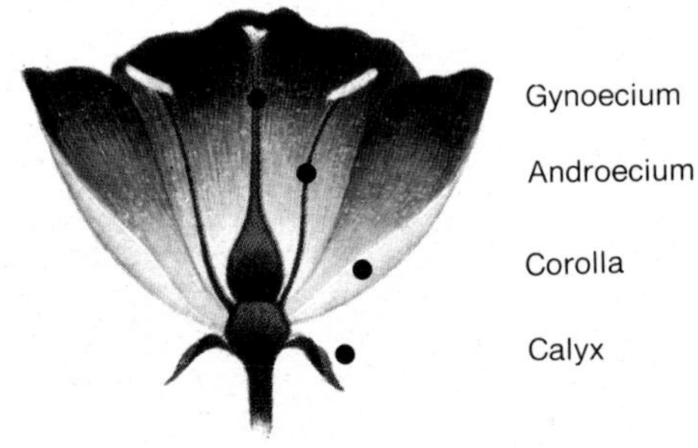

Whorls of a typical flower

AMENTUM A catkin; a spike of tiny, petal-less flowers of one sex, often pendulous.

AMERICAN ALOE
see AGAVE

AMERICAN CRESS
see WINTER CRESS

AMERICAN LAUREL *(Rhododendron maximum)*. A small tree or large shrub with 100–250mm long evergreen, ovate, LANCEOLATE leaves, CORYMBS of up to 24 bell-shaped 38mm wide rose or purple, green-spotted flowers.

AMITOSIS The division of a cell nucleus by simple constriction. *See also* CELL DIVISION.

● **AMORPHOPHALLUS** A genus of 100 species in the *Araceae* from the tropics of Asia and Africa. They have large corm-like RHIZOMES which produce solitary leaves on stout stalks. The leaf blades are SAGGITATE and deeply lobed. The inflorescence is arum-like, the spike of small flowers enclosed by a large coloured SPATHE. A carrion smell is given off when in bloom to attract pollinating flies. Some species are cultivated as ornamentals. *A. bulbifer*, has a 60cm wide leaf on a 800mm olive and pale green mottled stalk. The spathes are pale greenish-white and red, with a large club-shaped SPADIX. *A. titanum* is the largest AROID in the world, having a pale green spathe 600–900mm long, and a leaf up to 4.5m wide on a stalk of equal length.

AMPHIBIOUS Growing on land or in water, e.g. water pepper.

AMPHICARPIC [AMPHICARPUS] With two different kinds of fruit on the same plant, e.g. Pot marigold and *Aethionema*.

AMPLECTENS [AMPLEXICAUL|IS, -E] Stem-clasping; referring to the base of the leaf or leaf stalk.

AMPULLACEOUS Flask-shaped.

ANAEROBE BACTERIA that fail to grow in the presence of oxygen, e.g. those that cause tetanus.

ANANDROUS Without stamens.

ANAPHASE
see CELL DIVISION

ANASTOMOSE The linking up of the branches of a vein system in a leaf or bract.

ANATROPOUS
see OVULE

ANCEPS Two-edged, as in flattened, keeled stems; alternatively, doubtful or uncertain.

ANCHUSA A genus of 50 annuals, biennials and perennials in the *Boraginaceae* from Europe, Asia and N. Africa. They have LINEAR-LANCEOLATE or OBLONG leaves and funnel-shaped flowers in SCORPIOID RACEMES. A few species are cultivated for their bright blue flowers, notably *A. azurea* (sun. *A. italica*) a 0.9–1.5m perennial.

■

ANDRODIOECIOUS HERMAPHRODITE and male flowers on separate plants.

▲ **ANDROECIUM** The stamens in a flower.

ANDROGYNOUS Separate male and female flowers in one inflorescence.

ANDROMEDA A single-species genus in the *Ericaceae*, from the colder parts of the northern temperate zone. *A. polifolia* (bog rosemary) is a slender-stemmed, suckering shrub 100–300mm or more tall with linear to narrowly oblong leaves and nodding urn shaped 6mm long pink bells. The species varies in leaf width and some forms are GLAUCOUS beneath.

ANDROMONOECIOUS HERMAPHRODITE and male flowers on the same plant.

ANDROPHORE A stalk-like elongation of the RECEPTACLE of a flower bearing the stamens and pistil.

■ **ANDROSACE** A genus of 100 species of annual and perennial plants in the *Primulaceae* from the mountains of Europe, Asia and N. America. They are small, tufted or cushion-forming plants with LINEAR to LANCEOLATE or OVATE leaves and UMBELS of tiny, primrose-like flowers, tubular with five petal-lobes. Several species are cultivated as ornamentals: *A. lanuginosa* is mat-forming with silvery hairy leaves and 8mm wide, red-eyed pink flowers; *A. primuloides* has woolly winter rosettes and larger green summer ones with lanceolate leaves and pink flowers.

ANEMONE A genus of 150 species in the *Ranunculaceae*, mainly small, perennial plants with a world-wide distribution. Some have fleshy, tuber-like RHIZOMES and most have long stalked, deeply lobed or dissected leaves. The flowers have 5–20 petal-like sepals and are borne above a WHORL of three, leaf-like bracts. A number of species are grown as ornamentals. *A. apennina* is 15cm tall with 38mm wide, blue, daisy-like flowers; *A. coronaria* is 225mm tall with cup-shaped flowers of red or purple; *A. hupehensis japonica* is 600–900mm with wide, cup-shaped flowers of purple-rose or white; *A. nemorosa* (wood anemone) is 150mm, with 6-sepalled flowers 25–38mm wide, usually white flushed pink, sometimes lavender.

ANEMOPHILOUS Applied to wind-pollinated flowers.

ANGELICA A genus of 80 species of perennial or MONOCARPIC plants in the *Umbelliferae* from the northern hemisphere and New Zealand. They have bi- or tripinnate leaves of triangular outline and erect, branched stems bearing flat umbels of tiny flowers. *A. archangelica* provides the angelica of commerce. It is a 0.9–1.5m tall plant with greenish-white flowers.

ANGIOSPERMAE [ANGIOSPERMS] The division of the plant kingdom which contains all the true flowering plants. They are typified by the CARPEL which carries the OVULES folding in to form the OVARY.

ANGOSTURA BARK A fever cure obtained from the bark of *Cusparia febrifuga*, a small S. American evergreen tree with COMPOUND leaves.

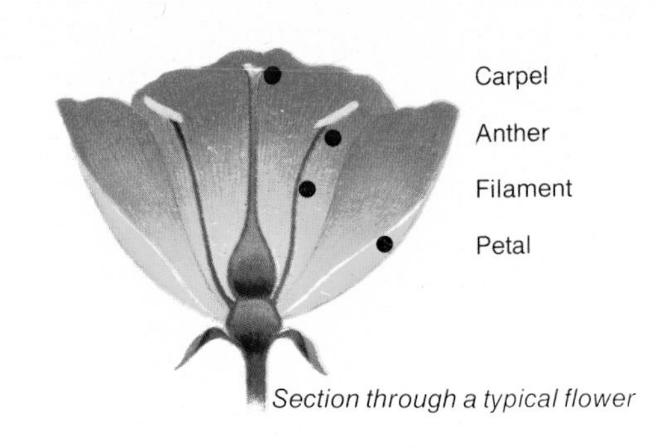

Section through a typical flower

Section through ripe anther

Pollen

Section through anther

Pollen sac

ANGULAR|IS, -E [ANGULATUS, ANGULOSUS] Angular or having angles, e.g. keeled stems or fruits.

ANGUSTAT|US, -A, -UM [ANGUST|US, -A, -UM] Narrow.

ANISE CAP *(Clitocybe odora)*. A toadstool belonging to the *Basidiomycetes* found in mixed woodland from late summer to autumn. It has a flattened cap and a STIPE which bulbs out at the base. The whole plant smells of anise and is a light blue-green.

ANISEED *(Pimpinella anisum)*. An annual plant in the *Umbelliferae* from Greece, but now much cultivated for its seeds. It has PINNATE leaves and compound umbels of tiny, yellowish flowers.

ANISOGAMETES GAMETES of unequal size as in the algae, where the males are much smaller and more active than the females.

ANNUAL A plant that grows from seed, flowers and fruits in one year.

ANNULAR|IS, -E Ring-shaped.

ANNULAT|US, -A, -UM Bearing rings.

ANNULUS In Basidiomycete fungi, a membranous ring of tissue on the stipe just below the cap. In ferns, a partial ring of thickened cells around the SPORANGIUM which straightens when dry, rupturing the sporangium wall and releasing the spores.

ANNU|US, -A, -UM Of annual duration.

● **ANTHER** The male part of a flower, consisting of a stalk or FILAMENT and usually two anther lobes which contain the pollen grains.

ANTHERIDIA Flask-like bodies within the antheridial cups.

▲ **ANTHERIDIAL CUPS** The so-called 'flowers' of mosses, surrounded by wide spreading leaves like petals. They produce antheridia which in turn release the male cells or spermatocytes.

ANTHEROZOIDS
see SPERM

ANTHESIS When the flower opens.

ANTHRACNOSE A fungal disease of plants causing blackish, depressed circular areas, usually on the fruits, e.g. bean and apple.

ANTHRACOBIA A small cup fungus belong to the *Ascomycetes*. It forms small, fleshy orange saucers about 6mm across, often on burnt ground.

ANTHUS A flower; often used as a suffix.

ANTIPODAL CELLS Three large cells in the EMBRYO SAC at the end furthest from the MICROPYLE. They are associated with the nutrition of the embryo sac.

▲

■ **ANTIRRHINUM** A genus of 42 species of annual, perennial and sub-shrubby plants in the *Scrophulariaceae*, all from the northern hemisphere. Some are erect, others of spreading habit, with entire LINEAR or LANCEOLATE to OVATE leaves. The tubular flowers having bulging COROLLA lobes formed like a closed mouth; only the bee pollinators are strong enough to push through. Several species are cultivated as ornamentals, in particular *A. majus* (snapdragon). This branching perennial reaches 600mm or more with lanceolate leaves and 38mm long flowers in shades of red, pink, purple, yellow and white.

APETALAE A group of plants within the *Dicotyledons* having flowers lacking petals.

APETAL|US, -A, -UM Without petals.

APHYLL|US, -A, -UM Without leaves.

APICAL The tip of an organ furthest from its points of attachment, e.g. the tip of a leaf or stem.

APIFER|A, -UM Bee-like, e.g. the bee orchid.

APLANOSPORE A unit of asexual reproduction found mainly in land forms of filamentous algae *(Vaucheria)*. An aplanospore is borne singly in a special cell known as an aplanosporagium.

APOCARPUS When the CARPELS in a flower are all separate, e.g. anemone, buttercup.

APOGAMY The origination of a young fern plant direct from a PROTHALLUS, thus bypassing the usual sexual processes.

APOGON A group of species in the genus *Iris* which lack beards on the lower petals.

APOMIXIS The production of seeds without the usual fertilisation process. Many of the blackberries and dandelions reproduce in this way.

APOPETALOUS With separate petals (*see also* POLYPETALAE).

APOPHYSIS The slightly thickened basal part of moss spore capsule.

APOSPORY The production of spores without the usual sexual processes.

APOTHECIA A type of fungal fruiting body characteristic of the *Ascomycetes*. It varies in size from a pin-head to several centimetres across, usually discoid, shallowly saucer-shaped or cup-like. *See* ANTHRACOBIA.

APPENDICULAT|US, -A, -UM Bearing small appendages.

APPLE The fruit of *Malus sylvestris* (syn. *M. pumila*), a deciduous 9m tree in the *Rosaceae* originally from S.E. Europe and S.W. Asia. All the different sorts of eating and cooking apples have been raised in gardens. The leaves are OVATE, 58–75mm long, PUBESCENT beneath. 5-petalled

● white, or pink and white flowers are borne as the leaves unfold. Fruits globose to ovoid, sunken at the stalk and flower end. The true fruit is the core and its seeds, *see* POME. Trees with small fruits, usually sour and less than 38mm across on long stalks with hairless leaves are known as crab apples or crabs. About thirty-five wild crab apples species are found in the northern temperature zone, several having colourful flowers and fruits and grown as ornamentals. *See also*: BALSAM, CUSTARD, KANGAROO, LOVE, MAMMEE, OAK, SUGAR, THORN apples.

APPLE MOSS [Bartramia pomiformis] A neat little tufted moss with pale green linear leaves and globular green spore CAPSULES like minute apples.

APPRESSED or **ADPRESSED [APPRESS|US, -A, -UM]** Lying flat, or almost so, e.g. the hairs on some leaves.

▲ **APRICOT** The fruit of *Prunus armenaica*, a deciduous 9m tree from China. It has broadly ovate leaves with rounded teeth and 25mm wide white or palest pink flowers before the leaves. The globular or ovoid 38mm long fruits are yellow, often flushed red. Several CULTIVARS have arisen in cultivation.

APTER|US, -A, -UM Without wings; as applied to plants which normally have keeled or winged stems or fruits.

■ **AQUATIC PLANTS** Many plants are adapted to living in water, some completely submerged, others floating or with their roots in water and their leaves above. They play an important part in the ECOLOGY of water life, the submerged ones providing both oxygen and shelter for many water creatures. A feature common to the submerged and some floating aquatics is the ability to absorb oxygen, carbon dioxide and some minerals through the leaves directly from the water. Another feature is the presence of AERENCHYMA.

AQUATILIS Submerged in water.

AQUIFOLIUM The classical name for holly.

ARABIS A genus of 120 species of annuals and perennials in the *Cruciferae*, from the northern hemisphere and the mountains of Africa. They are mainly low-growing, tufted or mat-forming plants with single, OVATE to OBOVATE or LANCEOLATE toothed leaves and RACEMES of 4-petalled flowers. These are usually white but occasionally pink, red, purple or yellow. A few species are cultivated as ornamentals: *A. caucasica* (syn. *A. albida*) is mat-forming with hoary, obovate leaves and white flowers on 150–225mm stems; *A. blepharophylla* is similar, with rose-purple flowers on 100mm stems.

ARACHNOIDES Covered with cobweb-like hairs as in cobweb houseleek *(Sempervivum)*.

ARAUJIA A genus of 2 or 3 species of evergreen, climbing plants in the *Asclepiadaceae* from S. America. They have twining stems and opposite pairs of ovate to lanceolate leaves. The flowers are tubular, inflated at the base and with five lobes at the mouth. One species is cultivated: *A. sericofera*, the cruel plant. This has ovate-oblong leaves 50–100mm long and 25mm wide, white flowers. These are followed by deeply grooved, 125mm long ovoid pods.

Moths visiting the flowers are sometimes trapped by their tongues.

ARBORESCENS Of tree-like form.

ARBOR-VITAE A named applied in a general way to the 5 species of *Thuja*, evergreen coniferous trees in the *Cupressaceae*, from China, Japan and N. America. They have tiny overlapping scale leaves and the stems are arranged in flattened, frond-like branchlets. They are grown as ornamentals or for timber. *T. orientalis* (Chinese arbor-vitae), usually a rather round-headed tree to 12.2m, the branchlets erect and 13–19mm long ovoid cones. *T. plicata* (western arbor-vitae, or western red cedar) a pyramidal tree to 60m with a buttressed trunk and 13mm long, narrow cones.

ARBOUR VINE
see MORNING GLORY

ARBUSCULUS A small shrub of tree-like form.

ARCHEGONIA In mosses and liverworts, the organs that contain the egg cells.

ARCHEGONIAL CUP Flower-bud-like structures at the tip of the moss shoots which contain the ARCHEGONIA.

ARCHICHLAMYDEAE A sub-class of the *Dicotyledons*, containing the two major groups *Polypetalae* and *Apetalae*.

ARCHIMYCETES Fungi frequently parasitic on other plants that function as single cells collectively or as naked masses of protoplasm. An example of the former is wart disease of potatoes; of the latter, finger and toe disease of brassicas.

ARCUAT|US, -A, -UM Arching or curving.

ARENARI|US, -A, -UM Frequenting sandy places.

AREOLE A tiny hump-like organ, found in the *Cactaceae*, which bears spines, hairs, wool or bristles. It arises in a leaf axil and probably represents a greatly modified shoot.

ARGENTE|US, -A, -UM Silvery

ARGILLACEOUS Whitish.

ARGUT|US, -A, -UM Sharply toothed or notched.

ARIL [ARILLUS] An extra covering layer around the seed, often fleshy and brightly coloured, e.g. spindle and yew.

ARISAEMA A genus of 150 species of perennial plants in the *Araceae* from E. Africa, tropical Asia and N. America. They have tuber-like RHIZOMES and long-stalked leaves cut into three lobes or sometimes PEDATE. The spikes of small, petal-less flowers are enclosed in a large spathe as in *Arum*. A few species are cultivated as ornamentals: *A. candidissimum* has broad, trifoliate leaves and a white striped, green spathe; *A. triphyllum* (Indian turnip, Jack-in-the-pulpit) has purple brown and green striped spathes.

▲ 

ARISTAT|US, -A, -UM Bearing an AWN or beard.

ARMAT|US, -A, -UM Armed, bearing an armature of thorns.

ARNICA A genus of 32 species of herbaceous perennials in the *Compositae* from northern temperate and arctic regions. They have clumps of OVATE to OBLANCEOLATE leaves and erect stems bearing yellow daisy flower heads. They yield tincture of arnica. *A. montana* (mountain tobacco) is sometimes cultivated. It has 300mm tall stems and 75–100mm wide flower heads.

AROID All members of the family *Araceae*.

ARROW-GRASS Several species of *Triglochin* in the *Juncaginaceae* which are grassy plants of wet places. They are rhizomatous perennials with linear leaves and slender spikes of tiny, greenish flowers: *T. palustris* (marsh arrow-grass) is slender, 150–450mm tall, the leaves deeply furrowed; *T. maritima* (see arrow-grass) is the same height but robust, the leaves not furrowed. Both are well distributed in the northern hemisphere.

ARROWHEAD Several species of *Sagittaria*, but primarily *S. sagittifolia*. A member of the *Alismataceae*, this water plant has long stalked, SAGITTATE leaves above the water, ovate floating ones and linear, submerged leaves. The 19mm wide, 3-petalled white flowers are borne in WHORLS.

● **ARROWROOT** This name covers several unrelated plants that yield a starchy powder from the tuberous roots. The main one is *Maranta arundinacea* (West Indian or Bermuda arrowroot) a 1.8m plant in the *Marantaceae* from S. America with slender, branching stems and lanceolate leaves; *Manihot utilissima* (Brazilian arrowroot) in the *Euphorbiaceae* is about 900mm with PEDATE leaves having 3–7 leaflets; *Curcuma angustifolia* (East Indian arrowroot), in the *Zingiberaceae* is also about 900mm with long-stalked, lanceolate leaves; *Tacca leontopetaloides* in the *Taccaceae* (East Indian arrowroot) is 600mm or more with trifoliate, PINNATIFID leaves. Formerly, *Arum maculatum* (Lords and ladies) was used as a source of starch under the name Portland arrowroot.

ARTEMISIA A genus of 400 species of shrubs, perennials and annuals in the *Compositae* from the northern temperate zone, S. America and S. Africa. They usually have deeply dissected or LINEAR leaves and tiny OVOID or rounded flower heads in RACEMES or PANICLES. Several species are cultivated as ornamentals: *A. abrotanum* (lad's love, old man or southernwood) is a 900mm shrub with hoary, sweetly aromatic foliage and panicles of dull yellow flower heads; *A. dracunculus* (tarragon), a 600mm herbaceous perennial with linear grey-green leaves and green-white flowers in panicles, is used for seasoning vinegar; *A. absinthium* (common wormwood) a 600–900mm perennial with a woody base and coarsely dissected, grey-green leaves. The yellow, globular flower heads are borne in leafy panicles. A medicinal plant and used to flavour absinth; *A. vulgaris* (mugwort) 0.6–1.2m perennial with PINNATISECT leaves, dark green above and greyish-white beneath; a wayside weed in northern temperate regions.

●

■

ARTICHOKE A name referring to three different plants all grown as vegetables. Globe artichoke *(Cynara scolymus)* is a large thistle-like plant from N. Africa, with grey-green lobed leaves and purple flower heads to 150mm wide. The latter are cooked when young and the fleshy bases of the scales sucked out; Jerusalem artichoke *(Helianthus tuberosus)* is a perennial sunflower from N. America with potato-like tubers that must be cooked before they are eaten; Chinese artichoke *(Stachys affinis)*, also known as crosnes, is a 300–450mm perennial with rough ovate leaves and tubular, pink flowers. The curious beaded RHIZOMES are cooked and eaten.

ARTICULAT|US, -A, -UM Jointed, or appearing to be so.

ARTILLERY PLANT *(Pilea microphylla* syn. *P. muscosa)*. A bushy, somewhat fleshy perennial to 225mm, in the *Urticaceae* from tropical America. It has tiny, crowded, obovate leaves and minute flowers, the stamens of which explode when touched, showering pollen in all directions.

▲ **ARUM LILY** *(Zantedeschia aethiopica* or *Richardia aethiopica)*. Also known as calla lily, this robust member of the *Araceae* has SAGITTATE leaves and stems to 900mm, with white, funnel-shaped SPATHES 100–150mm long.

ARUNDINACE|US, -A, -UM Like a reed. From ARUNDO.

ARUNDO A genus of 12 species of reeds in the *Gramineae* from the tropics and sub-tropics. They are mainly clump-forming plants with arching, narrow leaves and the flowering spikelets in large PANICLES: *A. donax* has broad, grey-green leaves and woody stems to 3.6m. The latter are used as walking sticks and fishing rods.

ARVENS|IS, -E Growing in cultivated fields.

ASAFOETIDA *Ferula asafoetida*, a 1.8–3.6m tall perennial in the *Umbelliferae* from S.W. Asia. It has finely divided leaves and UMBELS of small, greenish-yellow flowers. The roots yield gum asafoetida, a medicinal substance. *See also* FENNEL.

ASARABACCA *Asarum europeum*, a creeping, evergreen plant in the *Aristolochiaceae* from Europe. It has glossy, kidney-shaped leaves and solitary bell-shaped, 13mm long greenish-brown flowers with three triangular PERIANTH lobes.

ASCOMYCETES A class of fungi with SEPTATE HYPHAE typified by the mildews and *Penicillium*. The fruiting bodies are spherical or flask-shaped (PERITHECIA), or disk or cup-shaped (APOTHECIA), both of which carry spore-bearing ASCI.

ASCUS [ASCI] A SPORANGIA characteristic of the ASCOMYCETES. Asci are large, ovoid cells known as sacs produced by the fertile HYPHAE of certain mildews and moulds. Each one contains eight SPORES which are forcefully ejected when mature.

■ **ASH** A general name for all 70 species of *Fraxinus*. This genus of deciduous trees and shrubs from the northern temperate zone has PINNATE leaves in opposite pairs. A

● 

▲ 

group of species (Ornus or flowering ashes) has flowers bearing linear petals which are carried with the young leaves. The remaining species are without petals, the flowers appearing before the leaves. The flattened or spindle-shaped fruits (SAMARAS) are winged. Many ashes are good timber trees and some are grown as ornamentals. *F. americana* (white ash) grows to 30m, the leaves having 5–9 LANCEOLATE leaflets. The flowers are APETALOUS and the fruits 38–50mm long; *F. excelsior* (common ash) grows to 36m, the leaves with 7–11 leaflets. The flowers are apetalous. *F. ornus* (flowering or manna ash) grows to 15m, the leaves with 5–7 oblong leaflets. The scented flowers have white petals.

ASPARAGUS A genus of 300 species of perennials, some climbing or scrambling, in the *Liliaceae* from Europe and Asia. The apparently linear to ovate leaves are PHYLLOCLADES, the true leaves reduced to scales. The tiny flowers are bell-shaped or starry, the fruits berry-like. Several species are grown as vegetables or ornamentals: *A. medeoloides*, the florist's smilax, is a slender climber to 3m with ovate phylloclades; *A. officinalis* (common asparagus), an erect tufted plant with FILIFORM phylloclades, bell-shaped green-white flowers and red, pea-like berries. The young shoots are cooked and eaten; *A. plumosus* (syn. *A. setaceus*, asparagus fern) is tufted when young, with bristle-like phylloclades forming frond-like branchlets; a climber when older to 3m; *A. sprengeri* (now *A. densiflorus* 'Sprengeri') is a trailing species to 1.5m with linear phylloclades 25–38mm long, white starry flowers and red fruits.

ASPEN
see POPLAR

ASPER Rough to the touch.

● **ASPHODEL** Several different plants are grown under this name. The 12 species of *Asphodelus* have basal tufts of grassy leaves and branching racemes of 6-petalled, starry white or pink flowers. *Asphodeline lutea* is tufted, the 0.9–1.2m erect stems densely linear leafy and carrying yellow flowers. *Narthecium ossifragum* (bog asphodel) is a 100–300mm iris-like tufted plant with yellow starry flowers. *Tofieldia* (false or Scottish asphodel), is a 50–150mm iris-like plant with greenish-white flowers. All are in the *Liliaceae*.

ASPIDISTRA A genus of 8 species of evergreen perennials in the *Liliaceae* from E. Asia. They have stalked, leathery elliptic to lanceolate leaves and SESSILE erect, bell-shaped flowers, each with an umbrella-shaped style forming a lid. One species is popular as a house plant: *A. lurida* (syn. *A. elatior*), having 300–500mm long oblong to lanceolate leaves and dull purple flowers.

▲ **ASTER** A genus of 500 perennial and sub-shrubby species in the *Compositae* from the northern hemisphere and Africa. They have linear to broadly ovate entire leaves and starry, daisy-like flower heads either large and solitary or small and in PANICLES. Several are grown as ornamentals: *A. alpinus*, a small mat-forming plant with LANCEOLATE to SPATHULATE leaves and solitary, 25–50mm wide purple daisies on 140mm stems; *A. acris* a tufted, bushy species to

■ 

○ 

600mm with large, dense panicles of bright, pale, blue-purple flowers. *See also* MICHAELMAS DAISY; CHINA ASTER.

ASTERIAS Star-like.

ASTEROIDES Aster-like.

ASTILBE A genus of 25 perennial species in the *Saxifragaceae* from E. Asia and N. America. They have BI- or TRITERNATE leaves and plume-shaped panicles crowded with tiny florets. Several species and hybrids are grown as ornamentals, favouring moist soil. *A. X arendsii (A. X davidii X chinensis X japonica)* covers all the common cultivated varieties with flowers from white and pink to deep red.

■ **ASTRANTIA** A genus of 10 perennial species in the *Umbelliferae* from Europe and W. Asia. They have long stalked PALMATE or TRIFOLIATE leaves and erect, branched stems bearing rounded UMBELS of small flowers. Each umbel is surrounded by colourful petal-like bracts. A few species are cultivated as ornamentals. *A. major*, 450–600mm clump-forming plant with five lobed leaves and pinkish flowers, having white or pink flushed bracts.

ASYMMETRICAL Not symmetrical, e.g. the leaves of elm having one side larger than the other.

ATRAT|US, -A, -UM Blackish or blackened.

ATROPURPURE|US, -A, -UM Dark purple.

ATROVIRENS Rich or dark green.

ATTAR OF ROSES A fragrant oil distilled from the petals of musk and damask roses. Also apple geranium *(Pelargonium odoratissimum)*.

○ **AUBERGINE** *(Solanum melongena)* in the *Solanaceae* from tropical India. Also known as egg plant, this annual species is grown for its large egg-shaped purple or white fruits. The plant is a 0.6–1.2m annual, with 75–150mm long ovate, shallowly lobed leaves and 25mm wide, purple-blue ROTATE flowers.

AUBRIETA A genus of 15 evergreen, mat-forming plants in the *Cruciferae* from the mountains of Italy to Persia. The leaves are ovate to oblong, sometimes toothed, and the 4-petalled flowers are carried in RACEMES. One species is commonly grown as an ornamental: *A. deltoidea* (common aubrieta) has OBOVATE, toothed leaves and lilac to red-purple flowers on 100mm stems.

AUGUST|US, -A, -UM Notable, majestic, stately, tall.

AURANTIAC|US, -A, -UM Orange.

AURE|US, -A, -UM Golden-yellow.

AURICLE An ear-shaped organ or appendage, e.g. small rounded lobes at the base of a leaf.

AUSTRAL|IS, -E Southern, or south of.

AUTOECIOUS A term describing a rust fungus that has all its different sporing stages on the same plant.

● 

■ 

AUTOPOLYPLOID
see POLYPLOID

AUTOTROPHIC The normal means of plant nutrition: water, minerals in the soil, carbon dioxide, oxygen and PHOTOSYNTHESIS via the sun's energy. *See also* HETEROTROPHIC.

AUXANOMETER An apparatus for measuring the growth in height of a stem. A thread is attached to the growing point and to the base of a pivoted pointer, the tip of which moves down a graduated scale as the stem lengthens.

AVELLANA An old generic name for hazel.

● **AVENS** A name applied to a few species of *Geum* and *Dryas*, both in the *Rosaceae* Wood avens *(Geum urbanum)*, also known as herb bennet, is perennial to 600mm with basal PINNATE leaves and erect, 13mm wide 5-petalled, yellow flowers; Water avens *(G. rivale)* is about 450mm with nodding orange-pink flowers 19mm across. The seeds of all geums are ACHENES, each one bearing a stalked hook which aids distribution by animals. Mountain avens *(Dryas octopetala)* is a mat-forming species from Arctic regions and mountains further south, with 13–19mm long ovate, CRENATE leaves. The 7–10 petalled flowers are followed by long AWNED wind-dispersed achenes.

AVOCADO PEAR The fruit of *Persea americana* (syn. *P. gratissima*), an evergreen tree to 20m in the *Lauraceae* from Central America. It has elliptic, 100–150mm long leaves and 13mm wide green flowers. The pear-shaped fruits, also known as alligator pears, are 75–125mm long, yellow to deep green, sometimes flushed purple or brown.

AWLWORT *(Subularia aquatica)*. A small aquatic annual in the *Cruciferae* from the northern temperate zone. The leaves are SUBULATE, 25–63mm long and the 4-petalled, white flowers are in short RACEMES.

AWN A bristle-like projection, often hairy, that projects from the tips or backs of the lemmas of grasses, the tips of ACHENES or SEPALS, e.g. *Geranium*.

AXIL The point and angle where the leaf stalk joins the stem or a smaller stem joins a larger.

AXILLAR|IS, -E Borne in the leaf AXILS.

AXILLARY Growing out from leaf axils, the usual points of origin of lateral branches.

AZALEA
see RHODODENDRON

AZURE|US, -A, -UM Sky-blue or bright blue.

▲ **BABY BLUE EYES** *(Nemophila menziesii* syn. *N. insignis)*. An annual plant in the *Hydrophyllaceae* from California, sometimes known as Californian bluebell. It is a bushy, spreading plant to 100mm tall with narrow, PINNATIFID leaves and bowl-shaped sky-blue flowers 13–25mm across. White, purple, veined and spotted flowered forms occur.

▲

BACCAT|US, -A, -UM Having berries or berry-like fruits.

■ **BACHELOR'S BUTTONS** Applied to the pompon-like double-flowered forms of many plants, e.g. buttercup, campion, columbine, cornflower, herb robert, Jew's mallow, marsh marigold, periwinkle and scabious. In addition, the seed heads of burdock are so called.

BACILLUS Rod-shaped bacteria.

BACTERIA A group of single-celled fungi which multiply by splitting in two (binary fission). Some bacteria are parasitic and harmful disease organisms, e.g. tuberculosis, lock-jaw, typhus, cholera. Most of them are harmless SAPROPHYTES, and a few manufacture their own food substances using ammonia, sulphuretted hydrogen or iron as energy sources instead of sunlight. The various sorts of bacteria vary greatly in size and shape, being spherical, rod shaped or spiral. Some species live in chains, others in groups or packets.

BACTERIOPHAGE Extremely minute organisms that attack and destroy bacteria.

BALEARIC|US, -A, -UM From the Balearic Is. in the Mediterranean.

BALLOON FLOWER *(Platycodon grandiflorus)*. A 300–450mm herbaceous perennial in the *Campanulaceae* from China and Japan. It has toothed, OVATE-LANCEOLATE leaves, and terminal, broadly bell-shaped 50mm wide, purple-blue flowers that are balloon-like in bud.

BALM At least three different plants are known as balm, all in the *Labiatae*. Common or lemon-scented balm *(Melissa officinalis)* is a herbaceous perennial from Europe forming spreading clumps of erect stems to 600 or 900mm with OVATE to CORDATE leaves that smell of lemons when bruised. Small white, tubular flowers are borne in the upper leaf AXILS; Bastard balm *(Melittis melissophyllum)* is a tufted, 300–600mm perennial from Europe with wrinkled, ovate leaves and tubular, two-lipped, white flowers, spotted with rose-purple, each about 38mm long; Balm of Gilead *(Cedronella triphylla)* is a strong, sweetly aromatic sub-shrub from the Canary Isles, 0.9–1.2m in height with TRIFOLIATE leaves and TERMINAL spike-like clusters of small white or purplish tubular flowers. *See also* bee balm, under BERGAMOT.

BALSA The very light, corky wood of *Ochroma lagopus* an 18–24mm tree in the *Bombacaceae* from C. America. It has PUBESCENT, broadly ovate-cordate leaves and angular brown flowers. The oblong seed pod yields a silky down used as a substitute for kapok.

BALSAM This name is used for two different groups of plants: several members of the genus *Impatiens*, and various species of trees which produce a mixture of volatile oils and resins that form the medicinal balsam. *Impatiens balsamina*, common balsam, is an erect, fleshy-stemmed annual in the *Balsaminaceae* from S.E. Asia. It has oblong ovate, deeply toothed leaves and large ZYGOMORPHIC red flowers having rounded STANDARD and WING petals and an incurved SPUR. Fully double CULTIVARS in a variety of colours are frequently grown. For Canada balsam *see* FIR; for balsam poplar, *see* POPLAR.

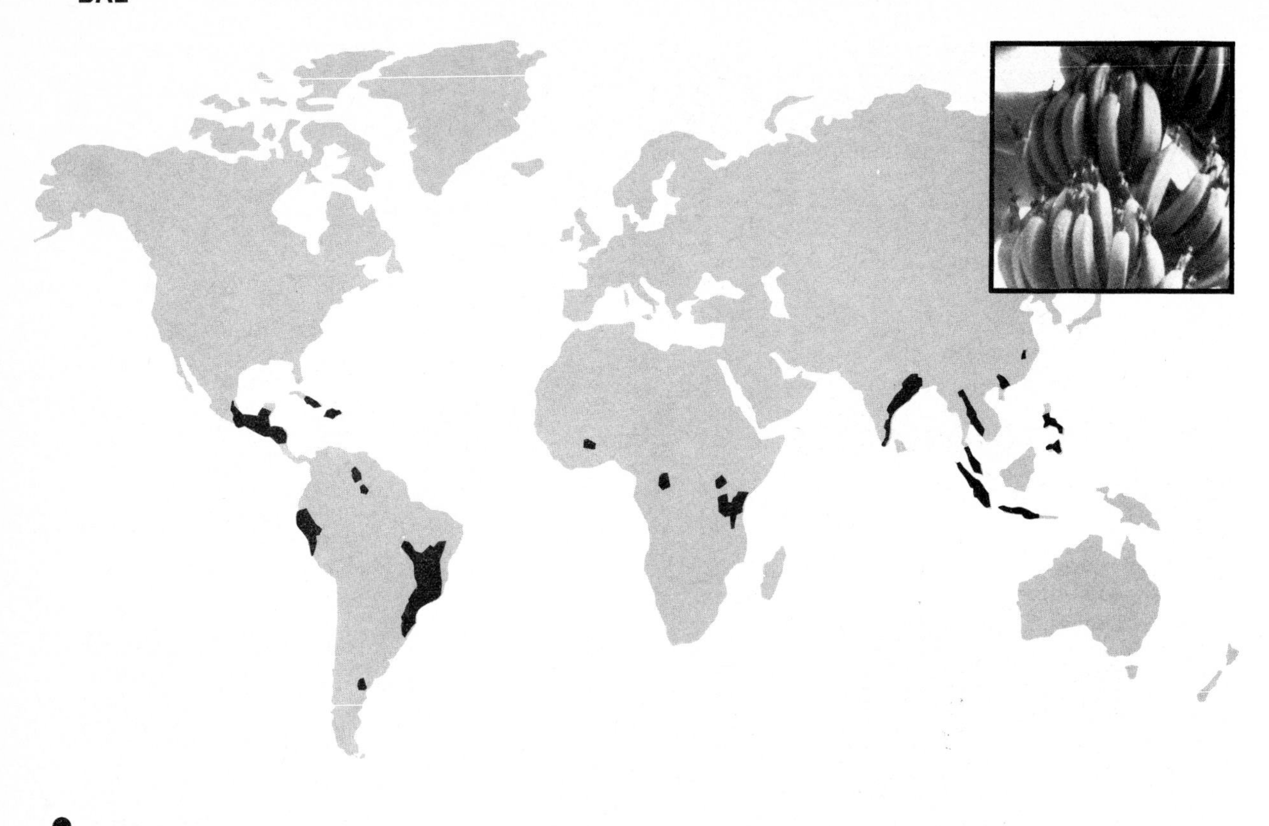

BALSAM APPLE *(Momordica balsamina)*. A climbing plant to 3m in the *Cucurbitaceae*, having PALMATE, deeply lobed, smooth leaves and brown-spotted yellow flowers which are followed by lantern-like, warted orange-yellow fruits to 75–100mm. These split suddenly when ripe to reveal oval seeds in fleshy, orange-red ARILS.

BALSAM PEAR *(Momordica charantia)*. Very similar to BALSAM APPLE, but larger in all its parts with hairy leaves and longer fruits, 100–200mm in length. The young fruits are the basis of Indian pickles.

BAMBOO A collective name for the stems of grasses in the *Bambuseae* group of the *Gramineae*. Bamboos are found around the world from temperate to tropical zones. They vary in height from under 1.2m in *Sasa* to over 30mm in *Dendrocalamus*. Some sorts are clump-forming, others spread by RHIZOMES into dense thickets. The stems are hollow between the joints and have a variety of uses: large ones are used for house building, piping and furniture, smaller ones for fishing rods and handles, while thin ones are woven into baskets, chair seats, etc. The leaves are basically LINEAR to OBLONG-LANCEOLATE, often attached to stem-embracing, sheathing stalks. The typical grass flowers are borne in small clusters to large PANICLES. Many bamboos flower very erratically, often only at intervals of several years, the stem that bears them dying after the seeds have ripened.

● **BANANA** The fruits of several species of *Musa*, large to gigantic perennial plants in the *Musaceae*. The bananas of commerce known as '*M. paradisiaca*' or '*M. paradisiaca sapientum*' are now known to be hybrids between *M. acuminata* and *M. balbisiana* from S.E. Asia. Many cultivars are grown in the tropics, but only a few are exported, the commonest in European shops being 'Gros Michel'. The dwarf Canary banana is a dwarf mutant of *M. acuminata* called 'Dwarf Cavendish'. The so-called banana tree is composed of 2.4m, oblong leaves, the broad, flattened stalks of which wrap closely around each other to form the 'trunk'. The tubular flowers comprise 3 sepals and 2 petals fused into a tube which splits when mature revealing a solitary petal and 5 stamens. They are borne in pendulous spikes, each flower cluster covered by a large, leathery, reddish bract. When the flower is pollinated, the young fruit bend upwards, each group finally forming a hand of bananas.

BANEBERRY *(Actaea spicata)*. A herbaceous perennial in the *Ranunculaceae* from N. Europe to Japan. It is clump-forming with BITERNATE or bipinnate leaves and erect stems to 600mm. Small whitish flowers are carried in erect TERMINAL SPIKES. Each floret has a single CARPEL which turns into a black, fleshy berry. There are 10 species of *Actaea* all from the cooler areas of the N. temperate zone: *A. pachypoda* has white berries, *A. rubra* red ones.

BANYAN *(Ficus benghalensis)*. A large, evergreen tree in the *Moraceae* from tropical India and Africa. The branches grow horizontally and bear aerial roots at intervals. These root into the ground and become trunk-like, acting as a support and supplementing the water and food supply. By such means one tree can become a small forest. The leaves are OVATE and the cherry-sized figs red.

▲

▲ **BAOBAB** *(Adansonia digitata)*. A deciduous tree in the *Bombacaceae* from arid regions of Central Africa. It has a bulbous trunk that stores water, and short stubby branches. Large specimens develop enormous trunks to 9m in diameter with a height of 9.0–12m. The leaves are PALMATE, deeply lobed and the 200–250mm long gourd-like fruits contain and edible floury pulp known as monkey bread.

BARBADOS GOOSEBERRY *(Peteskia aculeata)*. This is a climbing member of the *Cactaceae* with slender, woody, spiny stems and LANCEOLATE to OVATE fleshy leaves. White, pale yellow or pink flowers, 25mm across, are carried in large TERMINAL clusters and are followed by pale yellow, gooseberry-like fruits.

BARBADOS PRIDE *(Caesalpinia pulcherrima)*. A somewhat prickly shrub to 3.6m in the *Leguminosae* from the tropics. It has bipinnate leaves of 13mm long, OBLONG-OBOVATE leaflets and terminal RACEMES of 50mm wide, red, yellow-margined flowers with long red stamens.

BARBAT|US, -A, -UM Bearded with long tufts of hairs.

BARBERRY A name used for all members of the genus *Berberis*, which comprises 450 species of spiny shrubs. These are deciduous or evergreen from the north temperate zone and S. America. They are characterised by having LONG AND SHORT SHOOTS, TRIFID spines at each NODE, and yellow roots. The leaves vary from LINEAR-LANCEOLATE to OVATE, usually spiny margined. The flowers of the entire genus are yellow, orange or red tinted and they have sensitive stamens. When the tongue of an insect touches the base of a stamen it closes in rapidly, coating it with pollen. The berry fruits range from round to shortly cylindrical, black or red. Several species are cultivated as ornamentals: *B. darwinii*, a 2.4m evergreen species from Chile with small holly-like leaves, abundant orange flowers and blue-black fruits; *B. thunbergii*, a 1.2–1.8m deciduous species from Japan, with spineless obovate leaves, red flushed yellow flowers and waxy, red fruits; *B.* 'Atropurpurea' has purple leaves; *B. vulgaris* (common barberry), a 2.4m deciduous shrub from Europe, including Britain, with obovate, somewhat spiny leaves, yellow flowers and dull red fruits.

BARK The corky outer layer of the stems of woody plants, particularly trees. As the trunk of a tree expands, the outer TISSUES stretch and crack in characteristic patterns. In some trees the outer layer flakes off (e.g. plane), or peels in thin strips (e.g. birch), or it may become deeply fissured (e.g. elm and oak). As a response to this natural wounding, a cork CAMBIUM forms in the outer layers of the cortex to provide a protective, waterproof layer. In some trees this cambium functions for several years building up deep layers of bark (as in cork oak), but quite often the layer dies out and is replaced by another nearer to the surface. The barks of some trees have medicinal properties; *see* QUININE; WINTER'S BARK.

BARLEY Cereal-producing plants in the *Gramineae*. Two species of *Hordeum* are involved, both tufted, annual grasses with slender leaves and cylindrical flower heads with the long AWNED SPIKELETS arranged in groups of three: *H. distichon* (two row barley) has two rows of spikelets to each head, the centre flower only of each group producing a grain; *H. vulgare* (six row barley) has all three flowers producing grains.

BARLEY GRASS Various wild grasses in the genus *Hordeum*, mainly from Europe, Asia and N. America. They are tufted annuals or perennials of typical grassy appearance, the cylindrical flower spikes carrying many flexible AWNS and creating a squirrel's-tail effect. *H. murinum* (wall barley) is a common annual plant of roadsides and waste ground, with 300–600mm tall stems and green, bristly spikes of 100mm long. *H. jubatum* has arching heads to 150mm with silky, red-tinted awns.

BARREN Descriptive of plants that do not produce fruits or seeds, e.g. solitary plants of a DIOECIOUS species that do not have a male or female plant nearby to effect fertilisation.

● **BARTRAM, JOHN [1699–1777]** Born in Pennsylvania, a farmer and Quaker, he began collecting plants and seeds for customers in England in the 1730s, eventually devoting most of his time to the work. He was completely untrained in botany, but travelled widely in the eastern states of America as far south as Florida and in the Appalachians, Catskill and Blue Ridge mountains, collecting botanical and other natural history specimens. He created his own botanic garden in Philadelphia and among the many plants introduced into Europe by him are American laurel *(Rhododendron maximum)*, bergamot *(Monarda didyma)*, bugbane *(Cimicifuga racemosa)*, several lilies including *L. superbum* and the ostrich fern *(Matteucia struthiopteris)*. His son William was also a botanist of note.

BARTSIA A genus of 30 species of annuals and perennials in the *Scrophulariaceae* from the north temperate zone and tropical mountains. They are erect, bushy plants with OPPOSITE pairs of LANCEOLATE to OVATE toothed leaves and TERMINAL spikes of tubular, two-lipped flowers. They are partially parasitic on the roots of grasses. The common red bartsia is now placed in the genus *Odontites*. *B. alpina* is the most widespread. It is a perennial, 100–200mm tall with 19mm long purple flowers.

BASIDIA [BASIDIUM] Peg-like bodies growing from the hymenium of a basidiomycete fungus and which bear the basidiospores.

▲ **BASIDIOMYCETES** The class of fungi to which mushrooms and toadstools belong, these are the fruiting bodies of the fungus and typically comprise a stalk or STIPE and a cap or PILEUS. The underside of the cap has evenly spaced plates of tissue called GILLS (as in the mushroom and most toadstools) or angular pores (as in cep or *Boletus*). The surface of the gills or pores bears a fertile layer or hymenium which produces basidiospores.

BASIDIOSPORE The externally produced spores of the *Basidiomycetes*.

BASIFIXED ANTHERS ANTHER LOBES attached by their bases to the FILAMENT.

BASIL [SWEET BASIL] *(Ocimum basilicum)*. An annual plant in the *Labiatae* from tropical Africa and Asia with erect stems to 450mm, OVATE leaves and white tubular flowers with a four-lobed lip. The plant is used for flavouring food, and the extracted oil in perfumery and medicine.

BASS The inner, fibrous bark of various trees, particularly *Tilia* (lime). It is used as a twine for tying.

BASSWOOD
see LIME

BAST Now best known as PHLOEM.

BASTARD TOADFLAX *(Thesium humifusum)*. A slender, prostrate, semi-parasitic perennial in the *Santalaceae*, from south Britain to France and Spain. The yellow-green leaves are LINEAR, about 13mm long, and the minute, yellowish flowers are carried in CYMES.

BAUHINIA A genus of 300 species of woody climbers, trees and shrubs in the *Leguminosae*, named after John and Caspar Bauhin, botanists of the 16th century. Widely distributed over the tropics, where the climbing species provide some of the jungle LIANAS. These are noteworthy in having curiously flattened stems with the strengthening tissue down either side. The leaves are characteristically bilobed, in some cases appearing as if two OVATE leaves were fused side by side. The flowers are often large and showy, with stalked petals of differing sizes lending an orchid-like quality. Several species are cultivated as ornamentals: *B. variegata* (ebony wood) is a small tree with pink flowers, variegated red, purple or white, yellow in the centre. The bark is used for tanning and dying, and the flower buds and leaves for vegetables in India; *B. purpurea* (orchid tree) is a 1.8m shrub with flowers in shades of purple.

BAYBERRY
see SWEET GALE

BAY LAUREL *(Laurus nobilis)*. An evergreen tree in the *Lauraceae* from Mediterranean region, having aromatic, LANCEOLATE, leathery leaves used for flavouring. Pale yellowish flowers are borne in clusters in the leaf AXILS and are followed by black, OVOID fleshy fruits. Also known as sweet bay, the leaves were used to garland heroes in ancient Greece.

BEACH GRASS
see MARRAM GRASS

BEACH PEA
see SEA PEA

BEAD TREE
see MELIA

BEAN Many plants other than true beans go under this name, but most of them are members of the bean and pea family, the *Leguminosae*, and have edible seeds. *See* BROAD, BUTTER, HARICOT, KIDNEY, RUNNER, SOY beans; also BOG BEAN, which gets its name from the leaf shape.

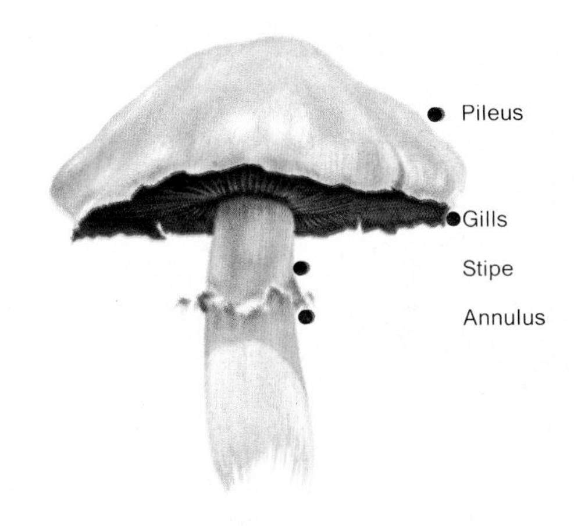

BEARBERRY *(Arctostaphylos uva-ursi)*. A prostrate mat-forming evergreen shrub in the *Ericaceae* from the north temperate zone. The 13–19mm leaves are OBOVATE-ELLIPTIC and the 9mm GLOBOSE bell flowers are white, tinged with pink and pendent.

BEARSFOOT
see HELLEBORE

BEDSTRAW Applied to several species of *Galium*, a genus of 400 species of cosmopolitan distribution in the *Rubiaceae*. They are annuals and perennials of slender growth with the LINEAR to LANCEOLATE leaves borne in WHORLS of 4–10 at each NODE. The stems are often sharply 4-angled and the tiny, tubular flowers have 4 spreading, pointed lobes. *G. mollugo* (hedge bedstraw) has a flexible stem pushing up through shrubs to 1.8m or more. The leaves are LINEAR to OBOVATE in whorls of 6–8 and the flowers white in a large, loose TERMINAL PANICLE. *G. verum* (lady's bedstraw) has linear leaves in whorls of 8–12 on reclining stems to 900mm or so; the flowers are bright yellow and fragrant. *See also* CLEAVERS (goose-grass), CROSSWORT, WOODRUFF.

BEECH Used of all 10 species of *Fagus*, deciduous timber trees in the *Fagaceae* from the north temperate zone and Mexico. They have ovate, often toothed leaves, and uni-sexual flowers, the tiny male FLORETS aggregated into rounded, pendulous heads. The larger female florets are in twos and threes, enclosed by four bracts that later become woody and enclose the triangular nutlets (MAST). *F. grandifolia* (American beech) is a 24.4m tree with 50–125mm long ACUMINATE toothed leaves; *F. sylvatica* (common beech) can attain 36.5m, with a smooth grey trunk and 50–75mm long, broadly OVATE leaves.

BEECH FERN *(Thelypteris phegopteris)*. A small de-ciduous fern in the *Thelypteridaceae* from the cooler, northern temperate zone. It has triangular PINNATE FRONDS 100–300mm high, with each PINNA deeply lobed. The tiny SORI are carried close to the margins of the pinnae.

BEE FLOWERS [BEE PLANTS] The flowers of certain plants produce abundant nectar or pollen and are good for bees, e.g. broad and field beans, blackberry, buck-wheat, clovers, dandelion, hawthorn, heather, lime *(Tilia)* mustard, sainfoin, sycamore *(Acer)*, willow-herb.

BEEFSTEAK FUNGUS *(Fistulina hepatica)*. A BASIDIO-MYCETE fungus attacking living trees, mainly ash and oak. The hoof-shaped fruiting bodies are up to 300mm wide and thrust out horizontally from the trunk. They are slimy, congealed, blood-red above and yellowish beneath with a red juice.

BEE ORCHID *(Ophrys apifera)*. A tuberous rooted, ground-dwelling species in the *Orchidaceae* from Europe and N. Africa. It has 38–75mm long elliptic-oval leaves and 150–450mm tall stems carrying 2–10, 25mm wide flowers. Each bloom has 3 broad, spreading pink or white sepals and a 13mm long semi-globose, velvety purple-brown LABELLUM like the body of a bumble bee.

BEETROOT Cultivated forms of wild beet *(Beta vulgaris)* with swollen roots and rounded or oval leaves, often waved and puckered. Garden beetroot have red, rarely yellow-fleshed roots that can be spherical, oval-oblong or long and tapered. The tall, branched flowering stems bear many globular clusters of 2–4 tiny green flowers. Later, each cluster enlarges and then turns brown and wrinkled, containing 1–4 seeds. *See also* SUGAR BEET, SPINACH BEET.

● **BEGONIA** A genus of about nine hundred species of mainly perennial plants in the *Begoniaceae* from the tropics and sub-tropics, particularly central and S. America. Some species have tuberous roots, others RHIZOMES; some are climbers or tall shrubs, others tufted and low growing. Most species are typified by having lop-sided, ear-shaped leaves, sometimes spotted or patterned with white, pink or red. The flowers are unisexual, the males often with two large and two small petals. The females have 4–5 more even-sized petals sitting on top of large and often winged and coloured OVARIES. Male and female flowers appear together in frequently pendent PANICLES. Many species are cultivated as ornamentals: *B. bolivensis* is tuberous rooted, about 600mm tall with LANCEOLATE leaves and scarlet, drooping fuchsia-like flowers; double-flowered HYBRIDS are commonly grown; *B. coccinea* has cane-like stems to 1.8m or more and large panicles of red flowers; *B. rex* has thick creeping rhizomes and large quilted leaves patterned red, pink, white and silver; *B. semperflorens* is fibrous rooted and bushy with rounded leaves, often copper or purple flushed. It has abundant small flowers in shades of white, pink and red, and is commonly grown as a summer annual.

BELLADONNA LILY
see AMARYLLIS

BELLBINE
see BINDWEED

▲ **BELL-FLOWER** *(Campanula)*. A genus of 300 species of annuals, perennials and shrubs in the *Campanulaceae* from the northern hemisphere. Most of them are tufted or clump-forming with ORBICULAR to LINEAR leaves and bell-shaped flowers, though some have starry blooms with widely expanded, pointed lobes. Bearded bell-flower, *(C. barbata)* is a perennial to 450mm with LANCEOLATE basal leaves and one-sided RACEMES of lilac-blue, nodding bells with a floss of white hairs in the mouth: Canterbury bell, *(C. medium)* is a biennial to 900mm with a coarsely hairy rosette of lanceolate 100–150mm long leaves and wide, 50mm long bells in shades of blue-purple, pink and white; peach-leaved bell-flower *(C. persicifolia)* is perennial with smooth, peach-like leaves and bowl-shaped flowers of lavender-blue or white; clustered bell-flower *(C. glomerata)* is perennial with narrow purple flowers in erect terminal clusters on 150–600mm stems.

BELL|US, -A, -UM Beautiful, handsome.

BENJAMINA From the Indian name benyan (banyan). *Ficus benjamina* is allied to the true banyan.

BENT GRASS Several kinds of *Agrostis*, a genus in the *Gramineae* containing at least 150 species from the temperate regions of the world. They are tufted in habit or

creep by means of rhizomes. The tiny flowering spikelets form loose or dense PANICLES on erect, wiry stems. *A. canina* (brown bent) may be tufted or RHIZOMATOUS with stems 300–600mm long, leaves with long, pointed LIGULES and loose panicles of shortly AWNED spikelets. *A. tenuis* (common bent) is similar but the spikelets are without visible awns and the leaves have collar-like ligules.

BERGAMOT Applied in a general way to several sorts of *Monarda*, a genus from N. America in the *Labiatae*. They are aromatic, clump-forming perennials with OVATE toothed leaves and crowded WHORLS of slender, tubular, hooded flowers: *M. fistulosa* (wild bergamot) has purple flowers on 600–900mm stems; *M. didyma* (oswego tea or bee balm) has larger, scarlet flowers. *See* bergamot mint under MINT, bergamot oil under CITRUS.

BERMUDA BUTTERCUP
see OXALIS

BERMUDA GRASS *(Cynodon dactylon)*. A mat-forming species in the *Gramineae* from tropical and warm temperate regions. It is fast growing with short, stiff leaves and inflorescences like 3–5 fingered hands, each finger a spike of crowded spikelets.

BERRY A fleshy or SUCCULENT fruit formed by the fusion of two or more CARPELS and containing a number of seeds, e.g. tomato, gooseberry and cranberry.

BETELNUT The seed of *Areca catechu*; *see* PALM.

■ **BETONY** *(Betonica officinalis)*. A herbaceous perennial in the *Labiatae* from Europe. It is clump-forming with short, woody RHIZOMES and erect stems carrying pairs of OVATE-OBLONG leaves with large rounded teeth (CRENATE). The 13–19mm long tubular flowers are two-lipped, bright red-purple, borne in clusters forming an oblong spike.

BETULIN|US, -A, -UM Like the birch *(Betula)*.

BICOLLATERAL BUNDLE
see VASCULAR BUNDLE

BIDENTATE Having two teeth.

BIENNIAL [BIENN|IS, -E] Plants which require 2 growing seasons to complete their life cycle, then die; i.e. a leafy plant forms during the first year, then flowers are formed and seeds the following year.

BIFID Cleft into two parts or lobes.

BIFURCATE Forked into two branches of about equal size.

○ **BIG TREE** *(Sequoiadendron giganteum)*; also Wellingtonia, or mammoth tree. An evergreen conifer in the *Taxodiaceae* from California. The largest coniferous tree (but not the tallest, *see* REDWOOD), with a trunk girth of up to 30m and 100m in height. The leaves are scale-like, spreading at the tips and 6mm long. Hard egg-shaped cones 38–50mm long hand from the shoot tips.

BILBERRY *(Vaccinium myrtillus)*. Blaeberry or whortleberry. A small deciduous shrub in the *Ericaceae* from the cooler parts of Europe and Asia. The leaves are OVATE, SERRULATE, on wiry, angled green twigs. Nodding, globular pale greenish-pink bell flowers give way to edible, black, pea-sized berries with a glaucous, waxy bloom.

BILLBERGIA A genus of 50 distinctive, evergreen perennials in the *Bromeliaceae* from Mexico to Brazil. They have LANCEOLATE to LINEAR leaves, often prickly toothed, arranged in narrow or tubular rosettes. The flowers are tubular with recurved petal tips borne with often colourful leaf-like bracts on a stem from the rosette centre. Several species are grown as ornamentals and make good house plants: *B. nutans* (angel's tears) is 300–375mm tall, forming tufts of tubular rosettes. The flower stem has pink bracts and terminates in pendulous flowers having reddish sepals and yellowish, blue-margined petals.

● **BINDWEED** Applied to several climbing plants, mainly species of *Convolvulus* and *Calystegia* in the *Convolvulaceae*. All are twining climbers with deeply buried, white RHIZOMES and ovate cordate to sagittate leaves. The flowers are funnel-shaped, white or pink and white: *Convolvulus arvensis* (bindweed or cornbine) has leaves 19–38mm long and widely funnel-shaped pink or whitish flowers about 19mm wide; *Calystegia sepium* (larger bindweed or bellbine) has leaves 75–150mm long and white or pink flowers 38–63mm wide; *Calystegia soldanella* (sea bindweed) grows by the sea and does not climb. The prostrate stems bear RENIFORM leaves and 25–38mm wide pink or pale purple flowers. *See also* BLACK BINDWEED.

BINOMIAL SYSTEM
see NOMENCLATURE

▲ **BIRCH** *(Betula)*. A genus of 60 deciduous trees in the *Betulaceae* from the northern temperate and arctic zones. They are graceful trees with usually slender and often pendulous branchlets. The leaves are mainly OVATE and toothed and the tiny, petal-less flowers have the sexes in separate catkins. The males are long and flexuous, usually yellowish, the females are much smaller and rigid, later enlarging to slender, cone-like structures. The tiny nutlets are winged. The bark is smooth and peeling, often white or yellowish. Several species are grown as ornamentals: *B. lutea* (yellow birch) is up to 30m with ovate-oblong leaves up to 100mm long; *B. papyrifera* (paper or canoe birch) is 18–21m with ovate leaves to 88mm long, and white bark; *B. pendula* (common or silver birch) is 15–18m with 25–50mm long, broadly ovate leaves and a white trunk.

BIRD CHERRY *(Prunus padus)*. A large shrub or deciduous tree to 14m in the *Rosaceae* from Europe, Asia and N. Africa. It has sharply toothed, 100mm long ELLIPTIC to OBOVATE leaves and 13mm white flowers in RACEMES. The pea-sized cherries are black and astringent.

■ **BIRD OF PARADISE FLOWER** *(Strelitzia reginae)*. A spectacular evergreen perennial in the *Musaceae* from South Africa. It is clump-forming with long-stalked, OBLONG-OVATE leaves to 900mm tall. The flowering stems are about the same height, bearing at the top a slender-pointed, boat-shaped bract, held at right angles. From the

base of this bract, erect purple-blue and orange flowers stand up, the overall effect being of a bird's crested head.

BIRD'S FOOT *(Ornithopus perpusillus)*. A small, prostrate annual in the *Leguminosae* from Europe. It has 19–32mm long PINNATE leaves with ELLIPTIC leaflets and terminal clusters of 3–6 red-veined, white pea flowers 4mm long. The slender pods are constricted between the seeds.

○ **BIRD'S FOOT TREFOIL** Several kinds of *Lotus*, a genus of 100 species of annuals, perennials and sub-shrubs in the *Leguminosae* from Europe, Asia, Africa and Australia. They have leaves of five, often obovate leaflets and AXILLARY stems bearing CYMES of yellow or red pea flowers. Pollen is shed into the tip of the KEEL petal and is ejected when an insect alights on it. *L. corniculatus* (also known as bacon and eggs) is a smooth prostrate perennial with heads of bright yellow, red streaked flowers and pods that stand out like birds' feet; *L. uliginosus* (greater birdsfoot) is erect to 300mm or more and usually PUBESCENT.

BIRD'S NEST ORCHID *(Neottia nidus-avis)*. A curious SAPROPHYTE in the *Orchidaceae* from Europe to Siberia. It has a mass of short, blunt roots like a bird's nest and erect, brownish stems to 300mm tall, covered with brownish scale leaves. The small flowers are of a similar colour with long LABELLUMS of dark brown. It grows in leaf mould beneath trees, especially beech.

BISERRATE [BISERRAT|US, -A, -UM] Doubly toothed, e.g. the larger teeth of a leaf again toothed or notched.

BISEXUAL
see HERMAPHRODITE

BITERNATE Twice TERNATE, e.g. a leaf with nine leaflets.

BITTER CRESS Several kinds of *Cardamine*, a genus of 150 species of annuals and perennials in the *Cruciferae* from the cooler parts of the world. They have basal tufts or rosettes of PINNATE leaves and RACEMES of small white or pink flowers. The SILIQUAE (seed pods) open suddenly when ripe, flinging out the seeds. *C. hirsuta* (hairy bitter cress) is a small annual weed 75–150mm tall with sparsely hairy leaves and tiny white flowers; *C. pratensis* (lady's smock or cuckoo flower) is a perennial with stems to 450mm with lilac flowers 13–19mm wide.

BITTERSWEET *(Solanum dulcamara)*. Also known as woody nightshade, this is a perennial, deciduous scrambler in the *Solanaceae* from Europe, Asia and N. Africa. It has OVATE leaves variously lobed or PINNATISECT at the base. The purple flowers have a rotate COROLLA and a central beak of yellow stamens. They are followed by pendulous clusters of translucent, OVOID red berries about 9mm long.

BLACKBERRY *(Rubus fruticosus)*. Often called bramble, this is a scrambling, thorny shrub in the *Rosaceae*, from Europe but naturalised elsewhere. It is an extremely variable plant and almost four hundred MICROSPECIES have been described. The leaves have 3–5 ovate or obovate leaflets with prickles on the veins. Whiter or pink 5-petalled flowers appear in terminal PANICLES. The typical black, glossy fruits are composed of many fleshy CARPELS.

BLACK BINDWEED *(Polygonum convolvulus)*. A prostrate or twining annual in the *Polygonaceae* from Europe, Asia and N. Africa, naturalised elsewhere. The leaves are OVATE, CORDATE to SAGITTATE, 25–50mm long. Tiny white flowers appear in small clusters and are followed by triangular seeds.

BLACK BOY
see GRASS TREE

BLACK BRYONY *(Tamus communis)*. A tuberous rooted, perennial climber in the *Dioscoriaceae* from Europe to the Middle East and N. Africa. It has strongly veined, broadly ovate, cordate leaves that taper to a fine point. The green flowers are 4mm across, the females in AXILLARY clusters, the males in RACEMES. The fruits are red, OVOID berries 13mm long.

BLACK GRASS *(Alopecurus myosuroides)*. Also known as slender foxtail, it is an annual member of the *Gramineae* from Europe and Asia, introduced in USA and elsewhere. It is tufted in habit, 300–750mm tall, with pointed leaf blades 75–150mm long. The flowering PANICLES are narrow and spike like 50–110mm long. A weed of cultivated land, very difficult to eradicate.

BLACK MUSTARD *(Brassica nigra)*. An annual plant in the *Cruciferae* from Europe, but much cultivated around the world for its seeds which produce the black mustard of commerce. It is a bristly, hairy plant to 600mm or more high with OVATE to LYRATE-PINNATIFID leaves and RACEMES of four petalled, bright yellow flowers about 13mm across.

BLACK SHIELDS *(Lecanora atra)*. A common LICHEN on rocks and walls especially near the sea. It forms pale grey, spreading crusts, the surface seamed with dark cracks. The spore-producing APOTHECIA (called shields by the early naturalists) are like black discs with thick rims.

BLACKTHORN *(Prunus spinosa)*. A deciduous thorny shrub or small tree to 4.5m, in the *Rosaceae* from Europe to Siberia. It has ELLIPTIC to OBLANCEOLATE leaves 19–38mm long with CRENATE-SERRATE margins. The white, 5-petalled flowers are about 13mm wide and borne before the leaves. They are followed by blue-black, astringent fruits 13–19mm long, known as sloes.

BLADDER CHERRY *(Physalis franchetii)*. A RHIZOMATOUS perennial in the *Solanaceae* from the Caucasus to China. It has erect, unbranched, 600mm stems with triangular-ovate 75–150mm long leaves and solitary white rotate flowers in the upper AXILS. Each cherry-like berry is entirely enclosed in an inflated, papery, orange-red CALYX 50–75mm long. The allied *P. alkekengi* is very similar but smaller in all its parts.

BLADDER FERN Several species of *Cystopteris*, ferns in the *Athyriaceae* of cosmopolitan distribution: *C. fragilis* (brittle bladder fern) is tufted, with 75–300mm long ovate-lanceolate bipinnate or tripinnate fronds. The sori are covered with convex, bladder-like INDUSIA; *Athyrium montana* (mountain bladder fern) is a spreading species with creeping RHIZOMES bearing 100–300mm long tri-pinnate leaves at intervals.

BLADDER NUT *(Staphylea pinnata)*. A deciduous large shrub or small tree to 4.5m in the *Staphyleaceae* from Europe and Turkey. It has 50–100mm PINNATE leaves of 5–7 OVATE leaflets and pendulous PANICLES of 13mm long, white flowers. The fruit is an inflated membranous CAPSULE.

BLADDER SENNA *(Colutea arborescens)*. A bushy, deciduous shrub about 3m tall in the *Leguminosae* from S. Europe and the Mediterranean. It has 75–150mm long pinnate leaves of 9–13, OBOVATE, notched leaflets. Bright yellow, 19mm long pea flowers are borne in AXILLARY RACEMES. The seed pods are inflated, bladder-like, often red flushed.

● **BLADDERWORT** *(Utricularia)*. Several species of aquatic, carnivorous plants in the *Lentibulariaceae*. Those mentioned here come from the north temperate zone, but most of the 120 species are tropical. They are submerged, rootless plants with finely divided leaves some of which bear small, bladder-like structures. Each bladder has a door with a sensitive trigger hair. Tiny water creatures such as *Daphnia* activate the trigger and are sucked into the rapidly expanded bladder. The products of their decay are utilised by the plant: *U. minor* (lesser bladderwort) has stems to 250mm long. Yellow, 8mm long ZYGOMORPHIC flowers are borne above the water on 50–150mm stems; *U. vulgaris* (greater bladderwort) has stems to 450mm, free-floating with numerous bladders. The 13–19mm long bright yellow flowers appear above the water on 100–200mm stems.

BLAEBERRY
see BILBERRY

BLAND|US, -A, -UM Pleasant, charming; not bitter.

BLANKET WEED
see SPIROGYRA

BLEEDING HEART *(Dicentra spectabilis* syn. *Dielytra spectabilis)*. Also known as Dutchman's breeches or lady in the bath, this is a fleshy-rooted perennial in the *Fumariaceae* from Siberia and Japan. The GLAUCOUS leaves are BITERNATE and from the arching stem tips hang heart-shaped, 25mm long rosy crimson flowers.

BLEWITS *(Tricholoma saevum* syn. *T. personatum, Lepista saeva)*. An edible fungus in the *Basidiomycetes* from the pastures of Europe and Asia. It has a blue-purple STIPE and a 63–138mm wide flattened, clay-coloured cap, often flushed blue-purple. The GILLS are flesh-coloured. Wood blewits *(T. nudum* syn. *Lepista nuda)* is similar in size and shape but is suffused purple to brownish-violet all over. It frequents woods.

BLIGHT Often used in a general way for fungi that cause wilting and death of foliage or whole plants. In a more specific sense it refers to *Phytophthora infestans*, a serious parasite of potatoes and tomatoes. This consists of long, colourless HYPHAE which grow into the leaf tissue and rob the cells. As a result the tissues become brown, then die. The hyphae can also attack potato tubers causing them to brown or rot. It freely produces CONIDIA which blow on to nearby plants spreading the disease. This fungus is native

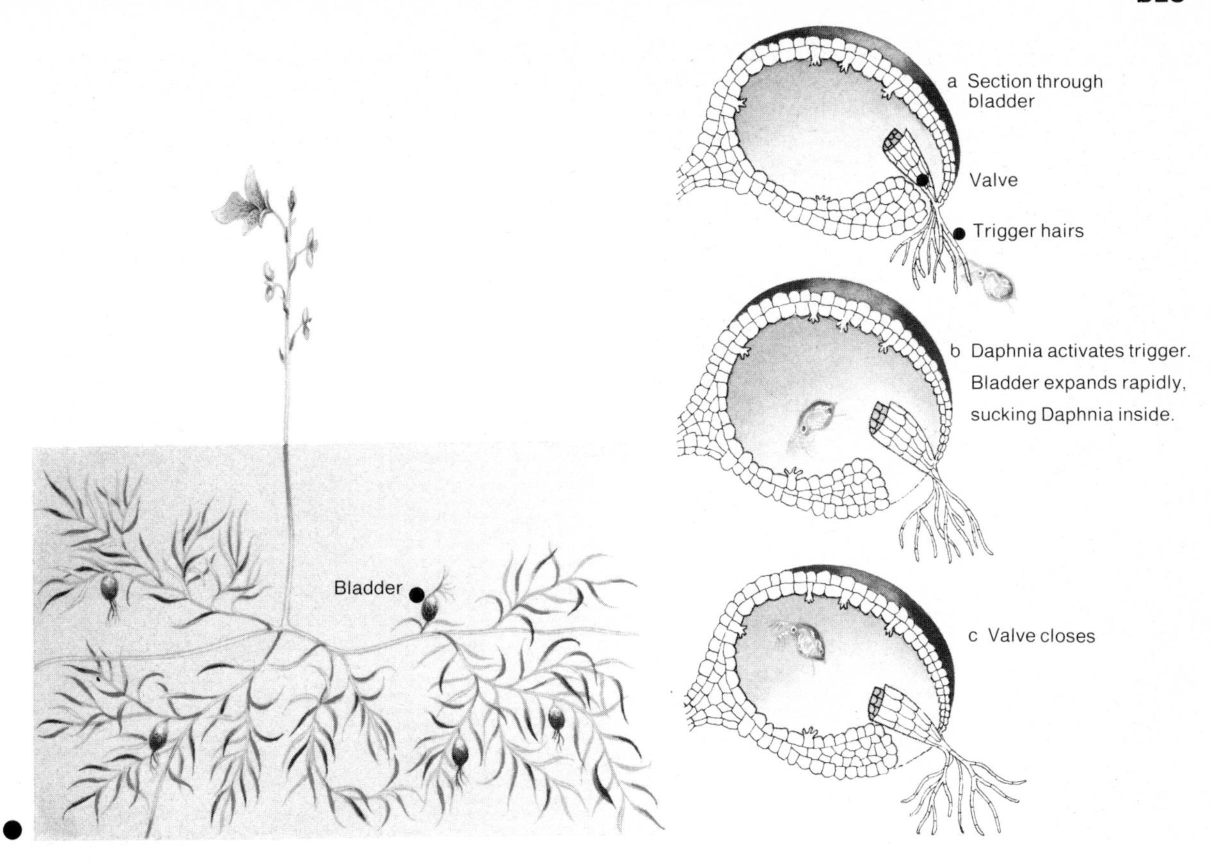

to N. America on wild species of *Solanum*.

BLINKS *(Montia fontana)*. A small amphibious annual or perennial in the *Portulacaceae* from temperate regions around the world. Stems vary from 25mm or so in land forms to 450mm under water. The SPATHULATE to OBOVATE leaves may only be 3mm in land forms, 19mm under water. The minute, white-petalled flowers are carried in terminal CYMES.

BLOOD LILY *(Haemanthus coccineus)*. A bulbous rooted perennial in the *Amaryllidaceae* from S. Africa. It has pairs of hairy, fleshy, tongue-like leaves to 450mm long, which lie flat on the ground. After they die away in summer, tulip-like inflorescences arise on red mottled stalks. The petals are really coloured BRACTS and surround a cluster of small florets with red stamens. Other species of *Haemanthus* are also sometimes called blood lily.

BLOOD ROOT *(Sanguinaria canadensis)*. Also called red puccoon, this is a herbaceous perennial in the *Papaveraceae* from N. America. It has fleshy RHIZOMES which exude a red sap. White, celandine-like flowers with 8–12 petals open on 100mm stalks as the leaves unfold. Later they expand to deeply lobed and scalloped grey-green blades on long stalks.

BLUEBELL In England and Wales, *Endymion non-scriptus*, a familiar bulbous plant in the *Liliaceae* from W. Europe. Also known as wild hyacinth it has strap-shaped leaves and nodding RACEMES of narrow, purple-blue bells. Spanish bluebell *(E. hispanica)* is similar, but with larger, wider bells. In Scotland, bluebell is *Campanula rotundifolia* (*see* HAREBELL).

BLUEBERRY Several species of *Vaccinium* often cultivated for their edible fruits, being deciduous shrubs in the *Ericaceae* from N. America: *V. angustifolium* (low bush blueberry) is a twiggy bush to 300mm in height with bristly SERRATE, LANCEOLATE leaves and narrowly bell-shaped flowers, white, streaked with red. The blue-black fruits are 8mm across; *V. corymbosum* (high bush or swamp blueberry) grows to 3m with elliptic leaves and OVOID pinkish bell flowers. The blue-black fruits are 8mm or more wide.

BLUE CUPIDONE *(Catananche caerulea)*. A rosette-forming, short-lived perennial in the *Compositae* from S. Europe. It has narrowly lanceolate, toothed leaves and wiry stems to 600mm or so bearing blue, scabious-like flower heads about 500mm wide.

BLUE-EYED GRASS *(Sisyrinchium angustifolium)*. A tufted, grass-like plant in the *Iridaceae* from N. America. It has narrow fans of flattened leaves like a slender iris and 6-petalled, 13mm wide starry, purple-blue flowers. The plant grown in gardens as *S. bermudiana* is the same species.

BLUE GRASS
see MEADOW GRASS

BLUE POPPY
see MECONOPSIS

BLUE WEED
see VIPER'S BUGLOSS

▲

■

BLUSHER *(Amanita rubescens)*. A fungus in the *Basidiomycetes*, from woodland in the temperate zone; edible when cooked. The 63–113mm wide cap is convex when young, expanding to almost flat. It is tan to reddish-brown with paler areas. When bruised or damaged, the flesh turns red. The STIPE is whitish with a skirt-like veil near the top.

BO TREE *(Ficus religiosa)*. An evergreen tree in the *Moraceae* from S.E. Asia and much venerated by the Buddhists and Hindus of India. It is about 8.5m tall, with broadly OVATE leaves drawn out to a very long, slender point (DRIP TIP). The fig-like inflorescences and purple fruits are borne in pairs.

● **BOG** A permanently wet area, the soil of which is acid and composed of partially decomposed plant remains, usually mosses. Bogs form in cool, humid climates where there is an excess of natural soil water, a lack of mineral salts, and poor aeration. Under such conditions dead plant remains do not decay properly and build up into layers known as peat. Characteristic plants are sphagnum mosses, heather, cross-leaved heath, purple moor grass, cotton grass and bog asphodel. *See also* FEN.

BOG BEAN *(Menyanthes trifoliata)*. A perennial bog or aquatic plant in the *Menyanthaceae* from the northern hemisphere. It has TRIFOLIATE leaves with OBOVATE to ELLIPTIC leaflets 38–63mm long. The tubular flowers are borne in RACEMES up to 300mm tall. Each flower is 13–19mm wide, pink in bud, opening white, the narrow petal lobed mossily covered and fringed with long hairs.

▲ **BOG MYRTLE** *(Myrica gale)*. Also called sweet gale. A deciduous shrub of wet soil in the *Myricaceae* from the cooler parts of the temperate zone. It has 19–38mm long OBLANCEOLATE, aromatic leaves and tiny petal-less DIOECIOUS, MONOECIOUS or HERMAPHRODITE flowers in short, stiff catkins.

BOG ROSEMARY
see ANDROMEDA

BOG RUSH *(Schoenus nigricans)*. A densely tufted grass-like perennial in the *Cyperaceae* from Europe, America and S. Africa. The very narrow, wiry leaves are overtopped by slender stems each tipped with a 13mm long, dense, OVOID head of pointed, black-brown spikelets.

BOG VIOLET
see BUTTERWORT; VIOLET

■ **BORAGE** *(Borago officinalis)*. A robust 300–600mm annual in the *Boraginaceae* from Europe. It has OBOVATE, bristly hairy leaves that smell of cucumber when bruised and sky blue, 19mm wide nodding, rotate flowers with pointed, reflexed lobes.

BOREAL|IS, -E Northern

▲ 

BOTRYOSE
see RACEMOSE

BOTRYOIDES Bunch-like; usually in the sense of a bunch of grapes.

BOTTLE BRUSH *(Callistemon)*. A genus of 25 species of evergreen shrubs in the *Myrtaceae* from Australia. They have LINEAR to LANCEOLATE leaves and dense spikes of small flowers with long red or yellow stamens like the bristles of a brush: *C. salignus* is a large shrub or small tree rarely to 9.1m, with yellow flowers and linear leaves; *C. speciosus* grows to 3m with rich scarlet brushes to 125mm.

BOTTLE GOURD
see CUCURBITS

BOUGAINVILLEA A genus of 18 species of shrubs, trees and climbers in the *Nyctaginaceae* from tropical S. America. The leaves are broadly lanceolate to ovate and the small tubular flowers are surrounded by brightly coloured, petal-like bracts. Some of the climbing species are much grown as ornamentals in greenhouses or in warmer countries: *B. glabra* has ovate leaves and ovate-cordate, rose-purple bracts in groups of 3. Crimson and orange hybrids are grown.

BOUTELOUA A genus of 40 distinctive species of annuals and perennials in the *Gramineae* from warm temperate and tropical America. They have typical grass foliage but the flowering spikelets are arranged in lateral arm-like spikes on the main stems, hence the name signal arm grass. They are also known as grama and some species are valuable as forage: *B. gracilis* (blue grama) is a densely tufted perennial 225–459mm tall with 1–3 signal arm spikes to each stem.

BOW WOOD
see OSAGE ORANGE

● **BOX** *(Buxus)*. A genus of 30 evergreen trees and shrubs in the *Buxaceae* from temperate Asia, W. Europe and America. They have opposite pairs of small, leathery leaves and axillary clusters of tiny, petal-less flowers: *B. sempervirens* ranges from 3–9m with angular twigs and OVATE to oblong leaves 9–25mm long. The flower clusters contain one female and several male flowers. The fruit is an OVOID, woody CAPSULE.

BRACHYPHYLL|US, -A, -UM Having short leaves.

BRACKEN *(Pteridium aquilinum)*. A fern in the *Dennstaedtiaceae* found throughout the world from the tropics to the Arctic circle. It has bipinnate fronds up to 1.8m or more which grow singly at intervals from deeply buried, branched RHIZOMES. Individual frond segments are OBLONG-LANCEOLATE, sometimes lobed, about 13mm long. The SORI forms a continuous band around these segments.

▲ **BRACKET FUNGI** A collective name for the fruiting bodies of certain species of fungi in the *Basidiomycetes.* They are PARASITIC or SAPROPHYTIC on trees, producing horse-shoe or bracket-like bodies at right angles to the trunk or branch. *See* BEEFSTEAK and SILVER LEAF fungi.

BRACT A modified leaf, usually associated with an inflorescence in the axils of which the flowers arise. Some bracts are scale-like and protective, others are brightly coloured and take the place of or supplement the petals.

BRACTEAT|US, -A, -UM Bearing bracts.

BRACTEOLE A little bract.

BRACTEOS|US, -A, -UM Having many bracts or large prominent ones.

BRAMBLE
see BLACKBERRY

BRANCH A lateral stem, e.g. one arising on the side of another and developing the same form.

BRANDY BOTTLE
see WATER LILY

BRASSICA A genus of about thirty species of annuals and perennials in the *Cruciferae* from Europe and Asia, particularly the Mediterranean region. They are often of rosette form with broad or PINNATIFID leaves and RACEMES of yellow or white, four-petalled flowers. Several food crops are members of this genus: BRUSSELS SPROUTS, CABBAGE, CAULIFLOWER, KALE, MUSTARD, RAPE, Sprouting Broccoli, SWEDE, TURNIP.

BRAZILIAN TEA
see MATÉ

BRAZIL NUT *(Bertholletia excelsa)*. An evergreen tree in the *Lecythidaceae* from S. America. It can attain 30m and more in height, with oblong-OVATE leather leaves and large, erect PANICLES of white flowers. The familiar brown nuts are botanically seeds, from 12–15 being neatly packed in a 100–150mm wide, woody CAPSULE like a small gourd. The nuts of commerce are still collected from wild trees in the Brazilian jungle.

■ **BREAD-FRUIT** *(Artocarpus altilis* syn. *A. incisa)*. An evergreen tree to 18.2m in the *Moraceae* from Malaysia and the Pacific Islands. It has 50–75mm long, leathery leaves, ovate and deeply cut into pointed lobes. The green flower clusters fuse together to form an OVOID, rough-skinned, multiple fruit 125–175mm long. This contains round, white seeds imbedded in pulp, though the cultivated sort is mainly seedless. Roasted, the pulp becomes like soft new bread.

BREADNUT *(Brosimum alicastrum)*. An evergreen shrub in the *Moraceae* from Jamaica. It grows to 1.8m, with OVATE LANCEOLATE leaves and greenish flower clusters that fuse to form a multiple fruit. This is cooked and the seeds eaten. The seeds of bread-fruit are also known as bread-nuts.

BRIAR
see ROSE

BROAD BEAN *(Vicia faba)*. An erect annual in the *Leguminosae*, the original home of which is unknown (probably in the eastern Mediterranean and West Asia).

●

It is a robust plant, the PINNATE grey-green leaves having 2–6 ELLIPTIC leaflets. The pea-shaped flowers are borne in AXILLARY clusters and have white STANDARD petals and velvety black KEELS. They are followed by long, thick pods, to 250mm long.

BROCCOLI
see CAULIFLOWER

BROME GRASS *(Bromus)*. A genus of 50 annuals or perennials in the *Gramineae* from the temperate regions of the world. They are mostly tufted grasses with slender leaves and the flowering spikelets aggregated into narrowly OVOID heads on slender stalks, creating an oat-like appearance: Barren brome *(B. sterilis)* is a hairy annual or biennial 300–600mm in height with AWNED spikelets hanging on flexible stems; *B. madritensis* is an annual to 600mm with a compact, erect inflorescence, the spikelets with prominent AWNS; *B. mollis* (soft brome) is a 300–600mm, softly hairy annual with broad spikelets having short awns.

BROOKLIME *(Veronica beccabunga)*. A perennial water plant in the *Scrophulariaceae* from Europe, N. Africa and Asia. It has creeping, then ascending, rather fleshy stems to 600mm with oval, CRENATE margined 25–50mm long leaves in pairs. The 8mm wide, ZYGOMORPHIC, blue flowers are carried in AXILLARY RACEMES.

BROOKWEED *(Samolus valerandi)*. A perennial plant of wet places, usually near the sea, in the *Primulaceae* from temperate climates around the world. It has tufts of OVATE to SPATHULATE leaves and slender, erect stems bearing RACEMES of tiny, 5-lobed, white flowers.

● **BROOM** A general name for several species of shrubs in the *Leguminosae*. The main one is *Sarothamnus scoparius* (syn. *Cytisus scoparius*), the common broom. This is a green, twiggy species to 2m or so, from Europe. The juvenile leaves are TRIFOLIATE, the adult simple and ELLIPTIC and soon falling. The bright yellow, 19mm pea-flowers are borne in AXILLARY clusters and the flattened 38mm pods expode on ripening. *S.s. maritimus* is a prostrate, maritime sub-species. *See* SPANISH BROOM, BUTCHERS' BROOM.

▲ **BROOMRAPE** *(Orobanche)*. Curious annual or perennial PARASITES in the *Orobanchaceae* which lack CHLOROPHYLL. They live on the roots of broom, gorse, ivy, yarrow, clover and many other plants less commonly, producing usually un-branched, brown-yellow or purplish stems bearing scale leaves and spikes of two-lipped, tubular flowers. The perennial species form underground tubers. There are 140 species of cosmopolitan distribution: Lesser broomrape *(O. minor)* is about 300mm high, with yellowish, purple-veined flowers 16mm long, growing chiefly on the roots of clover and other members of the *Leguminosae*.

BRUNNE|US, -A, -UM Brown.

BRUSSELS SPROUTS *(Brassica oleracea bullata)*. A form of the cabbage with an erect, leafy stem bearing a rosette at the top and plump, green buds in the leaf AXILS which are cooked and eaten as sprouts.

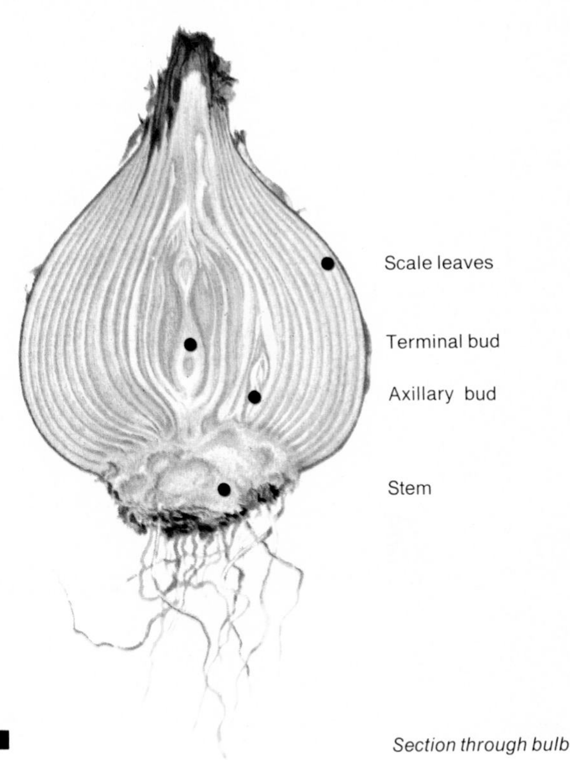

Section through bulb

BRYOIDES Of moss-like appearance.

BRYOPHYTA The group of primitive plants to which MOSSES and LIVERWORTS belong.

BUCKBEAN
see BOG BEAN

BUCKEYE
see HORSE CHESTNUT

BUCKLER FERNS Several species of *Dryopteris*, perennial ferns in the *Aspidiaceae* which die back in autumn. They are tufted plants with bi- or tripinnate, LANCEOLATE, arching fronds with the SPORANGIA attached to the veins of the PINNULES or final lobes of the fronds: Broad buckler fern *(D. dilata'ta)* has fronds to 1.2m long, triangular ovate to lanceolate in outline and tripinnate; Narrow buckler fern *(D. carthusiana)* is a plant of wet woods and marshy ground, with bipinnate, yellow-green lanceolate fronds, 300–900mm or more in height.

BUCKHORN *(Rhamnus catharticus)*. A 1.5–4.5m thorny, deciduous shrub in the *Rhamnaceae* from Europe, N. Africa and W. Asia. It has opposite pairs of OVATE to ELLIPTIC leaves and small clusters of greenish flowers, 5mm wide. The pea-sized purgative berries are black. Alder buckthorn *(Frangula alnus* syn. *Rhamnus frangula)* is very similar but without thorns and has OBOVATE, untoothed leaves. *See* SEA BUCKTHORN.

BUCKWHEAT *(Fagopyron esculentum)*. A sparsely branched annual in the *Polygonaceae* from Asia, but much cultivated in some countries as a cereal substitute, also for fodder. The 300–600mm stems carry long, pointed, SAGITTATE-OVATE leaves and a terminal PANICLE of tiny pink or white flowers. The fruit is a triangular nutlet about 8mm long.

BUD An immature and condensed shoot, containing in miniature a stem and leaves or flowers, the whole protected by special tough scale leaves. Buds are mainly produced by shrubs, trees and herbaceous plants.

BUFONI|US, -A, -UM The place of the toad; a plant growing in the same damp environment.

BUGLE *(Ajuga reptans)*. A mat-forming perennial in the *Labiatae* from Europe, S.W. Asia and N. Africa. It has OBOVATE basal leaves to 63mm long, and erect flowering stems 150–300mm in height. The tubular blue flowers are two-lipped, the lower the larger; and three-lobed.

BUGLOSS *(Anchusa arvensis* syn. *Lycopsis arvensis)*. An erect very bristly hairy annual or biennial in the *Boraginaceae* from Europe and Asia. It has LINEAR-OBLONG basal leaves to 150mm long with wavy margins and smaller, stem-clasping ones. The bright blue, tubular flowers are carried in simple or forked CYMES. *See also* VIPER'S BUGLOSS.

■ **BULB** A storage organ, usually deep in the soil, with bud-like structure. It is made up of a short, flattened stem called a base plate which produces roots on its lower surface and fleshy leaves or leaf bases on the upper. These are of two forms: tunicated where the leaves are broad

and wrap tightly round each other (e.g. daffodil); and scaly where the leaves are narrow and separate (e.g. lily). Within the middle of each dormant bulb is a shoot comprising leaves and flower buds in miniature. *See also* CORM.

BULBIFEROUS Possessing or producing bulbs.

BULBIL [BULBLET] Small bulbs, the former produced above ground as on the stems of certain lilies.

BULBOS|US, -A, -UM Possessing or producing bulbs.

BULLACE
see PLUM

BULLATE Puckered or blistered, e.g. the leaves of *Myrtus bullatus* with raised areas between the veins.

BULLOCK'S HEART *(Annona reticulata)*. A deciduous tree in the *Annonaceae* from tropical America, sometimes also known as CUSTARD APPLE. It grows to 7.5m, with oblong LANCEOLATE leaves and AXILLARY clusters of fleshy, green flowers purple blotched within. The OVOID to GLOBOSE fruits, 75–125mm long, have a netted, red-brown skin and a sweet pulp.

● **BULRUSH** *(Scirpus lacustris)*. A large, perennial rush-like water plant in the *Cyperaceae* of almost cosmopolitan distribution. It has branched, creeping RHIZOMES sometimes with narrow, submerged leaves. The stems can have an overall length of 3.0m, but are generally less, and bear a dense, usually branched head of tiny, red-brown flowering spikelets. The name bulrush is sometimes applied to REED-MACE *(Typha)*.

BUNYA-BUNYA PINE *(Araucaria bidwillii)*. An evergreen coniferous tree in the *Araucariaceae* from E. Australia. It grows to 45.7m in height with thick, peeling resinous bark and branches arranged in horizontal WHORLS. The leathery leaves are lanceolate, to 50mm long and the large cones can weight 4.5kg and be 300mm long.

BUR [BURR] Fruits or seed with barbed hairs that cling to passing animals, gaining dispersal in this way, e.g. burdock, bur marigold, cleavers.

BURDOCK Two species of *Arctium*, robust perennials in the *Compositae* from Europe to the Caucasus and N. Africa. They have rosettes of stalked, OVATE-CORDATE leaves 300mm or more long, grey-cottony hairy beneath, and branched stems to 1.3m in height. The flower heads are globular, the BRACTS tapering to stiff, hooked tips. The small tubular flowers are purple. When the heads ripen, each becomes a bur. Great Burdock *(A. lappa)* has flower heads 25–38min wide; Lesser burdock *(A. minus)* usually less than 25mm.

BUR MARIGOLD Several annual species of *Bidens*, water or wet-ground plants in the *Compositae*; those mentioned here are from Europe and Asia, though introduced elsewhere. The flower heads are mainly or entirely composed of tubular disk florets: Nodding bur marigold (*B. cernua)* grows to 600mm with LANCEOLATE leaves in slightly CONNATE pairs. The flower heads are nodding, up to 25mm across with leafy outer BRACTS;

▲

■

trifid bur marigold *(B. tripartita)* has semi-erect heads and leaves composed of 3–5 lanceolate leaflets.

BURNET SAXIFRAGE *(Pimpinella saxifraga)*. A slender 300–600mm perennial in the *Umbelliferae* from Europe to W. Asia. It has PINNATE or bipinnate lower leaves with OVATE to LINEAR-LANCEOLATE leaflets and terminal flat-topped 5–50mm UMBELS of tiny white flowers: Greater burnet saxifrage *(P. major)* is larger, 45–105cm tall, with umbels 32–63mm wide, the florets often pinkish.

▲ **BURNING BUSH** A name given to at least two very different plants. (1) *Dictamnus albus*, a herbaceous perennial in the *Rutaceae* from E. Europe and Asia. It has PINNATE leaves with 9–13 leaflets that smell of balsam when bruised and RACEMES of ZYGOMORPHIC, pale purple or white flowers. The plant gives off an inflammable, volatile oil which, during hot, still days can be ignited. (2) *Kochia scoparia*, an erect, bushy annual also called summer cypress, with LINEAR 50mm long leaves and minute GLOBOSE green flowers. In autumn the whole plant turns purple-crimson before dying.

■ **BUR REED** *(Sparganium)*. Several RHIZOMATOUS water plants in the *Sparganiaceae* from most countries except Africa and S. America. They have ELONGATE, LINEAR leaves either floating or erect, and small unisexual green flowers grouped into tight, globular heads: Common bur reed *(S. erectum)* has leaves triangular in cross section and stems 0.4–1.3m.

BUSY LIZZIE
see IMPATIENS

BUTCHER'S BROOM *(Ruscus aculeatus)*. A stiff, erect, very dark green DIOECIOUS shrub in the *Liliaceae* from Europe. The apparent ovate, spine-tipped leaves are really CLADODES, the true leaves being reduced to tiny scales. The minute 6-petalled, greenish white flowers are borne near to the middle of the cladodes, the female ones followed by red, fleshy fruits up to 13mm wide.

BUTTER-BEAN *(Phaseolus lunatus)*. Also known as Lima, sieva or duffin bean, this is an annual or perennial twining plant with TRIFOLIATE leaves and pale green and white pea flowers in AXILLARY RACEMES. They are followed by oblong, flattened 50–125mm pods which contain 2–4 flat white beans. Cultivars are known with red, brown, purple and black beans, and some are bush-like in form. Much grown in the tropics and subtropics, probably originally from Guatemala.

BUTTERBUR Several species of *Petasites*, RHIZOMATOUS perennials in the *Compositae* from the northern hemisphere. They are largely dioecious with creeping rhizomes and long stalked, rounded leaves. The racemes of flower heads appear before or with young leaves. Each head is functionally male or female, though the tubular florets of which it is composed are of both sexes: Common butterbur *(P. hybridus)* has ORBICULAR-CORDATE toothed leaves 300–900mm and 150–450mm tall flowering stems with pale redish-violet florets; White butterbur *(P. albus)* has leaves to 300mm wide and whitish florets.

BUTTERCUP Many sorts of *Ranunculus*, a genus of 400 species of annual and perennial plants of cosmopolitan distribution. The cup-shaped yellow or white flowers have 5 sepals and petals, occasionally more, often opening out flat. The fruits are hard, small ACHENES: Meadow buttercup *(R. acris)* is a tufted perennial to 900mm with PALMATELY 3–7 lobed basal leaves and 19–25mm wide yellow flowers; alpine buttercup *(R. alpestris)* is a tufted perennial 25–100mm tall with ORBICULAR, deeply 3-lobed, CRENATE leaves and 19mm wide, white flowers; Bulbous buttercup *(R. bulbosus)* is a tufted, tuberous rooted perennial with lobed, trifoliate leaves and 150–380mm stems with yellow flowers; glacier buttercup *(R. glacialis)* has TRIFOLIATE, lobed leaves and 38–250mm stems bearing white flowers often suffused with pale or deeper red-purple; creeping buttercup *(R. repens)* is a mat-forming, stoloniferous plant having BITERNATE leaves with lobed leaflets and yellow flowers on 150–450mm stems; corn buttercup *(R. arvensis)* is a slender annual with deeply, narrowly lobed leaves, small yellow flowers and large spine-covered achenes.

BUTTERFLY BUSH *(Buddleja davidii*, syn. *Buddleia davidii)*. A deciduous shrub to 3.0m or more, with LANCEOLATE-ACUMINATE, 100–250mm long leaves, white felted beneath. The terminal, slenderly pyramidal PANICLES of small, tubular, fragrant flowers are lilac or purple and attractive to butterflies.

BUTTERNUT *(Caryocar nucifera)*. An evergreen tree to 30m or more in the *Caryocaraceae* from Brazil. It has trifoliate leaves and purple flowers followed by 125–150mm wide globular fruits (DRUPES) which contain 2–4 hard-shelled nuts as big as a hen's egg and rather flattened. White WALNUT is also known as butternut.

BUTTERWORT *(Pinguicula)*. A genus of 46 CARNIVOROUS, insectivorous plants in the *Lentibulariaceae* from the north temperate zone and S. America. Leaves are borne in basal rosettes and are covered with glandular hairs that secrete a greasy fluid which traps small insects. The leaf edge rolls over on itself when insects are trapped, and when they die produces a ferment which digests the bodies and absorbs nitrogenous and other substances. The plants inhabit moist or wet places, often acid heaths and bogs which are deficient in soil nutrients. The flowers are violet-like, usually with a pointed spur: common butterwort *(P. vulgaris)* has yellow-green, OVATE-OBLONG leaves 27–75mm long, and 13–19mm purple flowers on stems 50–150mm in height; *P. grandiflora* is similar with larger flowers. Both species overwinter as rootless buds.

BUTTONWEED
see PLANE

● **BUTTRESS ROOTS** Aerial roots that radiate out from near the base of a stem or trunk. They grow obliquely down into the soil and serve as props, e.g. screw pines and several *Ficus*.

BUXIFOLI|US, -A, -UM Like the leaves of box *(Buxus)*.

CABBAGE *(Brassica oleracea)*. A biennial or perennial plant in the *Cruciferae*, the wild plant coming from the Mediterranean region and the coasts of western Europe.

■

Wild cabbage forms a rosette of large, rounded, glaucous, stalked leaves with wavy margins. The 4-petalled lemon-yellow flowers are borne in branched RACEMES to 600mm or more in height. The cabbages of commerce are cultivated mutants that form a conical or ball-shaped mass of concave, young leaves in the centre of the plant, known as a head. Botanically this is called *B.o. capitata.*

CABBAGE TREE *(Cordyline australis).* A somewhat palm-like tree in the *Agavaceae* from New Zealand. Reaching 12.2m in its homeland, this species has comparatively few erect branches, each topped by a tuft of arching, LINEAR leaves 300–900mm long. Large PANICLES of creamy-white fragrant flowers are borne terminally.

▲ **CACTUS [CACTI]** All those members of the *Cactaceae* with fleshy, succulent stems. They range from small, OVOID to globular plants, an inch or so across, to the massive tree-sized columns of the saguaro *(Carnegiea).* In between is a vast array of shapes and forms. The barrel cacti are rounded or oval, 0.6–1.8m tall and the prickly pear has flattened, pear-shaped stems called pads, joined end to end. ■ Cholla is also an *Opuntia,* but the stem sections are cylindrical, armed with barbed spines. Spines are a feature of many cacti, though a few are completely without, notably the mescal button *(Lophophora williamsii).* The main feature of the *Cactaceae* which distinguishes its members from other succulent plants is the AREOLE. Cacti come from arid and desert regions of N. and S. America; they are not restricted to tropical climates, many species growing as far north as Canada and high in the Andes. Although many are frost hardy, they need plenty of sunshine. A small section of the cacti is EPIPHYTIC, dwelling in the branches of trees. These are the so-called leaf or forest cacti. Generally they have flattened or much branched elongated, leaf-like stems and very small aeroles. True leaves are seldom found on true cacti, and when they do occur they are of short duration. Many seedlings show fleshy, tubular leaves, especially prickly pear *(Opuntia).* The flowers of cacti are often spectacular, funnel or trumpet shaped in shades of red, orange, yellow and white. They may be less than 25mm across or even over 300mm. The fruits of some species are fleshy and edible.

CADUCOUS Short-lived or falling early, e.g. the sepals of poppies.

CAERULE|US, -A, -UM Bright or sky-blue.

CAESI|US, -A, -UM Blue-grey or lavender.

CAESPITOSE [CAESPITOS|US, -A, -UM] Of tufted growth.

CALABASH TREE *(Crescentia cujete).* A spreading 9–12m evergreen tree in the *Bignoniaceae* from tropical America. It has broadly LANCEOLATE 100–150mm long leaves in clusters and solitary, somewhat bell-shaped flowers direct from the trunk or branches *(cauliflorus).* The fruit are large, gourd-like, woody berries 300mm or more long, formerly much used as vessels by native tribes.

CALAMINT Several species of *Calamintha,* perennial plants in the *Labiatae* from the northern temperate region. They are clump-forming plants with opposite pairs of OVATE leaves and stalked AXILLARY CYMES of 2-lipped, tubular flowers from the upper leaf AXILS: common calamint *(C. ascendens*, syn. *C. officinalis)* has erect stems 300–600mm tall and red-purple or lilac flowers.

CALCARAT|US, -A, -UM Spurred, e.g. the flowers of balsam.

CALCARE|US, -A, -UM Growing in limy or chalky soil.

● **CALCEOLARIA** A genus of more than 300 species of annuals, perennials, shrubs and scramblers in the *Scrophulariaceae* from S. America. They are all recognised by the 2-lipped flowers, the lower lip much inflated and fancifully like a slipper. The leaves are in opposite pairs and often RUGOSE. Several species and hybrids are grown as ornamentals. Most commonly seen is the hybrid race *C. X herbeohybrida*, bushy plants to 450mm, with 50mm wide kidney-shaped flowers in shades of orange, red and yellow, variously spotted. *C. polyrrhiza* is a mat-forming plant with LANCEOLATE leaves and slender, erect stems to 100mm with 25mm slipper flowers. The hybrid 'John Innes' is bigger and finer.

CALCICOL|US, -A, -UM Living on chalky or limy soil.

CALIFUGE Those plants, like rhododendron, that cannot grow on limy soil.

CALCIPHILOUS [CALCIPHILE] Plants that prefer to grow, or grow best, on limy and/or chalky soil.

CALICO BUSH *(Kalmia latifolia)*. A large evergreen shrub in the *Ericaceae* from N. America. It has oval to lanceolate leaves 50–125mm long and terminal CORYMBS of bright pink, cup-shaped flowers to 25mm wide. The 10 stamens are bent back into small pockets near the base of the flower and spring up when dislodged by an insect visitor, scattering the pollen.

CALIFORNIAN BLUEBELL
see BABY BLUE EYES

CALIFORNIAN LILAC A general name for the 55 species of *Ceanothus*, mainly evergreen shrubs in the *Rhamnaceae* from N. America, chiefly California. They have boldly veined, OVATE to OBOVATE or LANCEOLATE leaves and small blue flowers in AXILLARY RACEMES or TERMINAL PANICLES. Several species and hybrids are grown as ornamentals: *C. americana*, also called New Jersey tea, is one of the few deciduous species, 0.9–1.2m tall, with ovate 50–70mm long leaves and panicles of white flowers; *C. X veitchianus* is a 3m evergreen hybrid, much cultivated, with 13–25mm oval leaves and profuse panicles of deep blue flowers.

▲ **CALIFORNIAN POPPY** *(Eschscholzia californica)*. An annual or short lived perennial in the *Papaveraceae* from California, where it is aptly called 'copa de oro' (cup of gold). It grows to 300–450mm and has grey-green, TRIPINNATIFID leaves cut into slender segments and 60mm wide, 4-petalled orange-yellow flowers. Also sometimes called Californian poppy is *Romneya coulteri*, likewise in the *Papaveraceae* but a woody-based perennial to 2.4m with grey pinnatifid leaves and 100–125mm wide white flowers.

CALLIANTH|US, -A, -UM Bearing beautiful flowers.

CALLOS|US, -A, -UM Thickened or hardened.

CALLUS Tissue which forms over a wound giving a raised, protective layer. Roots and shoots can arise from this tissue: cuttings or slips a gardener takes heal in this way.

CALOPHYT|US, -A, -UM Beautiful plant.

CALTROPS Several species of *Tribulus*, creeping annual or perennial plants in the *Zygophyllaceae* from the tropics and sub-tropics. They have PINNATE leaves and 5-petalled flowers from the leaf AXILS. The fruiting CARPELS (MERICARPS) are triangular and woody, usually with three hard spines so arranged that, whichever way it lies, one stands erect. They stick into the feet of animals and are thus distributed. *See also* WATER CHESTNUT.

CALYCIN|US, -A, -UM Having a prominent CALYX.

CALYCULAT|US, -A, -UM Calyx-like; with bracts beneath the calyx forming an epicalyx.

CALYPTRA The pointed, hood-like sheath which covers the spore capsule of mosses.

CALYPTROGEN A self-renewing layer of cells on the root tip which wears away as the plant pushes through the soil, protecting the delicate growing point behind.

CALYX The outer whorl of the PERIANTH, usually green but sometimes coloured, which protects the petals and PISTIL within. The individual segments (SEPALS) may be fused into a tube, e.g. primrose, or separate and leaf or bract-like as in buttercup.

CAMASS *(Camassia quamash* syn. *C. esculenta)*. A bulbous plant in the *Liliaceae* from N. America, the bulbs formerly eaten by the local Indians. It has 300mm long LINEAR leaves and taller RACEMES of starry blue or white flowers with 6 narrow perianth segments.

CAMBIUM A layer of actively growing and dividing cells in stems and roots, and responsible for their increase in length and girth. In soft young stems it forms a zone between the XYLEM and PHLOEM of each VASCULAR BUNDLE. In the woody stems of DICOTYLEDONS it forms a continuous layer between the wood and the bark.

CAMBRIC|US, -A, -UM From Wales (Cambria).

CAMPANULAT|US, -A, -UM Bell-shaped.

CAMPESTR|IS, -E From field or flat places.

CAMPHOR TREE This name applies to several distinct tree species, the best known being *Cinnamomum camphora*, an evergreen species in the *Lauraceae* from Japan. It grows to 6m or more with ovate, strongly 3-nerved leaves and small green-white flowers. Sumatra, Borneo or Barus camphor *(Dryobalanops aromatica)* is a 30m evergreen tree with ELLIPTIC, leathery leaves and yellow flowers in the *Dipterocarpaceae* from Sumatra. It yields camphor oil and its crystalline solid by distillation.

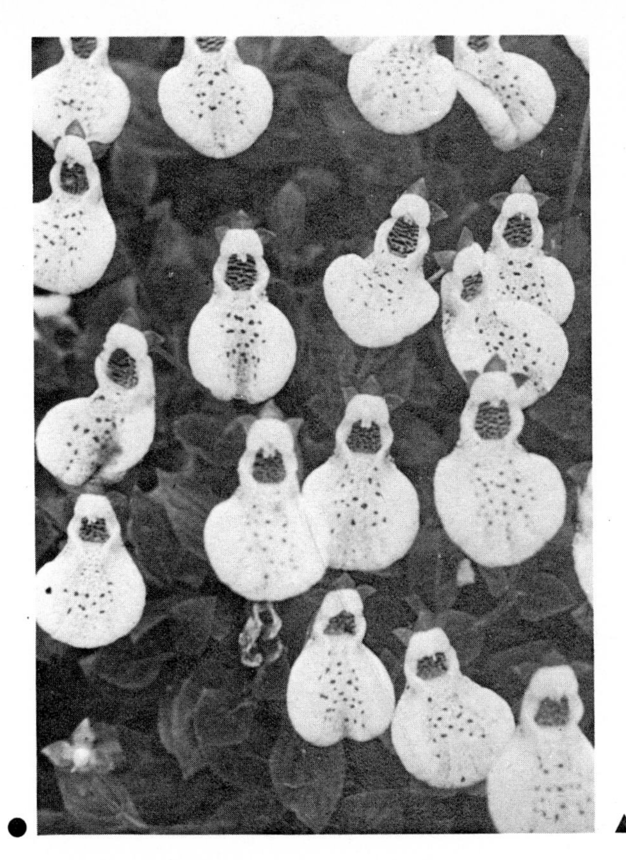

CAMPION Several species of *Silene*, annuals and perennials from temperate regions, in the *Caryophyllaceae*. They have opposite pairs of LINEAR to OVATE leaves and terminal CYMES of tubular flowers with 5 petals, each notched or bifid: bladder campion *(S. vulgaris* syn. *S. cucubalus)* is 300–900mm tall with GLAUCOUS, LANCEOLATE-ELLIPTIC leaves and nodding white flowers in an inflated bladder-like CALYX; moss campion *(S. acaulis)* forms moss-like cushions of linear leaves studded with almost stemless pink flowers; red campion *(S. dioica* syn. *Melandrium rubrum)* is 450–900mm, with ovate-oblong leaves and deep pink flowers; sea campion *(S. maritima)* is like a prostrate bladder campion with stems to 150mm; white campion *(S. alba* syn. *Melandrium album)* is similar to red campion, but with larger white flowers and lanceolate leaves.

CAMPYLOTROPUS
see OVULE

CANADA BALSAM *(Abies balsamea)*. An evergreen coniferous tree in the *Pinaceae* from N. America. It has LINEAR, leathery leaves to 25mm long with silvery STOMATA bands on the underside. The erect OVOID cones, 50–100mm long, contain winged seeds and disintegrate when ripe. The turpentine resin called Canada balsam is collected from the bark blisters.

CANADA RICE *(Zizania aquatica)*. Wild rice, Indian oats or water rice are other names for this annual plant in the *Gramineae* from N. America. It is an aquatic grass to 3m in height with wide, linear leaves and graceful PANICLES of unisexual flowering spikelets.

CANADIAN WATERWEED [CANADIAN PONDWEED] *(Elodea canadensis)*. A submerged, DIOECIOUS, aquatic perennial in the *Hydrocharitaceae*. From N. America but now widely naturalised in Europe. Stems are 300–600mm or more in length with oblong, LANCEOLATE leaves in WHORLS of three. The 5mm wide purple-green flowers are floating.

CANARY CREEPER *(Tropaeolum peregrinum)*. An annual or perennial climber in the *Tropaeolaceae* from Peru (not the Canary Islands as was once thought). It can reach 3m, with deeply 5-loped PALMATE leaves, the stalk of which acts as a TENDRIL. The spurred bright yellow flowers have only 2 fully developed petals which are deeply toothed and look like a tiny bird in flight.

CANARY GRASS *(Phalaris canariensis)*. An annual plant in the *Gramineae* from the Mediterranean region. Of tufted habit, it is 450–900mm tall with LINEAR, 150mm long leaves and OVOID heads of neatly overlapping green flowering spikelets. It is grown as a crop plant for bird seed.

CANCELLAT|US, -A, -UM Bearing a latticed pattern of veins or lines, usually on the petals.

CANDELABRUM Branched like an old-fashioned candlestick, e.g. the inflorescences of *Primula japonica*, one of the candelabra primulas.

CANDICANS Off-white or becoming white.

CANDID|US, -A, -UM Shining white.

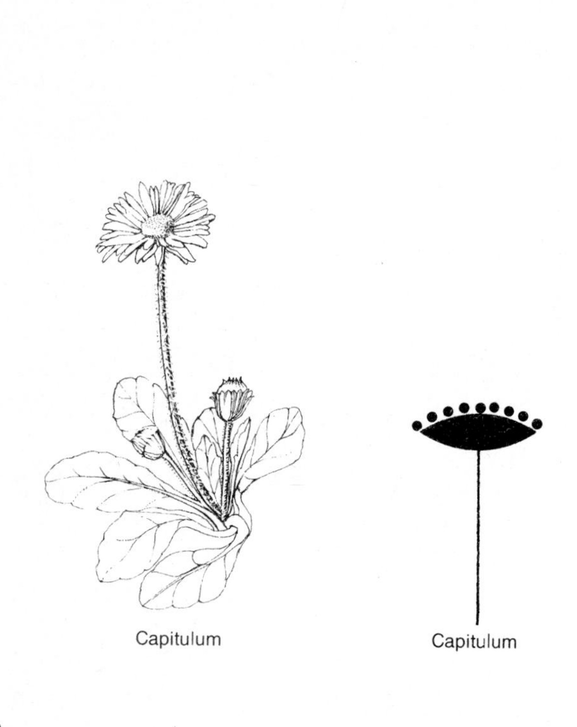

CANDLE NUT *(Aleurites moluccana)*. An evergreen tree in the *Euphorbiaceae* from Indonesia and the Pacific Islands. It grows to 12m with OVATE 3-lobed leaves and greenish, fleshy fruits that contain 2 hard-shelled, black, grooved nuts about 25mm wide. The latter contain very oily EMBRYOS which when pressed yield good lamp oil.

CANDYTUFT *(Iberis amara* and *I. umbellata)*. Bushy annuals in the *Cruciferae* from Europe. They have LANCEOLATE, toothed leaves on 150–300mm stems which are topped by CORYMBS of 4-petalled flowers that later elongate into RACEMES. Individual florets have the two outer petals larger than the inner: *I. amara* is white; *I. umbellata* purple, red or pink.

CANE In the strict sense, the stem of a bamboo, but applied to stems of grape vine, blackberry, raspberry, loganberry; also the PSEUDOBULBS of orchids such as DENDROBIUM.

CANESCENS Hoary or greyish.

CANNON-BALL TREE *(Coroupita guianensis)*. A remarkable evergreen tree to 15m or more in the *Lecythidaceae* from Guyana. It has ELLIPTIC to broadly LANCEOLATE leaves and fleshy, 50mm wide cup-shaped yellow, red and purple flowers. These are borne on short, pendulous branchlets direct from the trunk, and are followed by hard, brown globular fruits about 200mm across.

CANTERBURY BELL
see BELLFLOWER

CAN|US, -A, -UM Covered with grey-white hairs.

CAPE COWSLIP A general name for several species of *Lachenalia*, bulbous rooted plants in the *Liliaceae* from S. Africa. They have rounded bulbs, LINEAR, somewhat fleshy and often mottled leaves, and tubular or bell-shaped flowers in RACEMES. *L. aloides* has 300mm flowering stems with 25mm long pendulous narrow bells in shades of green, red and yellow.

CAPE GOOSEBERRY *(Physalis peruviana)*. An erect, branched, hairy annual or perennial in the *Solanaceae* from South America with OVATE-CORDATE leaves and axillary pendulous, white rotate flowers. The edible, cherry-sized yellow or purplish fruits are enclosed in enlarged CALYCES.

CAPE JASMINE
see GARDENIA

CAPE PONDWEED *(Aponogeton distachyus)*. A floating leaved aquatic plant in the *Aponogetonaceae* from S. Africa. Also known as water hawthorn, it has oblong-oval leaves about 150mm long and waxy white, fragrant flowers in two forked spikes held just above the water.

CAPE PRIMROSE *(Streptocarpus rexii)*. An evergreen tufted or rosette-forming plant in the *Gesneriaceae* from S. Africa. It has narrowly OBLONG to LINGUIFORM leaves to 250mm long, and wiry 150mm stems bearing ZYGOMORPHIC, obliquely trumpet-shaped blooms about 38–50mm long. Blue-purple is the usual colour, but cultivated hybrid kinds are also found in shades of red, pink and white.

CAPER *(Capparis spinosa)*. A somewhat straggling, spiny deciduous shrub in the *Capparidaceae* from S. Europe. It has OVATE leaves and white flowers, each with a central boss of long, purple stamens. The pickled flower buds are known as capers.

CAPILLAR|IS, E Slender; fine as a hair.

CAPILLUS-VENERIS Venus' hair.

CAPITAT|US, -A, -UM Growing in dense head-like clusters.

● **CAPITULUM** A flower head, like that of a daisy, where numerous tiny flowers are set close together on a flattened stem tip and surrounded by bracts, the overall effect being of one, large flower.

CAPREOLAT|US, -A, -UM Having tendrils for climbing.

CAPRE|US, -A, -UM Having associations with goats.

CAPSULAR|IS, -E Producing capsules.

▲ **CAPSULE** A dry, often box-like, fruit containing many seeds and opening by pores or slits, e.g. poppy and scarlet pimpernel. Some capsules open explosively, e.g. pansy.

CARAWAY *(Carum carvi)*. An erect 300–600mm biennial in the *Umbelliferae* from Europe and Asia. It has triangular bipinnate leaves and UMBELS of tiny, white flowers with deeply BIFID petals. The seeds (technically the two halves of a SCHIZOCARPIC fruit) are used for flavouring and perfumery.

CARDAMOMS *(Elettaria cardamomum)*. A large RHIZOMATOUS perennial in the *Zingiberaceae* from India. It forms thickets of erect stems to 3.6m in height clad with LINEAR-LANCEOLATE leaves to 600mm long. The tiny white flowers are marked with blue and yellow and borne in 300–600m long PANICLES near ground level. The OVOID 13–19mm CAPSULES are a main source of the cardamoms of commerce, used as curry spice and in medicine and confectionery.

■ **CARDOON** *(Cynara cardunculus)*. A member of the *Compositae* from the Mediterranean region and the Canary Isles. It is also widely naturalised. Looking very much like a globe ARTICHOKE, it has smaller flower heads with spine-tipped bracts. The leaves are blanched and used as a vegetable.

CARINAT|US, -A, -UM Keeled; e.g. pea flowers.

CARLINE THISTLE *(Carlina vulgaris)*. A small biennial thistle-like plant in the *Compositae* from Europe to Asia. It forms a rosette of OBLONG-LANCEOLATE PINNATIFID spiny leaves which die before the 100–300mm flowering stems fully develop. The 19–38mm wide flower heads are surrounded by slender straw coloured BRACTS which look like RAY florets.

CARMINE|US, -A, -UM The colour of carmine.

CARNATION *(Dianthus caryophyllus)*. A woody based evergreen perennial in the *Caryophyllaceae* from France. The glaucous leaves are LINEAR, closely borne in opposite pairs on sprawling stems. The fragrant 5-petalled, red flowers emerge from a tubular CALYX. The name carnation properly applies to the many large, variously coloured, double cultivars which have been favourite garden plants for centuries.

CARNE|US, -A, -UM Flesh-coloured or pinkish-red.

● **CARNIVOROUS [INSECTIVOROUS] PLANTS** A group of plant genera and species that have evolved the means of digesting animal tissues and absorbing essential food materials, mainly nitrogen. In most cases the leaf blade is modified into a trap for tiny animals, mainly insects. In *Sarracenia*, *Nepenthes* and *Darlingtonia* the leaves are pitcher-shaped. The sundews have sticky, long-stalked mobile hairs which move in to enmesh a caught insect. Butterworts have a sticky, greasy surface but catch only tiny prey. The most remarkable is Venus's fly trap, where the leaf is fashioned like a man trap and closes rapidly on any insect that touches the special trigger hairs.

CARNOS|US, -A, -UM Fleshy; usually of thick, succulent leaves.

CAROB [LOCUST] *(Ceratonia siliqua)*. An evergreen tree to 12.2m in the *Leguminosae* from S. Europe. It has leathery PINNATE leaves composed of 4–10 oval, wavy-edged leaflets 25–75mm long. The small RACEMES of flowers often grow direct from the larger branches. They are followed by narrowly oblong, thick pods, 125–300mm in length, which contain a sweet pulp. The hard seeds were the original carat weights of goldsmiths. The pods, known also as St. John's bread, are considered to be the locusts of the biblical story, eaten by John the Baptist. They are much used as stock feed.

CAROLINIAN|US, -A, -UM From Carolina, USA.

CARPEL A tiny, folded, specially modified leaf in the centre of a flower which contains the OVULES. It may either be separate or fused together and containing one to many ovules. After fertilisation it becomes the fruit.

CARPINIFOLI|US, -A, -UM With leaves like the hornbeam *(Carpinus)*.

CARRAGHEEN *(Chondrus crispus)*. A red seaweed of rocky shores in the *Gigartinaceae* from the N. hemisphere. It is a tufted species to 150mm or more in length, each frond much branched to form rather mossy masses. It has been a medicine for lung complaints and is still used in a limited way in the preparation of jellies and blancmanges.

▲ **CARRION FLOWERS** Flowers which give off an odour of carrion, and are often reddish-brown in colour. They are pollinated by insects that feed on or breed in carrion. *Amorphophallus*, *Stapelia* and Lords and Ladies are examples.

CARROT *(Daucus carota)*. A biennial in the *Umbelliferae* from Europe and Asia, much grown for its edible red root. It forms erect rosettes of deeply bipinnatisect leaves the first season, and a 300–600mm tall erect, branched

▲

flowering stem the following year. The tiny, whitish to purple flowers are carried in flattened UMBELS. The history of the garden carrot is not known for certain, but it appears to have been selected from the wild carrot which normally has thin, yellowish, rather woody roots. There are many cultivars varying in length of root and time of maturity.

CARTWHEEL FLOWER
see HOGWEED

CARUM
see CARAWAY

CARUNCLE A localised ARIL, usually in the form of a firm, warty outgrowth of the seed stalk and attached to the TESTA, e.g. the seed of castor oil plant.

CARYOPSIS An ENDOSPERMIC fruit formed when a CARPEL containing a single OVULE develops by fusing together, e.g. the grains of cereals.

CASCARA SAGRADA The bark of *Rhamnus purshiana*, a deciduous tree to 12m in the *Rhamnaceae* from N. America. It has oval to obovate, 50–150mm long leaves and tiny, green flowers in UMBELS. The 8mm wide berries are purple-black.

CASCARILLA BARK The barks of *Croton cascarilla* and *C. eluteria*, evergreen shrubs in the *Euphorbiaceae* from the Bahamas. They have OVATE to LANCEOLATE leaves, those of *C. cascarilla* being densely hairy beneath. The tiny greenish flowers are in spikes.

CASHEW NUT *(Anacardium occidentale)*. A small to medium sized evergreen tree in the *Anacardiaceae* from S. America, but much grown in India and Africa. It has oval to obovate leaves and small red flowers in terminal PANICLES. The fruit is a nut, shaped like a thick letter C, and hangs at the end of a pear-shaped, fleshy stalk, shaded yellow and red.

CASHMERIAN|US, -A, -UM From Kashmir.

CASSAVA *(Manihot utilissima)*. An evergreen shrub in the *Euphorbiaceae* from S. America and much cultivated elsewhere in the tropics as a food plant. It has thick, fleshy roots yielding starch and 900mm stems bearing PALMATE leaves composed of 7 lanceolate leaflets. The roots are also known as manioc, mandioica and tapioca and need to be cooked before they are eaten.

CASTANEOUS [CASTANE|US, -A, -UM] Chestnut-coloured *(Castanea)*, i.e., bright red-brown.

CASTOR OIL PLANT *(Ricinus communis)*. An evergreen shrub to 3m or so with large, deeply 5–12 lobed leaves, sometimes bronze or purple flushed. The separate male and female flowers are green and borne in terminal RACEMES. The dry fruits are softly spiny and explode to liberate three OVOID, nut-like seeds, the source of castor-oil. Japanese aralia is also sometimes known as castor-oil shrub.

CASUAL An alien plant that appears from time to time but never becomes naturalised.

CAT BRIER
see SMILAX

CATCHFLY A group of annual and perennial species in the genera *Silene* (*see also* CAMPION) and *Lychnis* (*see also* RAGGED ROBIN). They have sticky zones on the stems beneath each pair of leaves: Alpine catchfly *(Lychnis alpina)* is a tufted perennial to 150mm, having small, rich pink flowers in head-like clusters; sticky or red German catchfly *(L. viscaria* syn. *Viscaria vulgaris)* is very similar but much larger, the stem up to 600mm; Spanish or Breckland catchfly *(Silene otites)* is a slender, woody based, tufted perennial with LINEAR-LANCEOLATE leaves. The tiny flowers have bell-shaped calyces and greenish-white petals arranged in dense, cylindrical PANICLES.

CATHARINE'S MOSS *(Atrichium undulatum)*. A woodland moss from the north temperate zone. It forms tufted clumps or carpets of erect stems to about 25mm with densely arranged, spreading lanceolate leaves having wavy margins. Slender, beaked SPORE CAPSULES are carried at the arching tips of reddish stems.

CATHARTIC|US, -A, -UM For plants with purgative uses.

● **CATKINS** A dense spike of usually unisexual, petal-less flowers, often flexuous and pendent (e.g. hazel) but also erect (e.g. willow).

CAT MINT *(Nepeta cataria)*. An aromatic herbaceous perennial in the *Labiatae* from Europe and Asia; also naturalised in N. America and S. Africa. The stems are 450–900mm, clad with opposite pairs of OVATE-CORDATE, toothed leaves, white hairy beneath. The 13mm long tubular, 2-lipped flowers are white with purple dots and are carried in WHORLS in the upper leaf AXILS.

CAT'S CLAW VINE *(Bignonia unguis-cati)*. An evergreen climber in the *Bignoniaceae* from Argentina. It has leaves composed of two wavy-margined, LANCEOLATE leaves and between them a 3-toed, claw-like tendril that clings readily on any rough surface. The obliquely trumpet-shaped flowers are yellow.

CAT'S FOOT
see MOUNTAIN EVERLASTING

CAT'S TAIL or **CAT TAILS**
see REED-MACE

CAT'S TAIL GRASS Several species of *Phleum*, a genus of 10 species of annuals and perennials in the *Gramineae* from the temperate regions of the world. They are typified by having tiny flowering spikelets, densely borne in erect, tail-like PANICLES. Common or smaller cat's tail *(P. bertolonii)* is a slender, tufted perennial 150–450mm tall, with the sheathing bases of the leaves somewhat inflated. The flowering tails may vary from 13–63mm in length: *P. pratense* is best known as Timothy grass. It is a robust perennial to 900mm with panicles 50–150mm long, sometimes to 300mm.

CATTLEYA A genus of 60 perennial EPIPHYTES in the *Orchidaceae* from Mexico, central and South America. They have tough, creeping RHIZOMES from which erect,

narrowly club-shaped or stem-like PSEUDOBULBS arise. These bear 1–3 narrow, leathery leaves and a terminal flowering stem carrying 1–6 or more blooms. The latter have 3 spreading PETALOID SEPALS, 2 similar petals and a broad tongue-like LABELLUM, often waved or frilled. The large flowered hybrids and species are spectacular and much cultivated, the blossoms used for corsages. *C. labiata* has PSEUDOBULBS to 300mm tall with solitary leaves, 150–250mm long. The 150–175mm flowers have deep pink PETALS and a velvety-crimson labellum.

CAUDAT|US, -A, -UM Having a tail-like organ or extension; usually a leaf, bract or SPATHE tip, e.g. *Arisarum proboscideum*.

CAUDEX A low trunk or part of the rootstock above ground, often with leaves or herbaceous stems only at the summit.

CAULESCENS [CAULESCENT] Having a stem.

▲ **CAULIFLORY [CAULIFLOR|US, -A, -UM]** Of plants which produce the flowers direct from the trunk or branches, e.g. cocoa, Judas tree.

CAULIFLOWER [BROCCOLI] *(Brassica oleracea botrytis)*. A curious MUTATION of the CABBAGE which produces large, swollen, much branched heads of small white or purplish flower buds known as curds. They are surrounded by long, wavy-margined leaves. Of similar appearance and origins are white and purple sprouting broccoli and green calabrese with looser, leafy heads.

CAULINE Usually of leaves borne on a stem, as distinct from those in a basal tuft or rosette.

■ **CEDAR** In the strict sense applied to species of *Cedrus*, majestic, evergreen coniferous trees in the *Pinaceae* from N. Africa, the eastern Mediterranean and Western Himalaya. They are of pyramidal outline when young, broad and spreading when mature. The slender needle-like leaves are scattered along the leading shoots and arranged in terminal WHORLS on the short SPUR shoots (*see* LONG AND SHORT SHOOTS). The erect cones are barrel-shaped, 75–125mm long: Atlantic cedar (*C. atlantica)* grows to 36m and maintains its pyramidal shape longer than the almost identical Cedar of Lebanon *(C. libani)*, which eventually forms a massive broad head. Both are much grown as ornamentals, especially *C. atlantica* 'Glauca' with bright, blue-green foliage. The deodar or Indian cedar *(C. deodara)* is the tallest, to 60m with long needles (25–38mm) and pendulous branch tips. *See also* JAPANESE CEDAR, PENCIL CEDAR, RED CEDAR.

CEIBA
see KAPOK

CELANDINE *(Ranunculus ficaria)*. A tuberous rooted, herbaceous perennial, also known as pilewort and lesser celandine, in the *Ranunculaceae* from Europe and W. Asia. It has rosettes of long-stalked basal leaves, rounded or somewhat angular, CORDATE. The buttercup-like yellow flowers have 8–12 narrow petals.

CELERIAC *(Apum graveolens rapaceum)*. A form of CELERY grown for its swollen, turnip-like root.

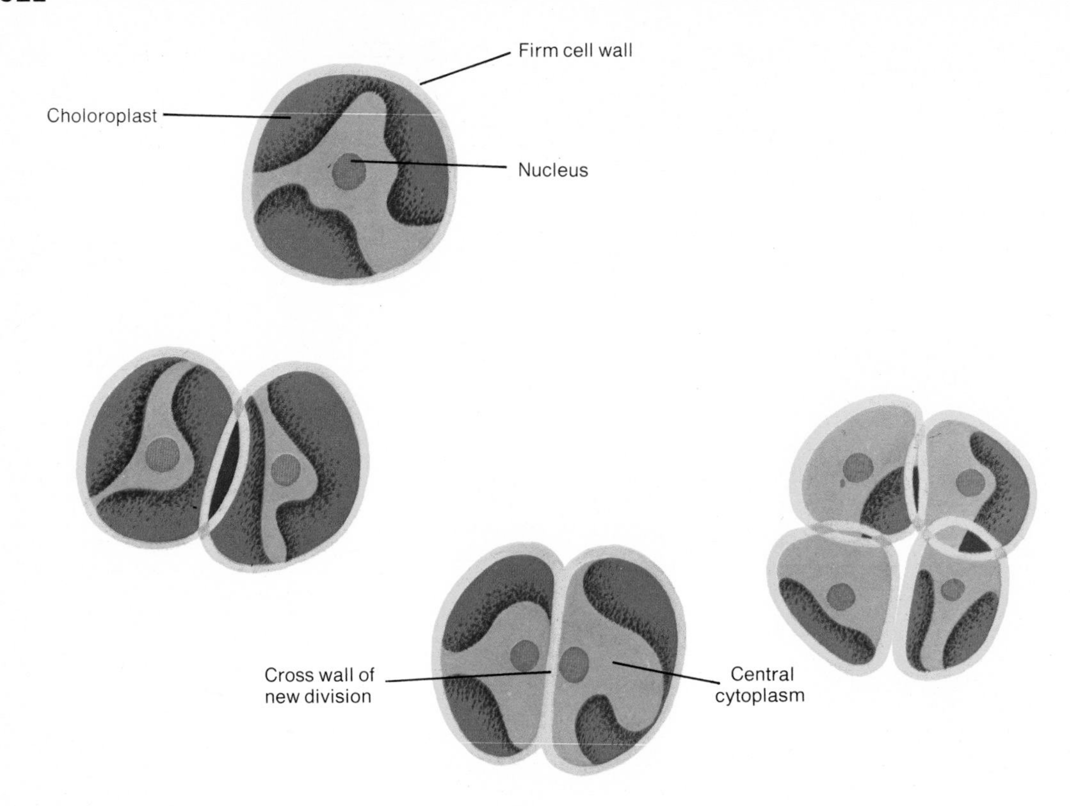

● *Cell division*

CELERY *(Apium graveolens)*. A biennial plant in the *Umbelliferae* from Europe, Asia, Africa and S. America, with edible leaf stalks. Wild celery usually grows near water and has PINNATE, basal leaves. The 300–600mm, branched flowering stems bear UMBELS of tiny, green-white flowers; garden celery *(A.g. dulce)* has larger, bipinnate leaves and thick fleshy stalks that are blanched before they are fit to eat. There are also newer green and golden-stalked cultivars that do not need blanching.

● **CELL DIVISION** In the meristematic tissues—those that are actively growing—the young cells are continually dividing by a process known as mitosis. This follows a set pattern in plants and can be looked at in a number of recognisable stages.

The entire dividing process is controlled by the nucleus, which is composed of a set number of long threads known as chromosomes. As the cell prepares to divide, these shorten and thicken and it can be seen that each one comprises two closely pressed, longitudinal halves. These are known as chromatids and are united at one central point called the centromere. This stage is known as prophase, and is followed by pro-metaphase which involves the formation of the spindle. Formed in the middle of the nucleus, the spindle is composed of gelatinous fibres that are joined at each end and bowed out in the middle. Metaphase follows as the chromosomes move to lie across the broad part of the spindle on the so-called equatorial plate, each one attached to a spindle fibre by its centromere. Anaphase quickly follows as the centromere parts and the two halves of the chromosome are pulled apart by the fibres, resulting in two complete sets of chromosomes which gather at the two poles of the nucleus. The final or telophase sees a membrane forming around each group and a prophase in reverse as they lengthen out. Two nuclei are now present and a new cell wall soon forms between them completing cell division.

The division of sex cells resulting in pollen grains and embryo sacs is somewhat different and is known as meiosis. Here the chromosomes are single and do not divide longitudinally. Instead, they pair up on the equatorial plate, and when anaphase takes place, one of each pair moves to either end of the spindle. Thus the two new nuclei and cells that form contain half the usual number of chromosomes. These two new cells then divide by the mitosis pattern, giving a group of four cells having only half the usual chromosome complement. If they are male then each of the four cells becomes a pollen grain; if female, each becomes a potential embryo sac.

CELLS The basic living units of all plants. Each cell may be likened to a microscopic box made of non-living cellulose and containing living protoplasm. The latter can be divided into two portions: the cytoplasm which makes up the bulk and the nucleoplasm which forms the small, ovoid nucleus. In the mature cell, the cytoplasm forms a layer in close contact with the cell wall (known as the plasma membrane) and with the nucleus. The remainder of the cytoplasm contains a number of spaces (vacuoles) which are filled with a watery solution of dissolved minerals. In addition to the nucleus, there are small protoplasmic bodies known as plastids. In photosynthetic cells, these become chloroplasts; in cells that store starch, leucoplasts; and in the coloured cells of petals and bracts, chromoplasts.

▲

CELLULOSE The substance of which the walls of plant cells are made. It is a carbohydrate allied to starch and is built up in the same way by glucose action.

CENSER MECHANISM The pepper-pot action of seed distribution. Capsules containing many small seeds, e.g. poppy and campion, are swung to and fro in strong winds and the seeds are flung out a few at a time through the small pores at the top.

CENTAURY Several species of *Centaurium*, mainly annuals in the *Gentianaceae* of cosmopolitan distribution: Common centaury *(C. erythraea* syn. *Erythraea centaurium)* is an erect plant to 300mm or more with a basal rosette of ELLIPTIC to OBOVATE 25–50mm long leaves. The stems carry opposite pairs of smaller, narrow leaves and are topped by dense CORYMB-like CYMES of tubular pink flowers with 5 flat petal lobes.

CENTURY PLANT
see AGAVE

CÉPE *(Boletus edulis)*. This is an edible *Basidiomycete* fungus distinguished from other toadstools and mushrooms by the spore-bearing underside of the cap being composed of narrow tubes, creating a smooth but perforated appearance. Cépe has a glossy brown convex cap to 60–200mm, whitish beneath. The stipe is thick and somewhat swollen at the base, pale brown with a paler raised network of veins.

CERASIFER|US, -A, -UM With cherry-like fruits.

▲ **CEREALS** All the grain-bearing grasses used as food by man. *See* WHEAT, OATS, BARLEY, RYE, MAIZE, RICE, MILLET.

CERIFER|US, -A, -UM Producing wax.

CERNU|US, -A, -UM Arching, nodding, drooping; usually of flowers.

CHALAZA The OVULE base, which bears an EMBRYO SAC surrounded by tiny scale-like structures (INTEGUMENTS).

CHALK GLANDS These are found at the tips or on the margins of leaves. They secrete excess water, often with dissolved lime which is left as a white deposit in evaporation, e.g. the so-called encrusted saxifrages.

CHAMAEPHYTE The second of a 7 category system of naming LIFE FORMS. Herbaceous or woody plants with winter resting buds above ground level but below 250mm in height.

CHAMOMILE Several species in three genera in the *Compositae*, but mainly *Chamaemelum nobile* syn. *Anthemis nobilis*. This is a sweetly aromatic perennial from Europe and N. Africa with creeping stems forming a mat and bi- or tripinnate leaves with LINEAR segments. The white, daisy-like flowers are about 25mm wide on 100–300mm stems. Corn chamomile *(Anthemis arvensis)* is an annual of looser, more erect growth to 450mm with white flower heads 19–32mm wide. *See also* MAYWEED.

CHAMPIGNON *(Marasmius oreades)*. An edible fungus in the *Basidiomycetes*, often forming fairy rings on lawns or closely grazed grassland. It has a 19–50mm wide light brown often pink-tinged cap flattened or slightly depressed in the middle with a central raised knob (UMBO). The spore-bearing gills and stipe are pale buff.

CHANNELLED WRACK
see WRACK

CHANTERELLE *(Cantherellus cibarius)*. An edible *Basidiomycete* fungus from woodlands in the northern temperate zone. It is wholly orange-yellow, smelling faintly of apricots. The funnel-shaped cap is 38–88mm across.

CHAPARRAL Scrub vegetation characteristic of dry, rocky hillsides in California, composed of dwarf oaks, california lilac, bearberry, buckeye, roses, *Baccharis*, *Garrya*, *Rhus*, etc.

CHARDS The branched shoots of globe ARTICHOKE. *See also* SPINACH BEET.

CHARLOCK *(Sinapis arvensis)* a roughly hairy annual, often a weed of fields, belonging to the *Cruciferae* from Europe, N. Africa, Asia and introduced elsewhere. The stems are 300–750mm tall with LYRATE leaves to 200mm long. The yellow, 4-petalled flowers are borne in RACEMES.

CHARTACEOUS Papery or parchment-like.

CHASMOPHYTE Descriptive of plants which usually grow in deep shady valleys, ravines or chasms.

CHECKERBERRY
see SERVICE TREE, WINTERGREEN

CHERIMOYER *(Annona cherimolia)*. A small evergreen tree to about 6m in the *Annonaceae* from Peru. It has OVATE-LANCEOLATE aromatic leaves, velvety-hairy beneath, and greenish-yellow, cup-shaped flowers 32mm long. The pleasantly acid fruits are roughly globular, about 100mm wide.

CHERRY Two species of deciduous trees in the *Rosaceae* from Europe and Asia. They have ELLIPTIC to OBOVATE toothed leaves, 50–125mm long, with slender points and 5-petalled, 19mm wide, white flowers in UMBELS of 2–6: wild cherry or gean *(Prunus avium)* can reach 24.4m with cupped flowers and dark red bitter or sweet cherries; sour cherry *(P. cerasus)* rarely exceeds 7.5m, has flat flowers and bright red, acid cherries. The sweet cherries of commerce are derived from both species, being largely of hybrid origin. There are many cultivars with large, fleshy cherries varying from black-red to yellow, flushed with red.

CHERRY LAUREL *(Prunus laurocerasus)*. An evergreen shrub or small tree to 6m in the *Rosaceae* from S.E. Europe and S.W. Asia. It has oblong, OBOVATE, distantly toothed, leathery leaves to 200mm long, and erect AXILLARY RACEMES of small, dull white flowers. The bitter fruits are like small, black cherries. Somewhat similar is Portugal laurel *(P. lusitanica)* from Spain and Portugal, but the leaves are smaller, SERRATE and OBLONG-OVATE.

CHERRY PIE
see HELIOTROPE

CHERRY PLUM
see PLUM

CHERVIL *(Anthriscus cerefolium)*. An erect, branched annual, 300–450mm tall in the *Umbelliferae* from E. Europe to Siberia and introduced to many other countries. It has fern-like tripinnate leaves with PINNATIFID leaflets and tiny, white flowers in 19–50mm wide UMBELS. The leaves taste and smell sweetly aromatic and are used for garnishing, seasoning and in salads. Rough chervil *(Chaerophyllum temulentum)* is a similar biennial species wild in hedgerows. It has rough purple-spotted stems and UMBELS that are pendulous in bud.

CHESTNUT
see HORSE CHESTNUT; SWEET CHESTNUT

CHICK PEA *(Cicer arietinum)*. An erect, well-branched annual in the *Leguminosae* probably from W. Asia. It is not known in the wild, but is much grown as a seed crop in India where it provides valuable protein. The leaves are PINNATE with about seven pairs of SERRATE, OVATE leaflets. The 8mm long pea-flowers are white, pink or bluish and followed by inflated, hairy pods 19–32mm in length containing one of two large furrowed peas which may be white, yellow, brown, red or almost black.

CHICKWEED *(Stellaria media)*. An annual weed of fields and gardens in the *Caryophyllaceae*, cosmopolitan in temperate regions. It forms prostrate mats of slender stems with pairs of 6–19mm ovate leaves and tiny flowers, having 5 tiny white bifid petals almost hidden by green sepals.

CHICLE [CHICELE] The milky sap of SAPODILLA which forms the basis of chewing gum.

CHICORY *(Cichorium intybus)*. An erect biennial in the *Compositae* from Europe, but introduced into many other countries. It forms basal rosettes of OBLANCEOLATE, RUNCINATE leaves, followed by 45–120cm branched stems bearing small clusters of 25–38mm wide flower heads composed of LIGULATE florets like sky-blue dandelions. The young leaves are blanched and used in salads and the dried, ground roots form an adulterant of coffee.

CHILEAN FIRE BUSH *(Embothrium coccineum)*. Despite its vernacular name, this is a semi-evergreen tree to above 18.2m in the *Proteaceae* from Chile. It has OVATE-LANCEOLATE, 50–100mm long leaves and profuse clusters of crimson-scarlet tubular flowers.

CHILE PINE
see MONKEY PUZZLE

CHIN In certain orchids, an outgrowth from the RECEPTACLE bearing the lateral sepals and LABELLUM in such a way that the sepals appear to grow out of the base of the labellum.

CHINA ASTER *(Callistephus chinensis)*. An erect annual in the *Compositae* from China. It varies from 150mm to 600mm in height, with coarsely toothed ovate leaves to 100mm long and terminal daisy-like flower heads 63–125mm wide. The wild type has dark blue-purple ray florets and yellow DISK florets. The cultivated forms come

in red, pink, purple, blue and white. Double-flowered sorts are much grown as ornamentals, the ostrich plume having very long, curled florets.

CHINESE GOOSEBERRY *(Actinidia chinensis)*. A vigorous, woody climber in the *Actinidiaceae* from China. Growing to 9m or more, it has broadly OVATE, CORDATE leaves 125–200mm long and AXILLARY clusters of 38mm bowl-shaped, cream or buff, 6-petalled flowers. The OBLONG to OVOID rusty, hairy fruits are likened to a gooseberry in flavour, but are more like a melon with a touch of strawberry. They are grown commercially in New Zealand and exported to Europe and elsewhere as Kiwi fruit.

CHINESE LANTERN
see BLADDER CHERRY

CHINESE SACRED LILY
see NARCISSUS

CHIVES *(Allium schoenoprasum)*. A tufted, herbaceous, perennial in the *Alliaceae* from the northern hemisphere. It has tubular, grassy leaves to 300mm long, smelling and tasting of onions, and spherical heads of starry pink or rose-purple flowers. It is much grown for flavouring salads, omelettes, etc.

CHLAMYDOMONAS A group of single-celled, motile algae which live mainly in still, fresh water. They are microscopic ovoid or spherical organisms with a cellulose wall that thickens to a colourless beak at one end. The cell contents are those usually found in a plant (*see* CELL), but with a single, large chloroplast in which is embedded a protein body or pyrenoid, associated with starch storage. At the apex of the cell are two fine thread-like outgrowths of the CYTOPLASM, known as CILIA or FLAGELLA. They make rapid backward strokes and pull the cell through the water. Sexual and asexual methods of reproduction take place, the former being the commoner. The contents of a cell divide mitotically into 2, 4 or 8 miniature cells called ZOOSPORES which swim away when the parental cell wall ruptures or dissolves. Some species of *Chlamydomonas* contain red pigment, and *C. nivalis* is sometimes so abundant it colours the snow red.

CHLOROPHYLL The green colouring matter found in all plants except fungi and a few PARASITIC flowering plants. It is usually located in special plastids in the cytoplasm of cells, known as chloroplasts. By means of the energy of sunlight, carbon-dioxide and water, chlorophyll builds up carbo-hydrates essential for the growth of the plant and much utilised by the animal kingdom.

CHLOROPLAST
see CHLOROPHYLL

CHLOROSIS A disorder or disease of plants showing as whitish areas in the leaves due to lack of chlorophyll. This may be caused by mineral deficiencies, viruses or mutation, the last two providing variegation in plants.

CHOCOLATE Derived from the seeds (beans) of COCOA.

CHRISTMAS CACTUS
see CACTUS

● **CHRISTMAS ROSE** *(Helleborus niger)*. An evergreen perennial in the *Ranunculaceae* from C. Europe to W. Asia. It has deeply divided, PEDATE leaves, each of the 7–9 lobes OVATE-CUNEATE, toothed at the tip. The large cup-shaped flowers are composed of 5 or 6 white sepals, 2 or 3 opening on 150–300mm stems.

CHRISTMAS TREE Several trees are known under this name, but in Europe the Norway spruce is the main one. *see* SPRUCE.

CHRIST'S THORN *(Paliurus spina-Christi)*. A thorny deciduous shrub to 3m from the E. Mediterranean. It has 19–38mm long OVATE leaves each with two spines at its base, one straight and the other hooked. Tiny greenish flowers are borne in AXILLARY UMBELS and followed by circular fruits rimmed with a broad wing.

CHROMOPLASTS Irregular plastids which contain colouring pigments mainly yellow to red shades, responsible for the colouring of petals, sepals and bracts.

CHROMOSOMES Thread-like bodies (except during cell division), which largely form the nucleus of a cell. Each chromosome is composed of a string of bead-like granules known as genes which control the development of the plant. The shape and number of chromosomes are constant for each species of plant. The basic number in each vegetative cell (leaves, stems, etc.) is of two sets, one from the male parent, one from the female, and is known as diploid. The single set found in sex cells is known as haploid. Many plants have multiples of the haploid or diploid number: 3 sets is triploid, 4 is tetraploid, 6 is hexaploid. All these are collectively known as polyploids.

CHRYSANTHEMUM A genus of 200 species in the *Compositae* from the northern hemisphere. They vary from low, mat-forming plants to tall perennials and sub-shrubs. The leaves are often variously lobed and the flowers of daisy form: *C. alpinus* is of sprawling, tufted growth to 150mm with PINNATISECT, LINEAR-lobed leaves and 19–32mm wide white flower heads. The popular florist's chrysanths, known in America as mums, are mainly double-flowered perennials of hybrid origin from China and Japan. They have broadly lobed, oak-like leaves and flower heads that range from 25–250mm in a wide selection of colours and forms.

CHRYSE|US, -A, -UM Golden yellow.

CHUFA
see GRASS NUT

CILIA [CILIUM] Thread-like outgrowths of the cytoplasm found in unicellular plants and motile spores. *See* CHLAMYDOMONAS; ZOOSPORES.

CINCINNUS A monochasial CYME, the successive lateral branches of which are borne on opposite sides of the stem, e.g. bugloss.

CINERARIA *(Senecio cruentus* syn. *Cineraria cruentus)*. A biennial or short-lived, sub-shrubby perennial to 300–900mm from the Canary Islands. It has ORBICULAR-CORDATE leaves with small, angular lobes and broad, terminal PANICLES of 19–38mm daisy flowers. Cultivars

▲

have red, purple, blue, cream and bi-coloured flowers.

CINERE|US, -A, -UM Grey or ashy-coloured.

CINNAMON *(Cinnamomum zeylanicum)*. An evergreen tree to 6m in the *Lauraceae* from S.E. Asia. It has OVATE-oblong leaves and small, greenish-white flowers followed by oval, berry-like fruits. Cinnamon is obtained from the bark of young trees.

CIRCINATE Rolled like a watch spring, e.g. the young fronds of ferns.

CIRCUMSCISSILE Of capsules that split around transversely, the top coming away like a lid.

CIRRHOS|US, -A, -UM, CIRRHAT|US, -A, -UM Bearing tendrils.

CITRIN|US, -A, -UM Lemon-coloured.

▲ **CITRUS** The generic name for all the citrus fruits, e.g. orange, lemon, grapefruit, tangerine, etc. All grow on evergreen, often spiny shrubs or trees in the *Rutaceae* from tropical Asia, but much grown elsewhere. The narrowly to broadly ovate leaves often have a winged leaf-like stalk, giving the appearance of two leaves end to end. The fragrant white flowers have 4–8 fleshy petals and are followed by the characteristic fruits, each of which is technically a berry known as a hesperidium. The citron *(C. medica)* has oblong, yellow fruits and stout spines, the leaf stalks are not winged; grapefruit *(C. paradisi)* has large, globular yellow or orange-yellow fruits; lemon *(C. limon)* has yellow fruits and stout spines; lime *(C. aurantifolia)* has small, almost globular, sour red fruits; tangerine, mandarin and satsuma *(C. reticulata)* have small, flattened, globose fruits, the mandarin with pale orange or yellow peel; pomelo (pummelo) or shaddock *(C. grandis)* is a spiny tree with large fruits up to 300mm across, either globose or pear-shaped, greenish or yellowish with sweet pinkish or yellow flesh. It is rarely seen away from the tropics; sweet orange (*C. sinensis)* is a variable species with small to large, almost globose to ovoid fruits having sweet pulp; sour or Seville orange *(C. aurantium)* has sour, bitter fruits cooked for marmalade. The peel is used medicinally and *C.a. bergamia* yields bergamot oil.

CLADODE A flattened leaf-like stem of one joint, e.g. butcher's broom.

■ **CLARKIA** A genus of 36 annual plants in the *Onagraceae* from N.W. America and Chile. They have linear to ovate oblong leaves and erect spikes of 4-petalled flowers. Some species are showy and grown as ornamentals: *C. pulchella* is 300–450mm tall with LANCEOLATE leaves and 19–38mm flowers in shades of lilac, violet, rose and white.

CLARY Several species of *Salvia*, perennial plants in the *Labiatae* from Europe. They have opposite pairs of ovate, RUGOSE leaves and RACEMES of tubular, prominently 2-lipped flowers: common clary *(S. sclarea)* grows to 1.2m with white or bluish flowers; wild clary *(Salvia horminoides)* is 300–900mm with toothed or PINNATELY lobed leaves and 13mm violet-blue flowers; some forms are CLEISTOGAMOUS.

CLASSIFICATION An arrangement of the plant kingdom according to the characteristics they have in common. It starts with the simplest or most primitive plants, like unicellular algae and finishes with the most highly evolved types such as orchids. Each class is further subdivided, e.g. the DICOTYLEDONS break down into orders, families, genera and species. The species is the lowest basic unit, though for convenience very variable species are further broken down into sub-species and varieties. *See also* NOMENCLATURE.

CLAVATE [CLAVAT|US, -A, -UM] Club-shaped.

CLAW The tapered base of a petal, the point of which is attached to the RECEPTACLE. It is best seen in members of the *Cruciferae*.

CLAYTONIA (syn. *Montia*). Annual and perennial plants in the *Portulacaceae* from eastern Siberia and N. America. They have fleshy, long-stalked ovate to elliptic leaves and 5-petalled flowers in RACEME-like CYMES. *M. perfoliata* (miners' lettuce in the USA) has white flowers which arise from a widely funnel-shaped structure formed of 2 fused bracts; *M. sibirica* (pink purslane) is similar, but has white or pinkish flowers and the 2 bracts are sessile but not fused.

CLEAVERS *(Galium aparine)*. An over-wintering prostrate or scrambling annual in the *Rubiaceae* from Europe and Asia, naturalised elsewhere. It has 4-angled stems with down-pointing prickles and WHORLS of 6–9 LINEAR-OBLANCEOLATE leaves with prickly margins. The tiny white, cross-shaped flowers give way to pairs of small, globular fruits covered with hooked bristles. Also known as goosegrass, sweethearts and sticky Willie.

CLEISTOGAMY Describing specially modified flowers which remain in the bud stage. They have ANTHERS and STIGMAS close together and pollination takes place within the bud. Violets are a good example.

CLEMATIS A genus of 230 species, mainly woody climbers in the *Ranunculaceae* from temperate regions of the world. The climbers usually have COMPOUND, often PINNATE leaves, the mid-rib of which acts as a tendril. The flowers have 4 coloured sepals, the petals being absent. Some species have PETALOIDS that look like true petals and hybrid cultivars often have 6 or more. Individual flowers may be bell-shaped *(C. tangutica)* or opened out flat *(C. X. jackmannii)*, both of which are much grown in gardens. *See* TRAVELLER'S JOY.

CLIMBING FERN Several species of *Lygodium*, ferns with tall twining fronds in the *Schizaceae*, widely distributed in the tropics and sub-tropics. The PINNAE are borne in alternate pairs, usually PINNATELY divided, and the SPORANGIA occur in spikes: *L. japonicum* grows to 1.8m or more with PINNAE to 200mm long.

CLIMBING PLANTS Usually fast growing plants with flexible stems that get into the sunlight by clambering on other plants. In many climbers the stems twist tightly around a support, others cling by sensitive leaf-stalks or tendrils which embrace any sufficiently thin object. Some climbers are really clamberers, pushing their stems through trees and bushes but not strictly holding on. Hooks prevent them from slipping back; e.g. roses. Some species hold on by their roots; ivy is an example of this.

CLONE A group of identical plants all raised vegetatively from one individual, e.g. all 'Cox's Orange' apples have been raised, by grafting, from the one original tree.

CLOUDBERRY *(Rubus chamaemorus)*. A RHIZOMATOUS perennial in the *Rosaceae* from circumpolar regions and northern mountains. It has 50–125mm herbaceous stems and almost ORBICULAR, CORDATE leaves PALMATELY 5–7 lobed. The white, 4–5-petalled flowers are followed by orange fruits made up of several large DRUPELETS.

CLOVER A general name for a number of species of *Trifolium*, small annual and perennial plants in the *Leguminosae* mainly from the northern hemisphere. They have TRIFOLIATE leaves and slender, small white, pink or red pea-flowers in crowded heads. The several yellow-flowered species are known as trefoils: Dutch or white clover *(T. repens)* is a prostrate, perennial plant with round heads of white or faintly pink-flushed flowers on erect stalks to 150mm or more; red clover *(T. pratense)* is a tufted, erect or semi-erect perennial plant with oval heads of pink to red-purple flowers; hop trefoil *(T. campestre)* is a semi-erect annual with rounded, yellow heads that turn pale brown and look like tiny hops when faded; lesser yellow trefoil, shamrock or suckling clover *(T. dubium)* is a slender annual with small, loose heads of yellow flowers; hare's foot clover *(T. arvense)* is an erect, softly hairy annual with white hairy OVOID to cylindrical flower heads and tiny pinkish florets.

CLOVES The dried flower buds of *Eugenia caryophyllus*, syn. *E. aromatica*, an evergreen tree to 12.2m in the *Myrtaceae* from Indonesia, but much cultivated elsewhere. The leaves are LANCEOLATE and the familiar flower buds are borne in small, terminal PANICLES. Left on the tree, they open to 100mm wide, 4-petalled, yellow flowers.

CLUBMOSS Moss-like plants classified in the same division as ferns *(Pteridophyta)* and belonging to two main genera of the *Lycopodiaceae*. The true clubmosses *(Lycopodium)* are erect or prostrate plants with tiny, LINEAR or LINEAR-LANCEOLATE overlapping leaves. The spores are borne in cones at the tips of the shoots: fir club moss *(L. selago)* is a branched, erect plant to 150mm or so from cool, northern temperate areas, mainly on mountains. Its SPORANGIA are borne in the leaf AXILS in zones on the stems rather than in proper, terminal cones; stagshorn or common clubmoss *(Lycopodium clavatum)* is a much branched, procumbent plant, 300–900mm long, the small, overlapping, LINEAR leaves ending in white flexuous hairs. The cylindrical sporing cones are 13–38mm long and borne erect on short stems; lesser clubmoss *(Selaginella selaginoides)* is a slender, prostrate plant with erect stems about 50mm high and cones about 13mm in length.

CLUSTER CUPS
see AECIDIAL CUPS

CLYPEAT|US, -A, -UM Shaped like a circular roman shield, the clipeus.

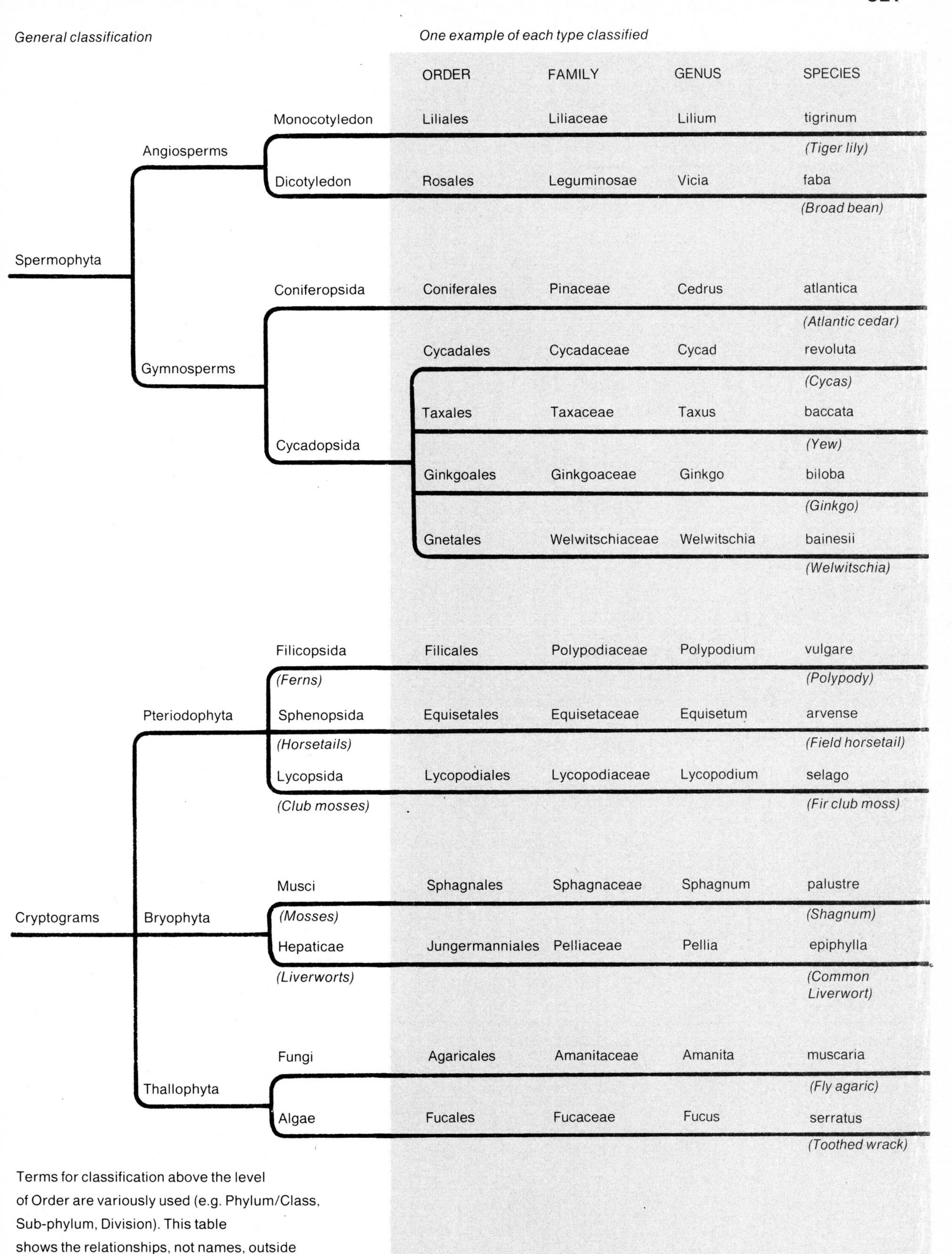

Terms for classification above the level of Order are variously used (e.g. Phylum/Class, Sub-phylum, Division). This table shows the relationships, not names, outside the tinted area.

COCAINE The pain-killing drug derived from *Erythroxylon coca*, an evergreen shrub in the *Erythroxylaceae* from S. America. It has OVAL to OBOVATE 50–75mm long leaves and AXILLARY clusters of greenish or yellowish 9mm wide flowers.

COCCINE|US, -A, -UM Scarlet.

COCCUS Spherical bacteria.

COCHLEAR|IS, -E Spoon-shaped, usually of leaves.

COCKLE-BUR *(Xanthium strumarium)*. A robust annual in the *Compositae* from the Americas and naturalised elsewhere. The 300–750mm stems bear triangular-OVATE, CORDATE leaves, PALMATELY lobed or toothed. The flower clusters are unisexual, males on the upper part of the plant, females below. The latter form hard, ovoid 13mm long fruits covered with hooked spines.

COCK'S COMB curious MUTANT of *Celosia argentea* in the *Amarantaceae* from tropical Asia. *C. argentea* is an erect annual with LANCEOLATE leaves and a dense terminal plume of tiny, white or red flowers; cock's comb *(C.a. cristata)* has the plume tightly condensed into a broad, arching crest.

COCK'S FOOT GRASS *(Dactylis glomerata)*. A robust species in the *Gramineae* from Europe, N. Africa and Asia, and introduced into other temperate countries. It has broad, rough leaves and stems to 900mm, topped by a densely lobed PANICLE of green or purplish spikelets.

● **COCOA** *(Theobroma cacao)*. A small evergreen tree to 4.5m in the *Sterculiaceae* from Central America and widely cultivated elsewhere. It has oblong, lanceolate leaves and clusters of 5-petalled, pink flowers direct from the trunk and branches. The grooved, ovoid pods are red or yellow, 150–200mm long and contain 50–100 cocoa beans imbedded in pulp. Beans and pulp are fermented, then dried before roasting and grinding.

COCO-DE-MER
see PALM

CODLINS AND CREAM
see WILLOW HERB

COELEST|IS, -E Sky blue.

COENOCYTE A filamentous alga without cell walls (non-septate), the filaments appearing like branched, hollow tubes. Cytoplasm with embedded nuclei and chloroplasts line the insides of the tubes, representing the combined contents of many cells.

COERULE|US, -A, -UM Blue.

▲ **COFFEE** *(Coffea arabica)*. An evergreen shrub to 4.5m in the *Rubiaceae* from Ethiopia, but much cultivated in Brazil and other tropical countries. It has opposite pairs of OBLONG-OVATE leaves and AXILLARY clusters of fragrant, 5-petalled white flowers. The cherry-like, red fleshy fruits contain two flattened ovoid seeds (the coffee beans of commerce) which are dried, roasted and ground.

COLA [GOORA NUT] *(Cola acuminata)*. An evergreen tree to 12.2m in the *Sterculiaceae* from tropical Africa. It has 100–150mm long OBLONG-OVATE, ACUMINATE leaves and PANICLES of 13mm wide, yellow flowers. The 125–150mm long woody fruits contain several glossy, brown nuts.

COLCHICUM A genus of 65 bulbous -rooted perennials in the *Liliaceae* from Europe, Central Asia and N. India. Ribbed LANCEOLATE to OVATE leaves arise in clusters of 3–8 from ovoid corms. The purple to lilac, rarely yellow, often chequered blooms open before the leaves from autumn to spring: *C. autumnale* has 100–150mm rose-lilac blooms and 150–250mm leaves: *C. speciosum* has 150–200mm rose-purple flowers and leaves 300mm or more long: *C.s.* 'Album' is pure white.

COLEOPTILE The outermost sheathing leaf of a very young grass seedling, which protects it as it pushes through the soil.

COLEORHIZA The protective sheath around the emerging root of a germinating grass seed.

COLLATERAL BUNDLE
see VASCULAR BUNDLE

COLLECTIVE FRUIT A tight or fused cluster or spike of small fruits forming a bigger one, e.g. pineapple, where a spike of flowers gives way to the well-known barrel fruit.

COLLENCHYMA Plant cells with a thickened strengthening of extra cellulose at the corners.

COLLIN|US, -A, -UM Growing on hills.

COLONIST [COLONISATION] The spread of a plant, usually by seeds into a new area, e.g. birch seedlings invading a burnt-over heath or mangroves extending into newly-formed mud banks.

COLTSFOOT *(Tussilago farfara)*. A vigorous, RHIZOMATOUS, herbaceous perennial in the *Compositae* from Europe, Asia and N. Africa. The solitary, yellow 19mm flower heads are borne on 75–150mm woolly purple-bracted stems in early spring before the leaves. These are 100–200mm long, rounded, toothed and 5–12 lobed.

■ **COLUMBINE** *(Aquilegia vulgaris)*. A tufted perennial in the *Ranunculaceae* from Europe, N. Africa and Asia and naturalised in N. America. It has BITERNATE basal leaves and erect 45–150cm stems bearing irregularly CYMOSE inflorescences of nodding flowers sometimes referred to as grannies' bonnets. Each purple-blue flower has 5-PETALOID sepals and 5-spurred tubular petals; red and white forms are known. Other species of *Aquilegia* are sometimes referred to as columbine.

COLUMN The central part of an orchid flower which bears the ovary, stigma and pollinia.

COMAT|US, -A, -UM, COMOS|US, -A, -UM Tufted, e.g. carrying tufts of hairs, bracts, sterile flowers.

COMFREY Several species of *Symphytum*, perennials in the *Boraginaceae* from Europe to W. Asia. They have fleshy roots, sometimes RHIZOMATOUS, with ovate to

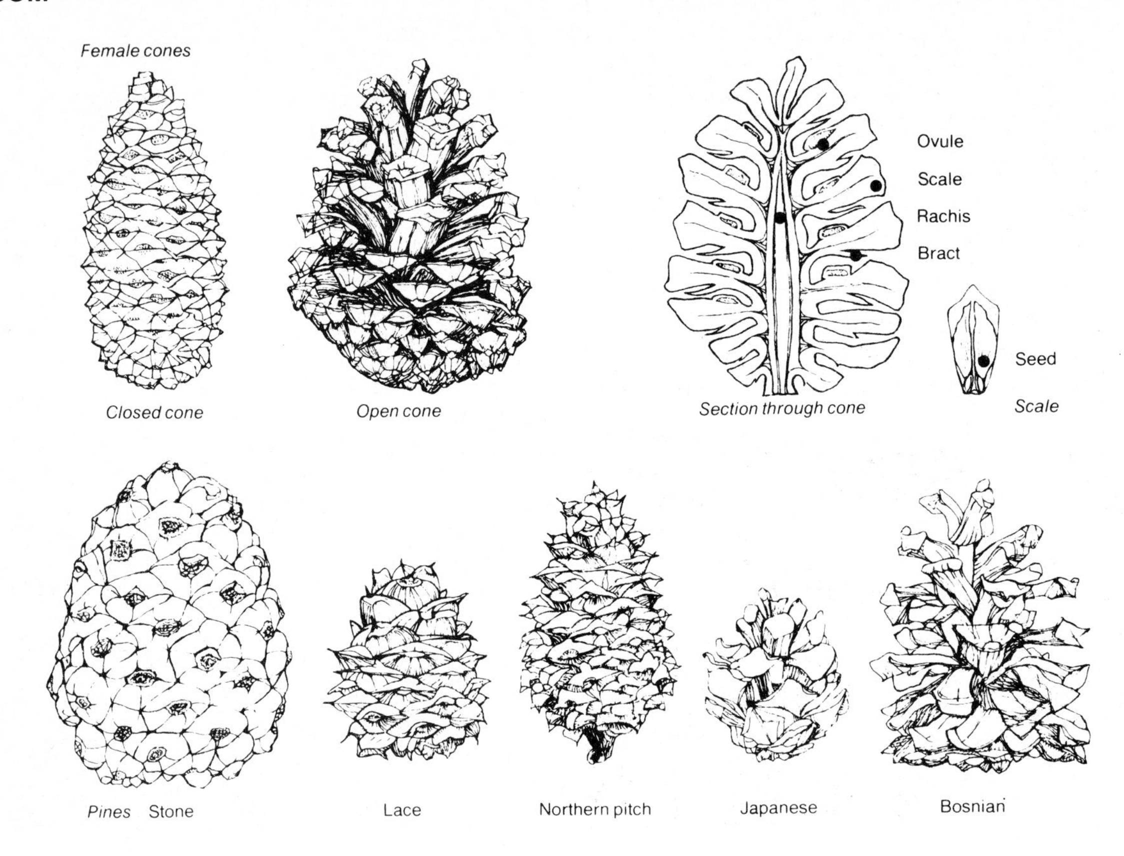

lanceolate leaves and tubular or funnel-shaped flowers in forked CYMES. Russian comfrey *(S. X uplandicum)* is commonly seen by roadsides in Great Britain. It is a variable, robust plant, with hispid, hairy, LANCEOLATE leaves to 300mm and stems 60–150cm with narrower DECURRENT leaves. The 13–19mm flowers can be blue or purple.

COMMUNIS, -E Common, occurring in numbers.

COMMUNITY A natural association of plants suited to a particular environment, e.g. a woodland with its backbone of trees and understory of shrubs with smaller plants growing on the soil beneath.

COMPASS PLANT *(Silphium laciniatum)*. A herbaceous perennial in the *Compositae* from N. America. It has erect stems 1.2–3.0m tall, bearing PINNATELY lobed leaves and daisy-like flower heads with yellow ray florets. The basal leaves are OVATE entire and held vertically with their edges aligned north-south. PRICKLY LETTUCE carries its stem leaves in the same way.

COMPLANAT|US, -A, -UM With a flattened surface.

COMPLICAT|US, -A, -UM Complicated, e.g. intricately folded petals.

COMPOUND Used for parts of a plant that are multiple, e.g. PINNATE and bipinnate leaves and double UMBELS of flowers.

CONCEPTACLE Minute flask-shaped cavities in the surface of seaweed fronds, e.g. bladder wrack, opening by a pore-like hole known as an ostiole. They contain ANTHERIDIA and/or oogonia.

CONCINN|US, -A, -UM Neat, elegant or pretty.

CONCOLOR Of uniform colour and shade.

CONDUPLICATE Folded lengthwise.

● **CONE** In the popular sense the fruit of pines, firs and other conifers, composed of woody central stems with overlapping scales that bear the seeds. Botanically a cone or strobilus is a spike-like structure that bears either spores (clubmosses and horsetails) or seeds.

CONFERT|US, -A, -UM Crowded together or compact.

CONFUS|US, -A, -UM Confused, e.g. one plant for another.

CONGLOMERAT|US, -A, -UM Clustered closely together.

CONIDIUM [CONIDIA] SPORANGIA which germinate directly and produce new HYPHAE, e.g. blight (potato).

CONIFER [CONIFEROUS] A colloquial name for members of the *Gymnospermae* division of plants, particularly *Coniferales*, i.e. pines, firs, spruces.

CONJUGATE [CONJUGATION] The fusion of male and female gametes that resemble each other in shape and

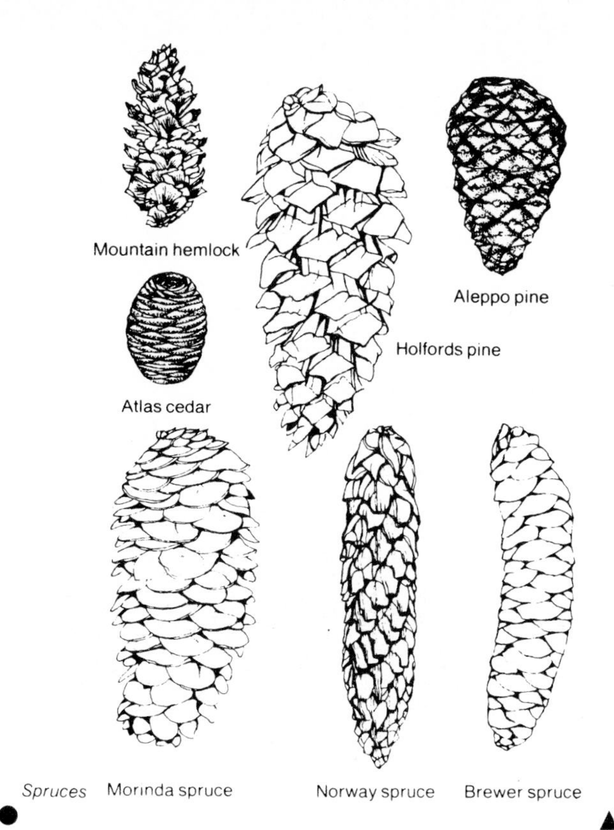

Spruces Morinda spruce Norway spruce Brewer spruce

●

▲

size. *See also* FERTILISATION, where the gametes are differentiated into a large OVUM and a small SPERMATOZOID.

CONNATE Joined together, e.g. the twin flowers of fly honeysuckle or the leaf bases of teasel.

CONNECTIVE The strip or web of tissue that joins the two anther lobes of a stamen and is in turn joined to the filament.

CONNIVENT Converging, e.g. leaf veins.

CONSIMIL|IS, -E Much resembling, e.g. another plant.

CONVOLUT|US, -A, -UM Rolled longitudinally, e.g. leaf margins.

COPAL A RESIN used in varnish-making, obtained from several tropical trees in the *Leguminosae*, in particular *Copaifera officinalis*, a small, evergreen tree from tropical America. It has PINNATE leaves comprising 4–10, unequally OVATE leaflets, and PANICLES of small, white flowers.

COPPICE [COPPICING] The regular cutting down of a tree or shrub to ground level. This results in a quantity of young shoots arising adventitiously. These are later cropped for various purposes, e.g. one year shoots of willow for basket-making; several year shoots or stems for fencing, bean poles etc. The past method of management for almost all our woodland.

▲ **COPSE** A wooded area of well-spaced, long-trunked (standard) trees with an undergrowth of coppiced trees or shrubs, e.g. oak trees with hazel or sweet chestnut, COPPICE.

CORAL BERRY
see SNOWBERRY

CORALROOT *(Corallorhiza trifida)*. A SAPROPHYTIC plant in the *Orchidaceae* from the northern temperate regions. Growing mainly in peaty, mossy woods, the 75–200mm erect stems are yellow-green, lacking chlorophyll. The flowers are yellow or olive, sometimes marked red, with narrow petals and sepals grouped round the oblong, 3-lobed lip.

CORALSPOT *(Nectria cinnabarina)*. A SAPROPHYTIC or PARASITIC fungus in the *Ascomycetes* easily recognised by the bright red or pinkish, pustule-like fruiting bodies. These are about 3mm wide and occur in large groups on dead stems, particularly on currant bushes.

CORD GRASS Several kinds of *Spartina*, perennial, RHIZOMATOUS plants in the *Gramineae*, native to mud flats by the sea in temperate regions: common cord grass (*S. X anglica)* is a fertile TETRAPLOID form of *S. X townsendii*, a hybrid between *S. alternifolia* and *S. maritima*. It is a robust plant, 30–120mm tall with greyish-green leaves and dense, erect PANICLES of 2–12 spikes. The small spikelets are closely overlapping. It is also much planted as a mud binder.

CORDATE [CORDIFORM|IS, -E] Heart-shaped; usually referring to an OVATE leaf with two basal rounded lobes.

CORIACE|US, -A, -UM With a stiff, leathery texture.

CORIANDER *(Coriandrum sativum)*. An erect annual from S. Europe in the *Umbelliferae*, grown for its aromatic seeds used to flavour liqueurs, confectionery and bread; also an ingredient fo curry powder. It grows to 300mm tall, the lower leaves bipinnate with lobed leaflets. The tiny, white or pinkish flowers are borne in 25mm wide UMBELS. The two-seeded fruits are globose to 9mm.

● **CORK** Layer of cell tissue that forms in the epidermis or near the surface of the cortex, or within the bark layer, and becomes coated with suberin, a waxy substance which makes it waterproof. Such tissue largely forms the bark layer of some trees and woody plants, e.g. cork OAK. Cork is formed by a special cambial layer called the phellogen.

CORM An underground storage organ derived from a stem base greatly swollen laterally. It is usually of annual duration, a new one forming on top of the old one each year. *Crocus* and *Gladiolus* are common examples.

CORMLET A tiny corm, usually formed at the base of a full sized one, but sometimes in the axils of the stem leaves.

CORN
see BARLEY, MAIZE, OATS, RYE, and WHEAT.

CORN COCKLE *(Agrostemma githago* syn. *Lychnis g.)*. A sparingly branched annual in the *Caryophyllaceae* probably from the Mediterranean area but widely naturalised in many temperate regions. It has opposite pairs of 50–125mm LINEAR-LANCEOLATE leaves and terminal salver-shaped, 25–38mm red-purple flowers.

CORN CROWFOOT
see BUTTERCUP

CORNELIAN CHERRY *(Cornus mas)*. A deciduous shrub or small tree to 7.5m in the *Cornaceae* from Central and Southern Europe. It has opposite pairs of 38–63mm long OVATE leaves and small UMBELS of 4-petalled yellow flowers before the leaves. The 13–19mm ovoid, glossy red fruits are edible.

CORN FLAG
see GLADIOLUS

CORNFLOWER *(Centaurea cyanus)*. An erect annual in the *Compositae* from Europe and W. Asia but naturalised elsewhere. It has lanceolate leaves, the lowest ones often pinnatifid, distantly toothed and grey, cottony hairy. The 25–38mm wide flower heads are composed of numerous slender, tubular florets of bright blue and red purple. Pink, red and white forms are grown.

CORN MARIGOLD *(Chrysanthemum segetum)*. A branched or single stemmed annual in the *Compositae*, probably from S. Europe and W. Asia but naturalised in other countries. It has GLAUCOUS leaves, the lowest OBLONG-CUNEATE, coarsely toothed or PINNATIFID. The 32–57mm wide flower heads are golden yellow.

CORNUT|US, -A, -UM [CORNICULAT|US, -A, -UM] Bearing a horn-like appendage, e.g. the spurs of the flowers of violets.

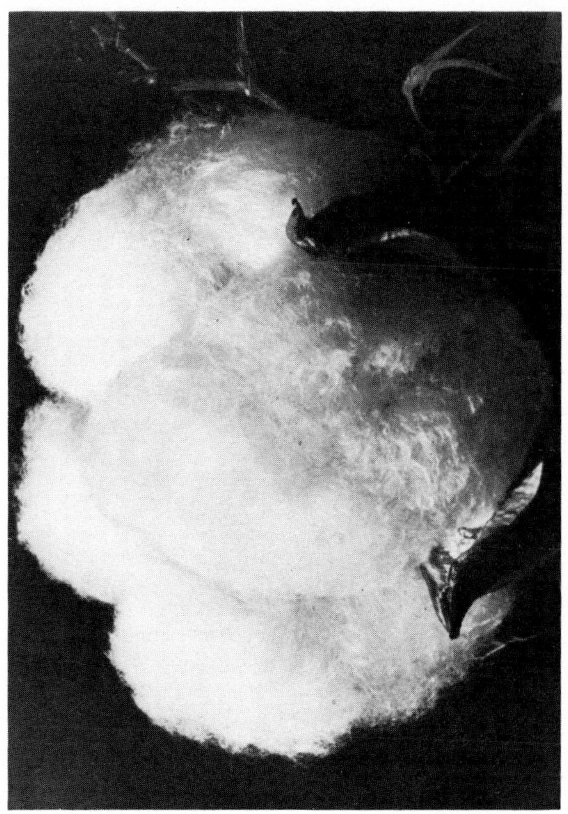

■ **COROLLA** The usually colourful part of a flower, comprising the petals which may be separate as in buttercup, or fused into a tube or trumpet, e.g. morning glory. With the SEPALS, they protect the vital sexual organs in the bud stage. Coloured petals also serve to attract pollinating insects.

CORONA An outgrowth of petal or perianth tissue forming a cup or trumpet, e.g. daffodil or narcissus. A false corona is formed in the sea-daffodil and its allies by the fusion of greatly flattened stamen filaments. The ring of thread-like outgrowths found in PASSION FLOWERS is also called a corona.

CORONARI|US, -A, -UM Crown-like or forming a crown.

CORONAT|US, -A, -UM Crowned or wreath-like.

CORPSE PLANT
see YELLOW BIRD'S-NEST

▲ **CORYMB [CORYMBOS|US, -A, -UM]** With flowers in a corymb. This is a branched inflorescence with the stalks of the lower flowers longer than the upper ones, forming a flattened or convex head, e.g. candytuft.

COSTATE [COSTATALIS, -E, COSTAT|US, -A, -UM] Fluted or ribbed.

■ **COTONEASTER** A genus of 50 species of deciduous and evergreen shrubs and trees in the *Rosaceae* from the north temperate zone. They vary from low, arching shrubs to tall shrubs and small trees, with white or pinkish, 5-petalled flowers in clusters (rarely singly). Several species and hybrids are cultivated as ornamentals for their showy, red, berry-like fruits (some species have black fruits): *C. franchetii* is an evergreen shrub 1.8–3.0m tall with 19–32mm long OVATE leaves, downy beneath and orange-red fruits; *C. frigidus* is deciduous, a large shrub or tree to 9.1m with elliptic leaves and globose red fruits. Its semi-evergreen hybrid, *C. X* 'Cornubia' is much grown for its abundant, bright fruits; *C. horizontalis* is deciduous, with rounded leaves to 13mm long and globose red fruits. It is much grown and easily recognised by the arching, fishbone habit of growth.

○ **COTTON** The long tangled white hairs or lint which covers the seeds of certain *Gossypium* species, annuals and perennials in the *Malvaceae* from the tropics and subtropics. Four species are involved, all with strong basic similarities; having CORDATE-PALMATE, lobed leaves and 50mm wide, funnel-shaped 5-petalled flowers with three conspicuous PALMATIFID calyx-like BRACTEOLES. The seed pods or bolls are rounded to ovoid-oblong: upland cotton *(G. hirsutum)*, is a shrub-like annual to 900mm with white to pale-yellow flowers ageing reddish; sea island cotton *(G. barbadense)* is similar with deep yellow flowers and long, lint hairs; tree cotton *(G. arboreum)* grows to 6m and has narrowly, deeply-lobed leaves and yellow to reddish flowers; India or Asiatic cotton *(G. herbaceum)* is smaller, annual or perennial with broader, lobed leaves.

COTTON GRASS [COTTON SEDGE] About twelve species of *Eriophorum*, tufted or creeping plants of boggy ground in the *Cyperaceae* from the colder parts of the north temperate zone: common cotton grass *(Eriophorum angustifolium)* has widely creeping RHIZOMES and erect stems and leaves to 450mm or more. The tiny flowers are carried in 3–7 ovoid heads, forming a terminal UMBEL. After flowering, the bristle-like PERIANTH elongates to white, cottony hairs.

COTTON LAVENDER *(Santolina chamaecyparissoides)*. A low, silvery grey evergreen shrub in the *Compositae* from S.W. Europe. It has 25–38mm long LINEAR-PINNATISECT woolly leaves and small button-like yellow flower heads.

COTTON WOOD
see POPLAR

COTYLEDON Seedling leaves which are present in miniature form within the seeds. The number of these leaves varies, MONOCOTYLEDONS have one, DICOTYLEDONS two, some *Gymnosperms* have 6 or more. In some plants, e.g. broad bean and maize, the cotyledons remain below ground, serving as food stores, a situation known as hypogeal germination. In epigeal germination, the seed leaves expand and become green, growing above ground. Frequently they are of different shape to that of the adult leaf.

● **COUCH GRASS** Several species of *Agropyron*, RHIZOMATOUS perennials in the *Gramineae* from temperate regions. They are typical grasses with wheat-like flower spikes formed of two rows of flattened spikelets: common couch grass *(Agropyron repens)* has somewhat dull, GLAUCOUS leaves and erect stems of 900mm with spikes 75–150mm long. Its wiry, invasive rhizomes make it a serious weed of farms and gardens: see couch grass *(A. pungens)* is similar, with shorter, broader spikes and is restricted to sand dunes and salt marshes.

COWBERRY *(Vaccinium vitis-idaea)*. Also known as red whortleberry, it is a small, spreading shrub in the *Ericaceae* from the cooler parts of the north temperate zone. The OBOVATE leaves are dark green above, paler beneath, 13–25mm long. Pendent, terminal clusters of small, white pink-tinged, bell-flowers give way to edible, globose red berries.

▲ **COW PARSLEY** *(Anthriscus sylvestris)*. Sometimes called Keck or Queen Anne's lace, this is an erect biennial or perennial in the *Umbelliferae* from Europe, Asia and N. Africa, also naturalised in N. America. The basal leaves are bi- or tripinnate about 300mm long, stalked. The tiny white flowers are borne in terminal UMBELS about 50mm wide.

COW PARSNIP
see HOGWEED

COW PEA *(Vigna unguiculata* syn. *V. sinensis)*. A sprawling annual in the *Leguminosae* from tropical Asia and Africa. It has TRIFOLIATE leaves and whitish, yellowish or purplish pea-flowers followed by slender pods to 450mm long, containing rounded seeds of various shades, often black-eyed. A climbing form of this plant, often known as *V. sesquipedalis*, is the snake or yard long bean, grown for its edible pods which can attain 900mm.

COWSLIP *(Primula veris)*. Also called paigle, this is a tufted perennial with somewhat rugose, ovate-oblong, stalked leaves 50–150mm long. Erect stems 100–300mm long bear UMBELS of cup-shaped, orange yellow flowers which are pendent and have ribbed, tubular CALYCES.

COW-TREE *(Brosimum glactodendron)*. Also milk-tree, this is an evergreen tree to 40m in the *Moraceae* from Venezuela, which yields a pleasant milk-like sap when incisions are made in the trunk. The narrowly elliptic leaves are up to 300mm long and the small, greenish-white flowers are carried in spherical clusters.

COW-WHEAT Several species of *Melampyrum*, annual plants in the *Scrophulariaceae* from the temperate northern hemisphere. They have opposite pairs of narrow leaves and terminal spikes of slender, 2-lipped, tubular flowers: common cow-wheat *(M. pratense)* is a slender branched plant to 450mm or so with linear-lanceolate leaves and yellow flowers.

CRAB'S EYE VINE *(Abrus precatorius)*. Also called weather plant, this is a climbing plant in the *Leguminosae* from the tropics. It has PINNATE leaves with numerous pairs of oblong leaflets, and small, pale purple, pink or white pea flowers in dense RACEMES. The pods that follow contain large, hard, shiny red seeds with a black eye, used in India as beads. The leaves fold in dull weather and the root is used as a liquorice substitute.

CRAMP BALLS *(Daldinia concentrica)*. Also known as King Alfred's cakes, this is an *Ascomycete* fungus which lives mainly on dead ash wood. The hard fruiting bodies are rounded, 25–50mm across and a dull, glossy black when mature. Cut open, they show a pattern of pale, concentric rings. They were formerly thought to prevent cramp or ague.

CRANBERRY *(Vaccinium oxycoccus)*. A slender prostrate evergreen shrub in the *Ericaceae* from bogs in the colder parts of the north temperate zone. The OBLONG-OVATE 6mm leaves are dark green above, GLAUCOUS beneath. The small pink flowers have 4 petals and are followed by edible globose to ovoid pale berries, often spotted red or brown.

CRANESBILL A vernacular name for all 300 species of *Geranium*, annual and perennial plants in the *Geraniaceae* from the temperate regions of the world. They have rounded, CORDATE leaves, deeply to shallowly lobed and 5-petalled flowers followed by beaked fruits: dove's foot cranesbill *(G. molle)* is a spreading annual plant with 7-lobed, softly hairy leaves 13–38mm wide and small rosy-purple flowers with notched petals; meadow cranesbill *(G. pratense)* is an erect, tufted perennial plant 300–600mm tall having long stalked, polygonal leaves deeply 5–7 lobed, each lobe PINNATIFID. The violet-blue flowers open widely to 25mm or more; herb Robert *(G. robertianum)* is a 150–450mm annual or biennial, with a rosette of red stemmed, PALMATE leaves having 3–5 deeply PINNATISECT leaflets. The 9–13mm wide flowers are rich pink.

CRASS|US, -A, -UM [CRASSIFOLI|US, -A, -UM] Thick or fleshy. The second term specifically of leaves.

CRATERIFORM|IS, -E Goblet, cup or bowl-shaped.

CREEPING CINQUEFOIL *(Potentilla reptans)*. A fleshy-rooted perennial in the *Rosaceae* from Europe and Asia, introduced into N. and S. America. It has slender stems to 800mm long rooting at the nodes, PALMATE leaves with 5 OBOVATE leaflets and solitary, 19–25mm wide, 5-petalled yellow flowers in the leaf AXILS.

CRENATE [CRENAT|US, -A, -UM] Having shallow, rounded teeth.

CRESS Various members of a number of genera in the *Cruciferae*, e.g. BITTER, GARDEN, HOARY, PENNY, WINTER CRESS.

CREST A ridge or flange of tissue, sometimes toothed or notched, e.g. the outgrowths on the styles of *Iris*.

CRESTED DOG'S TAIL *(Cynosurus cristatus)*. A tufted perennial in the *Gramineae* from Europe to Turkey. The erect stems bear narrow leaves and dense, narrowly oblong 25–75mm PANICLES of small, crested spikelets.

CRINIT|US, -A, -UM Bearing long weak hairs.

CRISPAT|US, -A, -UM [CRISPUS] Finely waved, e.g. leaf margins.

CRISTAT|US, -A, -UM Crested; e.g. crest-like FASCIATIONS.

CROCUS A genus of 75 bulbous-rooted perennials in the *Iridaceae* from Europe, N. Africa, to Central Asia, with tufts of grassy leaves, each with a pale or silvery central stripe. The goblet-shaped flowers open flat in the sun, each have 6 TEPALS, 3 STAMENS and a branched STIGMA. The leaves and flowers arise directly from a round, flattened corm often deep in the ground. The OVARY of each bloom stays buried until it becomes a ripe CAPSULE; the 'stalk' of each flower being in fact a corolla tube. Species of crocus flower from late summer through the autumn until spring, most of those flowering in the autumn, opening before the leaves appear. The dried stigmas of *C. sativus*, an autumn species with lilac flowers, provides saffron, an orange-yellow dye used for colouring and for flavouring food and confectionery; *C. speciosus* has 150mm, bright lilac-purple flowers before the leaves in autumn; *C. tomasinianus* has 75–100mm lavender flowers with short leaves in late winter; *C. vernus* has 100–150mm white to rich purple flowers in spring with the leaves; *C. flavus* (syn. *C. aureus*) is similar but with orange to yellow flowers. All are commonly cultivated.

CROSSWORT *(Cruciata chersonensis* syn. *Galium cruciata)*. A prostrate or sprawling perennial in the *Rubiaceae* from Europe and Asia. It has hairy, 4-angled stem with WHORLS of 4 elliptic leaves at each node. The small cross-shaped flowers are carried in AXILLARY CYMES.

CROWBERRY *(Empetrum nigrum)*. A small, 150–300mm heath-like evergreen shrub in the *Empetraceae* from the north temperate zone. The 13mm LINEAR-oblong leaves are crowded in branched stems which bear tiny, pink flowers. The flattened globose berries are black.

CROWFOOT Certain species of *Ranunculus*, most of which are aquatic with slender stems, having finely dissected submerged leaves. Some have PALMATE floating leaves and all have white, buttercup-like flowers held above the water. *See* BUTTERCUP.

CROW GARLIC *(Allium vineale)*. A bulbous, grassy plant smelling strongly of garlic in the *Alliaceae* from Europe, N. Africa and the Middle East. It has 300–600mm, grooved, hollow leaves and rounded UMBELS of small greenish-white to pinkish bells mixed with tiny BULBILS. *A.v. compactum* has bulbils only and can be a noxious weed.

CROWN IMPERIAL
see FRITILLARY

CRUEL PLANT
see ARAUJIA

CRUENT|US, -A, -UM The colour of blood.

CRYPTOGAM All those plants that reproduce by spores and do not bear obvious flowers, e.g. ferns, mosses, seaweeds and fungi.

CUCULLAT|US, -A, -UM [CUCULLARIS, -E] Hooded or hood-like.

CUCUMBER [CUCUMIS, CUCURBITA]
see CUCURBITS

● **CUCURBITS** Members of the *Cucurbitaceae* which includes gourds, cucumber, marrow, melon, water melon, squash, and pumpkin. All are prostrate or climbing plants with tendrils and PALMATE-CORDATE leaves often shallowly to deeply lobed. The DIOECIOUS flowers are widely funnel to trumpet-shaped, yellow or sometimes white. Female flowers are easily recognised by the prominent ovary beneath the COROLLA: wax gourd *(Benincasa hispida)* has 75–125mm yellow flowers and 300mm long waxy white marrows; all parts of the plant are eaten in tropical Asia; bottle gourd *(Lagenaria siceraria)* has 50–100mm white flowers and bottle, club or globular fruits with a woody shell much used for bowls, baskets and containers of many kinds, also for musical instruments; water melon *(Citrullus lanatus* syn. *C. vulgaris)* has very deeply lobed leaves, yellow 32mm wide flowers and ovoid fruits up to 600mm long; melon, including musk, canteloupe and casaba *(Cucumis melo)* has yellow, 19–32mm wide flowers (sometimes HERMAPHRODITE) and globular to ovoid fruits, often with a corky netting to the skin, 100–250mm long; cucumber *(Cucumis sativa)* has yellow flowers up to 25mm wide and shortly ovoid to truncheon-shaped 300–450mm fruits; marrow *(Cucurbita pepo)*, squash *(C. moschata)* and pumpkin *(C. maxima)* are rather similar, the last two also being known as winter squashes. Both trailing and bush forms are known, the yellow flowers to 125mm wide. The fruits may be globular or ovoid, smooth or warted, 300–900mm long or wide. Some of the squashes have tapered, crooked necks. *See also* BALSAM APPLE, LOOFAH, SNAKE GOURD.

CUDWEED Several species of the closely related *Filago* and *Gnaphalium*, annuals and perennials of almost cosmopolitan distribution. They have narrow, usually white hairy leaves and tiny tubular flowers in small woolly heads: common cudweed *(Filago germanica)* is a slender, erect, white woolly annual 50–250mm tall with yellow florets; wood cudweed *(Gnaphalium sylvaticum)* is a perennial 100–450mm tall with white, woolly stems and LANCEOLATE leaves smooth above, woolly beneath. The florets are pale brown.

CULM The hollow stems of grasses, particularly of the woody kinds, e.g. bamboos.

CULTIVAR Short for cultivated variety and referring to a distinct variant of a species or hybrid maintained in cultivation. Such a variety may be purposefully bred by man, or arise spontaneously as a mutation in gardens or in the wild, e.g. a plant with double flowers or variegated leaves. Such a plant does not usually come true from seed and is only maintained by vegetative propagation. *See* CLONE.

CULTRIFORM|IS, -E Shaped like a knife blade.

CUMIN [CUMMIN] *(Cuminum cyminum)*. A small slender annual in the *Umbelliferae*, from Europe to Asia and Africa, grown for its edible seeds used like caraway. The stems are about 150mm tall with 13–50mm long leaves cut into LINEAR segments. The tiny white or pinkish flowers are in small UMBELS and the ovoid, oblong bristly seeds are about 6mm long.

CUNEAT|US, -A, -UM [CUNEIFORMIS, -E] Wedge-shaped, tapering from a broad tip to a narrow base.

CUP AND SAUCER CREEPER *(Cobaea scandens)*. A vigorous evergreen climber in the *Cobaeaceae* from S. America. The pinnate leaves have 6, 75–100mm elliptic leaflets and much branched tendrils, each one ending in a tiny hook. The 75mm bell-shaped flowers open pale yellow-green and change to purple.

CUP FUNGI A group of fungi in the *Ascomycetes*, the fruiting bodies of which are neatly or roughly cup-shaped. They are fleshy or leathery in texture and the inner surface bears the spore producing asci. *See* ORANGE PEEL FUNGUS.

CUPRE|US, -A, -UM Copper-coloured.

CUPULE A cup-like structure holding the nut-like fruits of hazel, oak and beech. Such trees were formerly classified in a family known as the *Cupuliferae*.

CURRANT In a general way applied to the currant-like fruits of several different plants. The currants of commerce are the dried fruits of a particular sort of GRAPE. The red, black, white and flowering currants are species of *Ribes*, deciduous shrubs in the *Grossulariaceae* from the northern temperate zone and the mountains of S. America. They have PALMATE, often lobed leaves and erect or pendent RACEMES of small flowers: flowering currant *(R. sanguineum)* has showy, red, pendent flowers and purple berries covered with a white waxy bloom.

CURRY LEAF *(Murraya koenigii)*. A pungently aromatic evergreen shrub in the *Rutaceae* from India. It has PINNATE leaves with 10–20 oval leaflets and terminal clusters of racemes bearing small, palest yellow flowers. The leaves are used in curries, and the bark in a tonic.

CURRY PLANT *(Helichrysum angustifolium)*. A 300–600mm grey-white, bushy shrub in the *Compositae* from S. Europe. It has 25mm long, LINEAR leaves and 25–50mm wide flattened clusters of tiny yellow flower heads. The whole plant smells of curry powder.

CUSHION PLANTS Where the climate is cold and dry and windy, such as on mountain tops or exposed sea coasts, many plants have small, thick or hairy leaves on short stems formed into tight rounded hummocks, e.g. thrift, also many saxifrages. The cushion plant *(Pygmaea pulvinaris)* is a New Zealand alpine in the *Scrophulariaceae*. It forms rounded cushions of white, hairy, linear oblong leaves and bears tiny, 4-petalled white flowers.

CUSPIDATE [CUSPIDAT|US, -A, -UM] Suddenly narrowing to a slender point.

CUSTARD APPLE *(Annona squamosa)*. Also known as sugar apple or sweet sop, this is a deciduous tree to 6m in the *Annonaceae* from tropical America. It has LANCEOLATE leaves and 6-petalled greenish-yellow nodding flowers. The 75–100mm heart-shaped yellow-green fruits have a tubercled, segmented appearance and contain a sweet custard-like pulp.

CUTICULE [CUTIN] A thin, waxy layer over the surface of leaves and stems which prevents excessive water-loss by evaporation.

CYANE|US, -A, -UM Blue.

CYATHIUM A type of inflorescence peculiar to *Euphorbia*, consisting of 5 BRACTS which mimic a CALYX, a central female flower reduced to a PISTIL surrounded by several male flowers each reduced to one stamen.

CYBELE A flora of an area or country.

● **CYCAD** A semi-popular term for all members of the *Cycadaceae*, a primitive family of *Gymnosperms* representing one of the earliest groups of seed-bearing plants, much commoner in prehistoric times. They are palm-like plants with usually leathery, PINNATE leaves and woody trunks sometimes very short. The flowers are borne in cone-like structures. *See* SAGO.

CYCLAMEN A genus of 15 species of perennials in the *Primulaceae* from S. Europe, W. Asia and N. Africa. They have broadly OVATE, CORDATE leaves, sometimes entire but often shallowly lobed. They grow directly from a rounded, flattened corm. Quite often the upper leaf surface has a silvery marbling or pattern. The nodding, solitary flowers are tubular at the base, with five petals that REFLEX back to form a shuttlecock shape. In most species the seeding stem coils tightly, bringing the CAPSULE down to ground level. Ants distribute the sticky seeds: *C. persicum* is one species where this does not happen. It is the popular pot plant, with rounded, often handsomely marbled leaves and

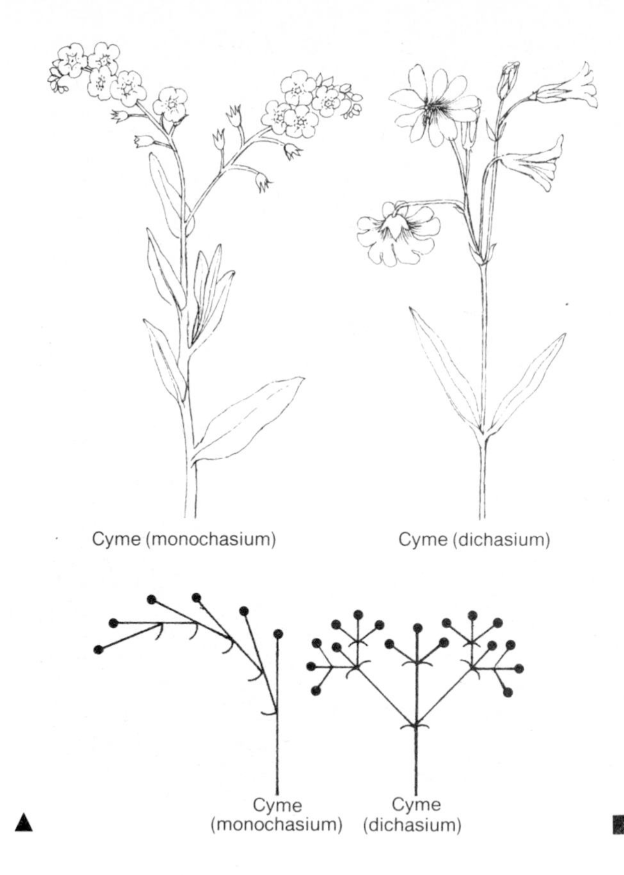

▲

■

large pink, red, purple or white flowers; *C. hederifolium* (syn. *C. neapolitanum*) has angular or lobed leaves patterned with silver and slender, 25–38mm white or pink flowers in autumn, before the leaves fully expand.

CYMBIDIUM A genus of 40 species of EPIPHYTIC and terrestrial evergreen plants in the *Orchidaceae* from tropical Asia and Australia. They are clump-forming with globose to conical PSEUDOBULBS sheathed by the leaf bases. The sepals and two petals are similar, surrounding a jutting 3-lobed labellum (lip). Several species and many hybrid cultivars are grown as ornamentals, the flowers being popular for corsages and for buttonholes. Characteristic are: *C. eburneum* with LINEAR, arching leaves and 75–100mm wide, white fragrant flowers, the labellum marked with yellow; *C. lowianum* with 100–125mm wide, pale green fragrant blooms having a pale yellow labellum bordered with crimson.

CYMBIFORMIS, -E Boat-shaped, e.g. of floral bracts.

▲ **CYME** A type of inflorescence made up by repeated lateral branching. In the monochasial cyme, each branch ends in a flower bud and one lateral branch. This in turn ends in a flower bud opposite a branch, and so on. In the dischasial cyme, each branch ends in a flower bud between two opposite branches.

CYMOS|US, -A, -UM Having flowers in CYMES.

■ **CYPRESS** Used in a general way for many trees that resemble the true cypresses, evergreen species of *Cupressus* and *Chamaecyparis* in the *Cupressaceae*. They come from S. Europe, Asia and N. America, and have tiny scale or awl-shaped, overlapping leaves. The much branched stems are arranged in plume-like sprays *(Cupressus)* and flattened frond-like sprays *(Chamaecyparis)*. In form the trees are narrowly conical or columnar, but some of the cupressus species become broad-headed with age. The ovoid or globular-shaped cones have umbrella-shaped scales with several seeds. The seedlings have soft, spreading LINEAR leaves at first, then develop the adult foliage: Monterey cypress *(Cupressus macrocarpa)* is broad-headed with age and grows to 21.3m or more, with bright green foliage and ovoid, 25–38mm cones; Mediterranean cypress *(Cupressus sempervirens)* reaches 30m and can be pyramidal or finger-like, the latter form a conspicuous feature around the Mediterranean where it is often called the Funeral cypress; Lawson cypress *(Chamaecyparis lawsoniana)* reaches 60m and is of columnar form with flattened sprays of grey-green foliage and globular 8mm wide cones; yellow or Nootka cypress *(C. nootkatensis)*, sometimes also known as yellow cedar, is narrowly pyramidal with dark green leaves and pendulous branchlets. The cones have a stiff point to each scale. *See also* SWAMP CYPRESS.

CYTOLOGY The study of cells and their functions.

CYTOPLASM
see PROTOPLASM

DACTYLOIDES Resembling fingers, e.g. leaf lobes.

DAFFODIL Species and CULTIVARS of NARCISSUS with trumpet-shaped CORONAS as long as the petals.

● 

DAHLIA A genus of 27 species of tuberous rooted, robust perennials in the *Compositae* from Mexico. They have erect stems and PINNATE or BIPINNATE leaves and large daisy-like flower heads in shades of red, orange, purple, yellow and white. Many hybrid cultivars are grown, the blooms varying from under 50 to 300mm across. Those with quilled petals are known as cactus dahlias. Most of the garden sorts are derived from *D. variabilis* (including *D. rosea*), a 0.9–1.5m species.

● **DAISY** *(Bellis perennis)*. An evergreen perennial in the *Compositae* from Europe and W. Asia. Has a basal rosette of OBOVATE, CRENATE toothed leaves and 13–25mm wide white daisies, often pink in bud, borne singly on 38–125mm stems. A common plant of short grassland.

DAISY BUSH Applied in a general way to *Olearia*, a genus of 100 species of evergreen shrubs from New Zealand, Australia and New Guinea. They have LINEAR to OVATE leathery leaves and profusely borne terminal PANICLES of small, usually white, daisy flowers. Several species are cultivated as ornamentals: *O. avicennifolia* grows to 2.4m or more with ovate leaves, buff to white felted beneath, and 6mm fragrant flowers; *O. macrodonta* grows to 6m with 50–75mm long ovate leaves having softly spine-toothed margins and a white felting beneath. The 13mm white flowers have reddish DISK florets.

DAME'S VIOLET *(Hesperis matronalis)*. A biennial or perennial plant in the *Cruciferae* from Europe to Asia. It has erect, 450–900mm stems and lanceolate, rough hairy and finely toothed leaves. The fragrant 4-petalled, violet or white flowers are borne in long RACEMES.

DAMSON
see PLUM

▲ **DANDELION** Several species of *Taraxacum*, rosette-forming perennials in the *Compositae* from the northern hemisphere but naturalised elsewhere. They have OBLANCEOLATE leaves, usually PINNATELY toothed or lobed and yellow daisy-like flower heads composed entirely of LIGULATE florets: common dandelion *(T. officinale)* has boldly RUNCINATE-PINNATIFID leaves and 32–57mm wide flower heads; lesser dandelion *(T. laevigatum)* is a miniature plant from dry, sandy or limy soils; marsh dandelion *(T. palustre)* grows in wet places with entire to sinuately-lobed leaves and 25–38mm wide flower heads.

DAPHNE A genus of 70 evergreen and deciduous shrubs in the *Thymelaceae* from Europe, Asia and the Pacific. They have linear to ovate entire leaves and clusters of tubular flowers with 4 spreading lobes. The fleshy fruits are one-seeded DRUPES. Several species are cultivated for their attractive, scented flowers and *D. papyracea* is used for paper-making in N. India. *D. mezereum* (mezereon) grows to 1.2m or so with deciduous lanceolate leaves and fragrant pink to red-purple or white flowers before the leaves. *D. laureola*, spurge laurel, is 0.6–1.2m with OBOVATE-LANCEOLATE evergreen leaves and night fragrant green-yellow flowers.

DARNEL
see RYE GRASS

● **DARWIN, CHARLES ROBERT [1809–82]** British biologist, born at Shrewsbury and educated there and at Cambridge. He showed an early interest in natural history and in 1831 was appointed naturalist to H.M.S. 'Beagle', a ship fitted out to explore the Pacific area. Sharply observant, Darwin kept a daily journal during the five years the explorations lasted and published a popular account of the voyages upon his return, and later a scientific account. His observations and investigations led him, early on, to suspect that species and forms were evolved from existing ones by natural selection and were not the result of a single creation by God as was believed at the time. By 1858, his theories were well backed by observations and researches. At this time he received a paper from Alfred Russel Wallace which presented theories practically the same as his, and soon afterwards he read both Wallace's paper and his own to a meeting of the Linnean Society in London. The following year his book *Origin of Species* was published, creating consternation and excitement in scientific, lay and religious circles. In 1871, *The Descent of Man* appeared and caused even more of a furore. His researches into many aspects of plant and animal life resulted in a series of books, those on botanical matters being *Insectivorous Plants* (1875), *The Power of Movement in Plants* (1880), *Climbing Plants* (1875), and *Fertilisation of Orchids* (1862). (*See also* EVOLUTION.)

DATE
see PALM

DAWN REDWOOD *(Metasequoia glyptostroboides)*. A deciduous tree in the *Taxodiaceae* from China. Also known as water fir or water larch, this conifer was only discovered in 1941 in Szechwan, and was introduced to the western world in 1946. In the wild it forms a broad-headed tree to 45m when mature; young specimens are conical. The leaves are LINEAR, 13–38mm long, carried in flattened sprays on short, deciduous branchlets. The cones are globose to ovoid, up to 25mm long.

▲ **DAVID, Père ARMAND [1826–1900]** A French missionary who joined the *Missions Étrangères* in Peking in 1862. He was primarily a zoologist and is probably best known for saving from extinction the deer which bears his name. He was also an excellent botanist and all-round naturalist. He made three extensive journeys into the relatively unexplored areas of China, the most important the second one from 1868–70 when he stayed on the Tibetan border and made many new discoveries in this rich area. His third journey (1872–74) was marred by ill-health and the loss of many specimens when his boat capsized. He returned to France the same year, where in Paris he founded his own museum and lived for the rest of his life. Among his introductions from China were the white flowered peach *(Prunus davidiana)*, the pocket ■ handkerchief tree *(Davidia involucrata)*, and *Clematis armandii, Lilium davidii* and *Rhododendron davidii*, all named for him from specimens sent home to Paris.

DAY LILY Several species of *Hemerocallis*, perennials in the *Liliaceae* from Asia. They are clump-forming with LINEAR, arching leaves and funnel-shaped flowers of 6 tepals which last one or two days only; *H. flava* has 900mm

stems bearing several lemon-yellow 88mm wide flowers which are eaten by the Chinese; *H. fulva* reaches 1.2m and has orange-red, 100mm flowers. There are many hybrids and CULTIVARS grown in gardens.

○ **DEADLY NIGHTSHADE** *(Atropa belladonna)*. Also called dwale, this is a robust perennial in the *Solanaceae* from Europe, W. Asia and N. Africa. It has OVATE-ACUMINATE leaves, sometimes in unequal pairs, the largest to 200mm long, and 25mm bell-shaped, lurid purple or greenish, nodding flowers. The glossy, black berries are 13–19mm across. The whole plant is poisonous and contains the drug belladonna.

DEAD NETTLE Several species of *Lamium*, annual and perennial plants in the *Labiatae* from Europe, Asia and N. Africa, also naturalised elsewhere. Their stems are square in cross-section (true nettles are rounded), and have tubular, two-lipped flowers, the upper lip hooded, borne in VERTICILLASTERS from upper leaf AXILS: red dead nettle, *L. purpureum* is a spreading annual with OVATE, CORDATE, CRENATE leaves and rose-purple, 13mm long flowers; white dead nettle *(L. album)* is perennial with erect stems to 450mm and larger, white flowers; henbit *(L. amplexicaule)* is an annual, similar to a slender red dead nettle but with more rounded leaves, those beneath the flowers semi-amplexicaul.

DEAL A name for the timber of several coniferous trees, but usually Scots pine (yellow deal) and Norway spruce (white deal).

DEALBAT|US, -A, -UM Whitened, as if by white powder.

DEATH CAP
see AMANITA

DECANDR|US, -A, -UM Flowers with 10 stamens.

DECAPETAL|US, -A, -UM Having 10 petals.

DECIDUOUS [DECIDU|US, -A, -UM] Becoming leafless or almost so during inclement seasons.

DECIPENS Deceiving; of one plant resembling another.

DECLINATE [DECLINAT|US, -A, -UM] Curved downwards or forwards.

DECORTICANS [DECORTICAT|US, -A, -UM] Shedding bark.

DECUMBENS Prostrate stems with the tips erect.

DECURRENS [DECURRENT] Running downwards; e.g. when the edges of a leaf blade or the gills or a mushroom extend in a tapering fashion down the stem.

DECUSSATE [DECUSSAT|US, -A, -UM] Extending at right angles.

DEFLEXED [DEFLEX|US, -A, -UM] Bent downwards.

DEFOLIATE [DEFOLIATION] The shedding of leaves.

DEHISCENCE The opening of fruits or anther lobes to shed seeds or pollen.

DELIQUESCENT Melting or becoming fluid, e.g. ink cap fungus, the petals of day flower and others.

DELTOIDE|US, -A, -UM Triangular like the Greek letter delta (Δ).

DEMERS|US, -A, -UM Aquatic: under water.

DEMISS|US, -A, -UM Drooping, lowly or humble.

● **DENDROBIUM** A genus of 1400 EPIPHYTIC perennials in the *Orchidaceae* from Asia, the Pacific Islands and Australasia. The species vary greatly in habit and size but most have stem-like or club-shaped PSEUDOBULBS and comparatively short, narrow leaves. The flowers also vary greatly but usually have 3 similar, spreading sepals; 2 usually broader, flared petals and a forward thrusting, often shell-like labellum (lip) which may be fringed. Many species and CULTIVARS are grown as ornamentals, the flowers used for buttonholes and corsages: *D. nobile* is the best known species with 450–900mm long pseudobulbs and large, pink-tipped, white flowers with a velvet crimson blotched labellum. There are many varieties.

DENDROID [DENDROIDE|US, -A, -UM] Having a tree-like form.

DENS-CANIS Like a dog's tooth.

DENTATE [DENTAT|US, -A, -UM] Toothed: with outwards pointing teeth.

DENTICULAT|US, -A, -UM With very fine teeth.

DEPENDENS Hanging down.

DEPLASMOLYSIS
see PLASMOLYSIS

DEPRESS|US, -A, -UM Flattened; pressed down.

DERMATOGEN The outermost layer of cells of the MERISTEM in a root or stem tip.

DESERTORUM From desert regions.

DESTROYING ANGEL
see AMANITA

DEVIL-IN-A-BUSH
see LOVE-IN-A-MIST

DEVIL'S BIT SCABIOUS *(Succisa pratensis)*. A tufted perennial in the *Dipsacaceae* from Europe and N. Africa to Siberia. It has rosettes of OBOVATE-LANCEOLATE, sparsely hairy leaves to 150mm or more long, and erect 300–900mm branched stems. The hemispherical flower heads are composed of many tubular, 4-lobed florets which are mauve to purple-blue.

DEWBERRY *(Rubus caesius)*. A prostrate shrub in the *Rosaceae* from Europe and Asia. It is much like the blackberry but more slender, the leaves always of three leaflets and the stem covered with a waxy-white patina. The white or pinkish flowers are followed by fruits composed of few, large drupelets bearing a white, waxy covering.

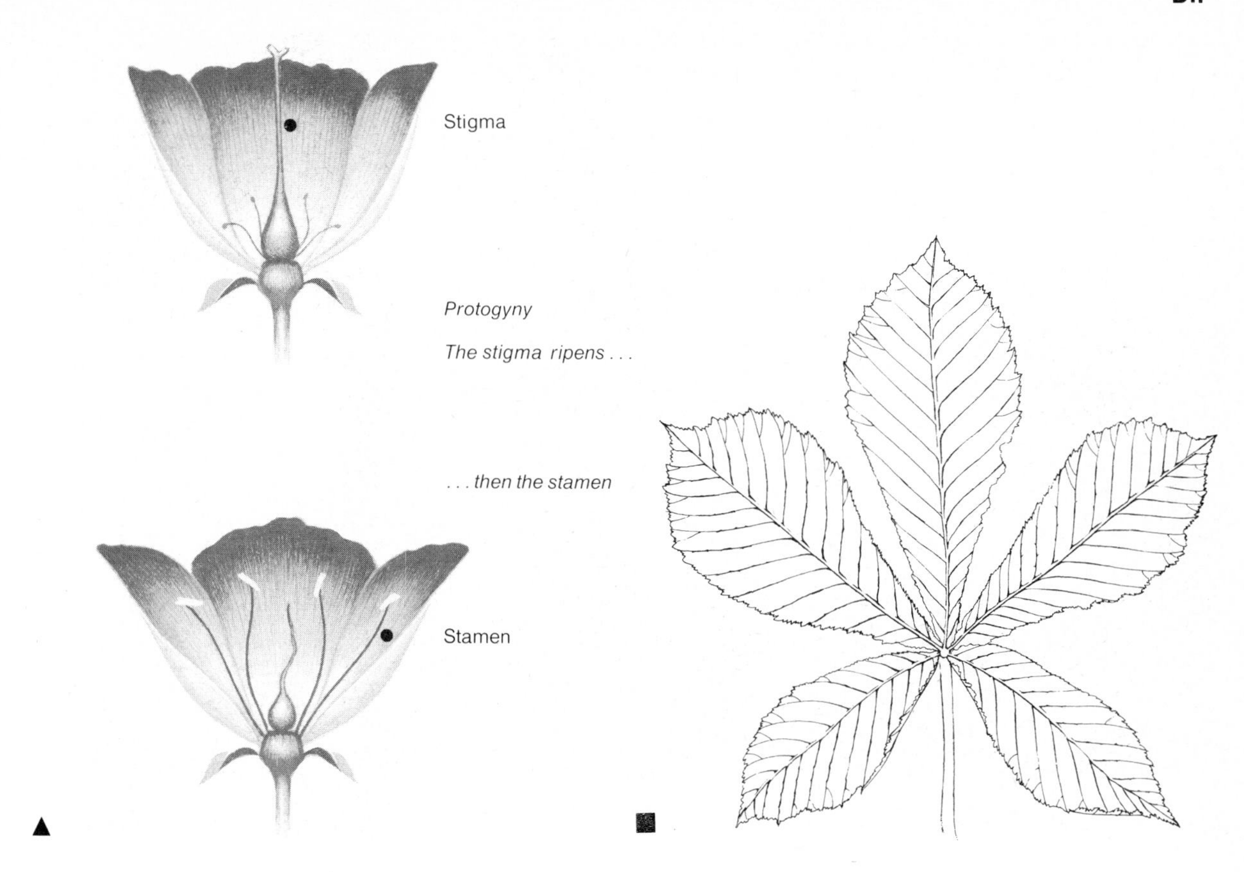

DIADELPHOUS Two groups of stamens (joined by their FILAMENTS).

DIAGEOTROP|IC, ISM
see GEOTROPISM

DIANDR|US, -A, -UM Having two stamens only.

DIAPHOTOTROP|IC, -ISM
see PHOTOTROPISM

DIARCH Of roots containing two groups of PROTOXYLEM vessels: triarch have three, tetrarch four, and pentarch five.

DICHLAMYDEOUS Of flowers with a distinct calyx and corolla.

▲ **DICHOGAMY [DICHOGAMOUS]** Having the stamens and stigmas maturing at different times, thus preventing self-pollination. In protandrous flowers the stamens ripen first, in protogynous, the stigma and ovary are first.

DICHOTOMOUS [DICHOTOM|US, -A, -UM] Stems that repeatedly fork into two at the ends, e.g. lilac.

DICOTYLEDONS One of the two classes that make up the ANGIOSPERMAE. It is characterised by the seedlings having two COTYLEDONS and the flowers having their parts in fours or fives. *See also* MONOCOTYLEDONS.

DIDYM|US, -A, -UM Divided into two, or twinned.

DIFFORM|IS, -E Of unusual form.

DIFFUS|US, -A, -UM Loosely or widely spreading.

■ **DIGITATE [DIGITAT|US, -A, -UM]** A compound PALMATE leaf, with the leaflets radiating out like the fingers of a hand.

DIGYN|US, -A, -UM Having two CARPELS or twin fruits.

DILATAT|US, -A, -UM Expanded or broadened out.

DILL *(Peucedanum graveolens)*. A 600mm annual in the *Umbelliferae* from Europe, but naturalised elsewhere. It has finely dissected, tri-PINNATE somewhat blue-green leaves and tiny yellow flowers in terminal UMBELS. The leaves and seeds are used for flavouring.

DIMORPHIC [DIMORPHISM] Having two different sorts of flowers or leaves on the same plant.

DIOECIOUS Having single-sexed flowers on separate plants.

DIOIC|US, -A, -UM With dioecious flowers.

DIPHYLL|US, -A, -UM Having two leaves only.

DIPLOID Two sets of CHROMOSOMES in each cell: the basic normal number.

DIPTER|US, -A, -UM Having two wings; usually of seeds.

DISCOID [DISCOIDE|US, -A, -UM] Round or flattened like a coin.

DISCOLOR Two different or contrasting colours on the same plant.

DISK (1) A flattened RECEPTACLE, usually around the ovary, e.g. flowers of rue and cow parsley. (2) The flattened central portion of a daisy flower, composed of tubular disk florets (as distinct from strap-shaped ray florets).

DISSECTED [DISSECT|US, -A, -UM] Leaves, bracts and sometimes sepals and petals that are cut into several deep lobes.

DISTAL Furthest from the axis, i.e. stem tip.

DISTANS Separated widely, e.g. flowers on a stem.

DISTICHOUS [DISTICH|US, -A, -UM] Having leaves or flowers in two opposite ranks.

DISTRIBUTION The geographic area within which a particular genus or species is found naturally.

DITTANY
see BURNING BUSH *(Dictamnus)*

DIURNAL Flowers opening during the day, closing at night.

DIVARICAT|US, -A, -UM Widely spreading.

DIVERGENS Divergent; branching at a wide angle.

DOCK Several species of *Rumex*, erect perennials in the *Polygonaceae* from Europe, Africa and W. Asia, naturalised elsewhere. They have long-stalked, basal leaves and narrow pyramidal PANICLES of tiny flowers with 3 sepals and 3 petals. The fruits are triangular with 3 winged angles: curled dock *(R. crispus)* is a common weed with 600–900mm stems and LANCEOLATE leaves to 300mm having crimped and undulate margins; broad-leaved dock *(R. obtusifolius)* has OVATE-OBLONG, CORDATE leaves to 250mm and stems to 900mm; water dock *(R. hydrolapathum)* grows in wet places with lanceolate to ovate leaves to 900mm and flowering stems often over 1.8m. *See* SORREL.

● **DODDER** *(Cuscuta)*. A genus of 170 PARASITIC species in the *Cuscutaceae* well distributed in temperate and tropical regions. They are slender twining plants with reddish or yellowish stems bearing tiny scale leaves and small, 5-lobed, bell-shaped flowers in heads or spikes. The seed germinates normally, but as soon as the stem finds a host plant, it twines around it and produces a HAUSTORIA, a sucker which penetrates the tissue and draws off food materials. Its own root then dies. As the dodder grows and branches it produces more haustoria and can seriously weaken the host plant. Some species are pests of clover, flax and other crops: common dodder *(C. epithymum)* has almost thread-like red stems which form tangled masses on clover, gorse, heather, etc. The fragrant pinkish flowers are borne in dense clusters 8.0mm across.

DOG DAISY
see OX-EYE DAISY

▲

DOG LICHEN
see LICHENS

DOG'S TOOTH VIOLET *(Erythronium dens-canis)*. A perennial plant in the *Liliaceae* from Europe and Asia. It has stalked, broadly-ovate leaves marbled with purple-brown, which spring direct from a tooth-shaped corm. The nodding 50mm wide rose-purple or white flowers have 6 narrow, reflexed TEPALS and are carried singly on 100–150mm stems.

DOGWOOD *(Cornus sanguinea* syn. *Swida* and *Thelycrania sanguinea)*. A deciduous shrub in the *Cornaceae* from Europe to S.W. Asia. It has opposite pairs of OVATE-ACUMINATE 38–75mm leaves and flat-topped CYMES of small white, 4-petalled flowers. The pea-sized fruits are black.

DOMESTIC|US, -A, -UM Used for or grown in the home.

DOMINANT GENE
see RECESSIVE GENE

▲ **DOMINANTS** One or several species in a natural association of plants that have the greatest effect on the community as a whole, e.g. in an oakwood the trees dominate all other plants with their shade, and the other components of this association must be plants which can tolerate this.

DOOB GRASS
see BERMUDA GRASS

DORMANT A resting period, enabling plants better to survive bad weather conditions, e.g. deciduous trees and shrubs stop growing and lose their leaves before the cold of winter. Many herbaceous perennials die back to ground level. Bulbous plants usually lose both leaves and roots during the hottest and driest months. This sort of dormancy is broken when warmer, longer days arrive, or in the case of many bulbs when the autumn rains arrive. Dormancy can also refer to buds and seeds. Buds of woody plants may remain dormant for years, only breaking into growth when the plant's main shoots get broken by a storm or are cut back by man. At least some seeds of many plants stay dormant and require certain conditions before they will grow. Sometimes the testa of the seed is very hard and takes many months or even years to rot. Quite often it contains a chemical inhibitor, and only after successive soakings by rain will it be leached out and germination take place. The seeds of many alpines, hardy trees and shrubs need chilling before they will grow and others, usually from very hot areas, need a heat shock.

DORSIFIXED Used of anthers which are attached to the FILAMENT by their backs.

DOUBLE COCONUT
see PALM

DOUBLE FLOWERS Flowers with more petals than normal, filling the centre of each bloom. Quite often a MUTATION occurs and most if not all of the stamens are transformed into petal-like structures (PETALOIDS), e.g. as in roses. More rarely the CARPELS also turn into petals, forming a densely double flower, e.g. as in bachelor's

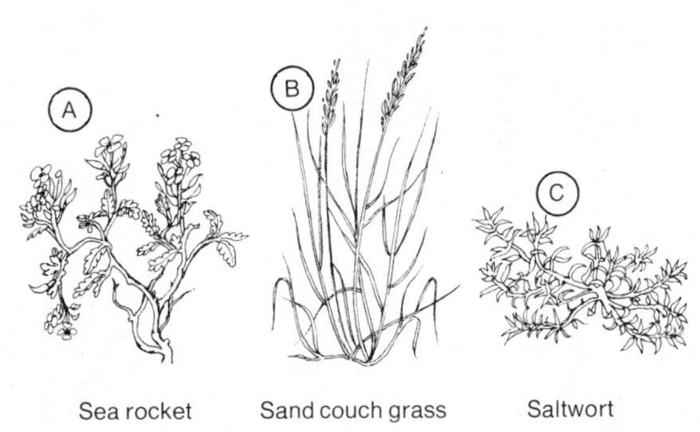

Sea rocket Sand couch grass Saltwort

buttons (*Ranunculis acris* 'Flore-pleno'). This condition is known as petaloidy. Such flowers are either completely sterile or male sterile. A different form of doubling occurs among members of the daisy family where the tubular DISK florets become strap-shaped RAY FLORETS.

DOUGLAS FIR *(Pseudotsuga menziesii* syn. *P. douglasii)*. Also known as Douglas spruce, Oregon pine and red fir, this fine evergreen conifer in the *Pinaceae* was introduced into Europe from its native north-western America by David Douglas in 1827. In the wild it can reach over 90m forming a spire of greenery. The 25–32mm needle-like leaves are densely borne in two ranks along the twigs and have an aromatic fragrance when bruised. The pendulous 75–100mm ovoid cones are formed of light brown, thin scales from between which protrude narrow, 3-lobed bracts like forked tongues. It is an important timber tree in the USA and much planted as an ornamental elsewhere.

DOUM PALM
see PALM

DOVE'S FOOT CRANESBILL
see CRANESBILL

DOVE TREE *(Davidia involucrata)*. Also known as ghost tree and pocket handkerchief tree, this deciduous tree in the *Davidiaceae* comes from China where it was first found by Père DAVID. It grows to 18.2m in the wild, with broadly-OVATE, coarsley toothed, 75–150mm long leaves, silky hairy beneath. The tiny petal-less flowers are borne in dense globose heads 19mm wide; one central female surrounded by many males. Each head is protected by two large white ovate bracts which hang down and suggest its vernacular names. The ovoid fruits are like hard green plums on long green stalks. *D. i. vilmoriniana*, a form with the leaves hairless beneath, is usually seen in gardens.

DOWN TREE
see BALSA

DRAGON TREE *(Dracaena draco)*. An evergreen tree in the *Agavaceae* from the Canary Islands. Growing to 18.2m or more, it has crowded branches each tipped by a cluster of 450–600mm, GLAUCOUS, LINEAR-LANCEOLATE leaves. Small greenish-white flowers are carried in large PANICLES. The reddish resin exuded by the trunk is known as dragon's blood and was used in the varnish for violins. The famous huge specimen on Teneriffe was 21.3m tall with a girth of 13.7m when it was blown down in 1868. The tree was said to be 6000 years old.

DRIP TIP The long, slender leaf points found on trees in tropical rain forests, supposedly to drain the leaf surface quickly and allow rapid drying after rain, e.g. FIG.

DROPPER A young bulb produced at the end of a downwards-growing RHIZOME. Shallowly planted tulips often produce them.

DROPWORT
see MEADOWSWEET and WATER DROPWORT

DRUPACE|US, -A, -UM Bearing drupes.

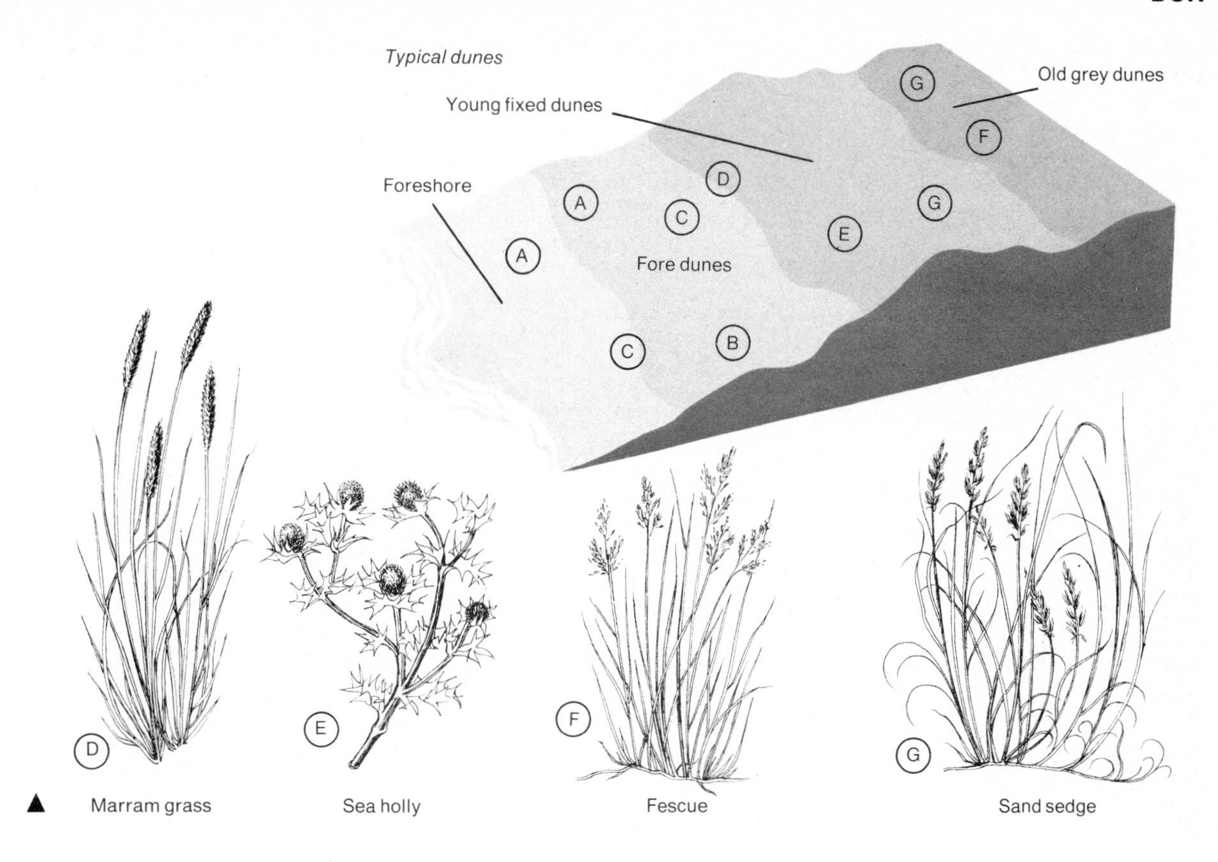

DRUPE A fruit with a fleshy PERICARP and usually a single stone-like seed, e.g. cherry, plum.

DUCKWEED The 10 species of *Lemna*, also known as ducks' meat, aquatic plants in the *Lemnaceae* widespread around the world. Each plant consists of a tiny, oval flattened stem with one to several roots, termed a thallus. This produces branches which break away and form new colonies. The minute flowers are borne in a pocket at the edge of the thallus: common duckweed *(L. minor)* is a floating species with a bright green, OBOVATE thallus 4.0mm long; ivy-leaved duckweed *(Lemna trisulca)* is submerged with translucent, stalked, OVATE thalli to 13mm long, in branched clusters.

DUFFIN BEAN
see BUTTER-BEAN

DUKE OF ARGYLL'S TEA TREE *(Lycium halimifolium)*. A deciduous shrub to 1.8m in the *Solanaceae* from S.E. Europe and W. Asia. It has arching, grey-white sometimes spiny stems and grey-green LANCEOLATE leaves to 50mm long. The rose-purple, 8mm wide flowers have 6 petal-lobes and are followed by ovoid, scarlet berries.

DULC|IS, -E Sweet.

DUMB CANE *(Dieffenbachia)*. A group of 30 species of evergreen perennials or soft shrubs in the *Araceae* from tropical America. They have OBLONG-OVATE, stalked leaves often variegated yellow or white and spikes of small petal-less flowers in an arum-like spathe: *D. picta* is a variable plant with erect, fleshy stems and dark-green leaves spotted with white and pale green; *D. seguine* is similar with stems to 1.8m and CUSPIDATE, undulate leaves with pellucid white spots. Both are frequently grown as house plants, but have acrid sap which can cause the tongue to swell, hence the vernacular name. The latter species was used to torture slaves in the West Indies.

DUMOSE [DUMOS|US, -A, -UM] Of low, bushy habit.

▲ **DUNE PLANT** Plants adapted to living in dry, sandy places usually near the sea. They have very extensive root systems and fleshy or rolled leaves with only a few STOMATA to cut down water loss. When covered with drifting sand they are soon able to reach the surface again and continue growing. Several of them help to stabilise shifting dunes with their extensive roots and RHIZOMES. Marram grass in particular being planted for this purpose. *See also* SEA BUCKTHORN, TAMARISK, LYME GRASS. Low-lying, moist or wet areas among dunes are known as dune slacks and are characterised by a different set of plants, e.g. marsh orchids and creeping willow.

DURAMEN The heartwood of a tree that has become hardened with mineral matter.

DURIAN *(Durio zibethinus)*. A notorious evergreen tree to 30m in the *Bombacaceae* from Malaysia. It has oblong-ovate leaves and large CAULIFLOROUS white or pink flowers followed by 200–250mm long, spiny, ovoid fruits weighing 2.7–3.6kg. It contains a custard-like pulp much esteemed in S.E. Asia but the subject of many vitriolic

adjectives by Europeans. The smell is described as a mixture of 'over-ripe cheese, rotten onions, turpentine and bad drains' and the flavour as 'French custard passed through a sewer'. Nevertheless, many Europeans, while agreeing about the smell, have pronounced it delicious. The seeds are roasted and eaten like chestnuts.

DURMAST OAK
see OAK

DUTCH CLOVER
see CLOVER

DUTCHMAN'S BREECHES
see BLEEDING HEART

DUTCHMAN'S PIPE *(Aristolochia marcrophylla* syn. *A. durior)*. A deciduous, woody climber in the *Aristolochiaceae* from N. America. It has RENIFORM to OVATE-CORDATE leaves 100–250mm long and tubular, siphon-shaped yellow-green flowers expanded at the mouth to 3 purple-bordered lobes.

DWALE
see DEADLY NIGHTSHADE

DWARF SHOOT Short or condensed shoots that bear flowers or leaves but do not grow out to form stems or branches, e.g. pines. The dwarf shoots of apple and cherry are called spurs.

DYE PLANTS Before they were synthesised or chemical substitutes were found, most of the dye colours came from plants, e.g. saff-flower (rouge), crocus (saffron-yellow), woad (blue), indigo (blue-violet).

DYER'S GREENWEED *(Genista tinctoria)*. A 300–600mm shrub in the *Leguminosae* from Europe to the Urals. It is an erect or ascending plant with green, striate twigs and oblong, LANCEOLATE leaves to 25mm long. The 13mm long yellow pea-flowers are borne in terminal RACEMES.

● **EAR FUNGUS** *(Auricularia auricula)*. A *Basidiomycete* fungus PARASITIC or SAPROPHYTIC on elder. It has liver-brown fruiting bodies of a rubbery, jelly-like texture rather like upside down, irregular-shaped ears.

▲ **EARLY PURPLE ORCHID** *(Orchis mascula)*. A tuberous rooted, ground dwelling species in the *Orchidaceae* from Europe, Asia and N. Africa. It has purple-spotted, OBLANCEOLATE 50–200mm long leaves and 150–600mm stems carrying a spike of crimson purple flowers. These have broad, 3-lobed LABELLUMS and upturned spurs.

EARTH NUT
see PIGNUT

EARTH STAR *(Geastrum triplex)*. A *Gasteromycete* fungus of temperate zone woodland, especially under beech. It is shaped like a pale-brown onion at first, then splits open into about 6 roughly pointed lobes creating a star-shape, the ovoid, spore-bearing structure in the middle.

EBONY Several species of *Diospyros*, but principally *D. ebenum*, an evergreen tree in the *Ebenaceae* from India and Sri Lanka. It has OVATE to oblong, 75–175mm leaves and small, tubular unisexual flowers followed by globose, 13–25mm fruits. Only the heart wood is black, the outer layer being softer and yellow-white.

EBRACTEAT|US, -A, -UM Without bracts (where they would normally occur).

EBURNE|US, -A, -UM Ivory-white, yellow-white.

ECALCARAT|US, -A, -UM Lacking spurs (where these are normally expected).

ECHINAT|US, -A, -UM Covered with prickles.

ECOLOGY The study of the relationship between plants and their natural environment.

EDAPHIC Factors associated with the soil in relation to plant growth, e.g. some plants need an acid soil, others a wet one.

● **EDELWEISS** In a general way all 30 species of *Leontopodium*, perennial plants in the *Compositae* from the mountains of Europe, Asia and S. America. They are tufted plants with LINEAR-LANCEOLATE, white woolly leaves and erect stems bearing compact groups of small flower heads surrounded by off-white to cream, felted, petal-like bracts: common edelweiss *(L. alpinum)* has linear leaves and 150mm flower stems with off-white bracts.

EDENTAT|US, -A, -UM Lacking teeth (where expected).

EDUL|IS, -E A plant, or part of one, used for food.

EEL GRASS Several species of *Zostera*, marine grass-like plants in the *Zosteraceae* from the coasts of the Mediterranean, Europe and N. America. They grow in sand and mud below the low water of spring tides, spreading by RHIZOMES. The leaves are strap-like, and the tiny, petalless flowers are pollinated under water. Common eel grass or grass wrack *(Z. marina)* has leaves to 450mm or more.

EFFUS|US, -A, -UM Spread out thinly.

EGLANDULAR Without glands.

EGYPTIAN LOTUS
see WATER LILY

ELATIOR Taller; the tallest species in a genus.

ELAT|US, -A, -UM Tall or very tall.

ELDER Several species of *Sambucus*, deciduous shrubs in the *Sambucaceae* and of cosmopolitan distribution. They have PINNATE leaves in opposite pairs and PANICLE-like CYMES of small, 5-petalled flowers followed by tiny black-purple, blue or red berries: common elder *(S. nigra)* grows to 6m or more with OVATE-LANCEOLATE, SERRATE leaflets, 25–75mm long, and flattened clusters of cream flowers and black berries; red berried elder *(S. racemosa)* grows to 3m, with round flower clusters and red berries.

▲

ELECAMPANE *(Inula helenium)*. A clump-forming perennial in the *Compositae* from Asia but naturalised in many temperate countries. It has erect, stout stems to 1.5m with OVATE-CORDATE, AMPLEXICAUL leaves and terminal daisy-like, bright yellow 63–75mm wide flower heads.

ELEGANTISSIM|US, -A, -UM Most graceful.

ELEPHANT'S EAR
see BEGONIA

ELLIPTIC|US, -A, -UM Elliptic; broadest in the middle, tapering equally to either end, e.g. of leaves.

▲ **ELM** The 45 species of *Ulmus*, deciduous trees in the *Ulmaceae* in the northern temperate zone. Characteristically they have coarsely toothed leaves with one side longer than the other. The tiny, greenish flowers open before the leaves and give rise to oval or circular winged fruits. Several species are planted as ornamentals and in the past the leaves were much used as animal feed: American elm *(U. americana)* grows to 30m, with a rounded head of branches and 50–125mm ovate to obovate leaves; slippery elm *(U. fulva)* is 12–18m with ovate-oblong, acuminate leaves 100–200mm long, the inner mucilaginous bark is nutritious and has medicinal value; English elm *(U. procera)* grows to 30m with a round head and several lower spreading branches, the leaves are sub-orbicular to ovate 45–88m long; wych elm *(U. glabra)* can attain 40m and has 75–150mm long obovate to elliptic leaves.

EMARGINAT|US, -A, -UM Notched at the tip, e.g. of petals.

EMASCULATE The removal of the (unripe) anthers.

EMBRYO The tiny rudimentary plant within the seed.

EMBRYO SAC
see OVULE

EMMER
see WHEAT

■ **ENCHANTER'S NIGHTSHADE** *(Circaea lutetiana)*. A stoloniferous perennial in the *Onagraceae* from Europe east to Iran, and N. Africa. It forms large colonies of erect stems 200–450mm tall with opposite pairs of OVATE-ACUMINATE, somewhat CORDATE-based 38–75mm long leaves. The tiny white flowers have 2 bilobed petals and are followed by 6mm long OBOVOID fruits covered with small, hooked bristles.

ENDEMIC A plant originating in one specific area or country and found nowhere else in a natural state.

ENDIVE *(Cichorium endiva)*. A biennial or annual salad plant in the *Compositae* from the Mediterranean region. It forms a dense rosette of OBLANCEOLATE leaves which may be PINNATELY lobed or SINUATE toothed. The stem grows to 900mm, with blue, 38mm wide flower heads. Several CULTIVARS are known, the curled types with deeply cut leaves being the most popular. They have to be blanched (kept in the dark) for one to two weeks before use.

ENDOCARP The inner layer of a DRUPE, i.e. the shell of a

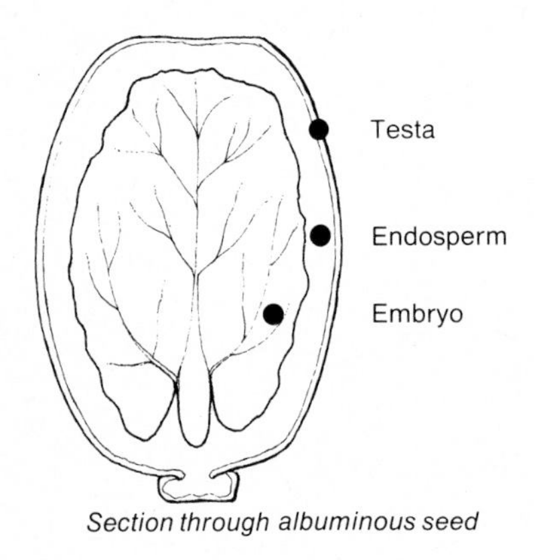

● *Section through albuminous seed*

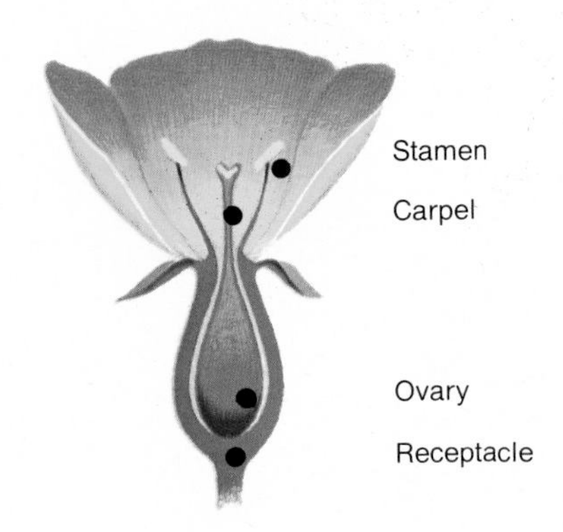

▲

plum or cherry stone. The fleshy layer is known as the mesocarp and the skin of the fruit the epicarp.

ENDODERMIS The innermost layer of the cortex in a stem or root. Within it lies the VASCULAR SYSTEM. In the stem, the cells that form it contain starch grains and it is sometimes called the starch sheath.

ENDOGENOUS Relating to the origin of lateral roots from internal tissue as distinct from the surface, or exogenous origin of lateral stems.

ENDOPARASITE
see PARASITE

ENDOSMOSIS
see OSMOSIS

● **ENDOSPERM** Nutritive tissue in seeds. In seeds such as those of castor oil plant, the embryo is embedded in endosperm, while in broad beans the embryo plant entirely fills the seeds. Those with endosperm are known as albuminous, those without as exalbuminous. *See also* GERMINATION; SEED.

ENDOTROPHIC
see MYCORRHIZA

ENSAT|US, -A, -UM Sword-like.

ENSIFORM|IS, -E Straight-sided, with a point, e.g. sword-like.

ENTIRE Smooth-edged, not cut or toothed.

ENTOMOPHILOUS (Insect-loving). Used of plants with flowers that are insect pollinated, providing nectar or surplus pollen as a reward.

EPHEMERAL Plants that complete their life cycle, growing from seed, flowering and fruiting several times in one season, as do many weeds. Also of flowers that last for one day or less.

EPICALYX A ring of bracts beneath a flower forming a second calyx, e.g. cinquefoil.

EPICARP
see ENDOCARP

EPIDERMIS The outermost layer of cells which forms the skin of a leaf or stem.

EPIGEAL
see COTYLEDON

▲ **EPIGYNOUS** Flowers with the ovary sunken in a hollow receptacle, the sepals, petals and stigma growing above it, e.g. apple and pear. Such ovaries are said to be inferior.

EPIPETALOUS Used of stamens which are attached to petals.

EPIPHYTE A plant that grows upon another but is not parasitic. In rain forests and jungles, such plants are common (notably orchids and bromeliads). They perch on

the bark of trees using special clinging roots and gain some nourishment from decayed plant remains in the crevices. Many have special water storage tissue, e.g. the PSEUDOBULBS of orchids or the cup-like structures formed by the leaf bases of bromeliads. Some of the orchids also have aerial roots, covered with special tissue (velamen) which can absorb water from the atmosphere; the leaves of bromeliads have scales and hairs modified to perform the same function.

EPITHET The specific (species) name of a plant, e.g. *acris* is the specific epithet of meadow buttercup, *Ranunculus acris*.

EQUATORIAL PLATE
see CELL DIVISION

EQUESTR|IS, -E Relating to horses.

ERECT|US, -A, -UM Upright in growth or habit.

ERICACEOUS All plants in the *Ericaceae*, e.g. heath, heather, rhododendron, bilberry.

ERICOID Like heath *(Erica)*.

EROSE [EROS|US, -A, -UM] With an irregularly notched margin, appearing as if gnawed.

ERUBESCENS Blushing; red or rosy.

ERYNGO
see SEA HOLLY

ESCAPE A plant growing outside a garden, but originating from it naturally either by seed or vegetative spread; usually of short duration.

ESCULENT|US, -A, -UM Edible; fit for food.

ESPARTO The leaves of several grasses used for paper making, but mainly *Stipa tenacissima*, a tufted perennial in the *Gramineae* from Spain and N. Africa, naturalised elsewhere. It has 600–900mm stems and narrow flowering PANICLES, the spikelets with 27–75mm feather AWNS.

ETIOLAT|E, -ION Plants kept in the dark producing long, weak, yellowish shoots.

■ **EVENING PRIMROSE** Several biennial species of *Oenothera* in the *Onagraceae* from N. and S. America, but naturalised elsewhere. They have rosettes of LINEAR-LANCEOLATE to OVATE leaves and terminal, often leafy, spikes of yellow, funnel-shaped flowers. Common evening primrose, *(O. biennis)* grows 450–900mm with narrowly OBLANCEOLATE basal leaves and flowers about 38mm long.

EVERGREEN Plants that stay leafy all the year round, with leaves that live for at least one full year before falling.

EVERLASTING The flowers of several plants that stay firm and colourful when dead. Most of them belong to the *Compositae* with heads that have petal-like bracts. The best known is *Helichrysum bracteatum* 'Monstrosum' an annual from Australia which is 0.6–1.2m tall with oblong, lanceolate leaves to 125mm long. The 38mm wide chaffy flower heads are red, pink, yellow or white.

EVERLASTING PEA Several species of *Lathyrus* in the *Leguminosae* from Europe, so-called because of their perennial nature (sweet and garden peas are annuals): broad-leaved everlasting pea *(L. latifolius)* is a climber to 1.8m or so with leaves composed of 2 OVATE to ELLIPTIC leaflets and large tendrils. The 25mm wide pink, pea flowers are carried in AXILLARY RACEMES; narrow-leaved everlasting pea *(L. sylvestris)* has LANCEOLATE leaflets and 13–19mm wide flowers.

EVOLUTION The theory that all plants are related, from the simplest alga to the largest tree or the most complex orchid. Although stated as a theory, there is now overwhelming evidence to show that new plant groups arise successionally from those before, gaining in complexity over the ages. The operative force for this is mutation, the spontaneous change in an organ or plant part caused by natural radiation striking the cell nucleus and its chromosomes. Such changes are mainly very small (though occasionally dramatically large), but become cumulative over the years when selected for, or against, by the environment: that is, the climate, soil, predators, etc. This is the process known as natural selection (first suggested by DARWIN in 1859) which covers both animals and plants.

EXALBUMINOUS
see ALBUMEN

EXALTAT|US, -A, -UM Lofty or very tall.

EXCELS|US, -A, -UM [EXCELSIOR] Tall. Most tall.

EXCIS|US, -A, -UM Cut out; e.g. division between petal lobes.

EXCURRENT A small point or vein-tip projecting beyond the end of a leaf.

EXFOLIATING Peeling off in thin layers, e.g. birch bark.

EXIGU|US, -A, -UM Very small, weak or feeble.

EXILIS, -E Very slender; thin or meagre.

EXINE
see POLLEN

EXODERMIS The outermost layer of the cortex, often of larger corky cells.

EXOGENOUS
see ENDOGENOUS

EXOSMOSIS
see OSMOSIS

EXOTIC Any plant introduced from another country which is not native. By wrong implication, often referring to something gorgeous or tropical.

EXSCAP|US, -A, -UM Stemless. (Without scape.)

EXSERTED Projecting; e.g. stamens beyond petals.

Name of family	Typical member	Name of family	Typical member	Name of family	Typical member
Acanthaceae	acanthus	*Cucurbitaceae*	melon	*Oxalidaceae*	oxalis
Aceraceae	maple	*Cupressaceae*	cypress	*Palmae*	date palm
Agavaceae	century plant	*Cyperaceae*	sedge	*Papaveraceae*	poppy
Aizoaceae	mesembryanthemum	*Dioscoriaceae*	yam	*Pinaceae*	scots pine
Alliaceae	onion	*Ericaceae*	heather	*Piperaceae*	pepper
Amarantaceae	love lies bleeding	*Euphorbiaceae*	spurge	*Polygonaceae*	knotweed
Amaryllidaceae	daffodil	*Fagaceae*	beech	*Portulacaceae*	purslane
Anacardiaceae	sumac	*Gentianaceae*	gentian	*Primulaceae*	primrose
Annonaceae	custard apple	*Gesneriaceae*	hot water plant	*Ranunculaceae*	buttercup
Apocynaceae	periwinkle	*Gramineae*	grass	*Rhamnaceae*	buckthorn
Aquifoliaceae	holly	*Iridaceae*	Iris	*Rosaceae*	rose
Araceae	lords and ladies	*Juglandaceae*	walnut	*Rubiaceae*	coffee
Araliaceae	ivy	*Juncaceae*	rush	*Rutaceae*	orange
Aristolochiaceae	dutchman's pipe	*Labiatae*	dead nettle	*Salicaceae*	willow
Begoniaceae	begonia	*Lauraceae*	bay-laurel	*Saxifragaceae*	saxifrage
Berberidaceae	barberry	*Leguminosae*	pea	*Scrophulariaceae*	figwort
Betulacaceae	birch	*Liliaceae*	lily	*Solanaceae*	potato
Boraginaceae	forget-me-not	*Magnoliaceae*	magnolia	*Theaceae*	camellia
Bromeliaceae	urn plant	*Malvaceae*	mallow	*Thymelaeaceae*	daphne
Cactaceae	cactus	*Moraceae*	fig	*Tiliaceae*	lime
Campanulaceae	harebell	*Musaceae*	banana	*Tropaeoleaceae*	nasturtium
Caryophyllaceae	pinks	*Myrtaceae*	myrtle	*Ulmaceae*	elm
Compositae	daisy	*Nyctaginaceae*	bouganvillea	*Umbelliferae*	carrot
Cistaceae	rock rose	*Oleaceae*	olive	*Urticaceae*	nettle
Cornaceae	dogwood	*Onagraceae*	evening primrose	*Violaceae*	violet
Crassulaceae	stonecrop	*Orchidaceae*	orchid	*Vitidaceae*	grape vine
Cruciferae	cabbage	*Orobanchaceae*	broomrape	*Zingiberaceae*	ginger

▲

EXSTIPULATE Without stipules.

EXTRORSE Used of stamens which shed their pollen to the outside of the flower away from the pistil.

EYE The centre of a flower, particularly when of a contrasting colour.

● **EYEBRIGHT** Many species of *Euphrasia*, a genus of 200 annuals and perennials in the *Scrophulariaceae* from temperate regions. They are mainly small plants, gaining some of their nourishment from the roots of grasses and other nearby plants. The leaves are OVATE with several lobe-like teeth. The tubular flowers expand to 5 notched lobes, the 3 lower large and spreading, the upper 2 erect. They have spined stamens which catch against visiting insects, covering them with pollen. Common eyebright includes many species formerly classified together under *E. officinalis*. One of the most widespread in Britain is *E. nemorosa*, an erect, branched species 100–200mm tall, with spikes of white, 6.0mm flowers, yellow in the throat and sometimes blue-tinted on the upper petals.

FAIR MAIDS OF FRANCE [FAIR MAIDS OF KENT] *(Ranunculus aconitifolius)*. A perennial plant in the *Ranunculaceae* from Europe. It has PALMATE leaves divided into 5 deeply toothed lobes and 300–600mm branched stems bearing white buttercup flowers. The pompon-like double-flowered form is known as bachelor's buttons.

FAIRY RING A darker green, often irregular ring in grassland caused by several species of fungi, notably champignon. The original fungus plant, growing from a chance spore, sends HYPHAE radiating in all directions, like the spokes of a wheel. Later the inner MYCELIUM dies, leaving a slowly expanding band like the perimeter of a wheel. It appears that the more vigorous dark green grass is a result of additional nitrogen derived from the decomposition of the dead mycelium, left behind.

FALCATE [FALCAT|US, -A, -UM] Sickle-shaped, usually of leaves.

FALSE ACACIA *(Robinia pseudacacia)*. A deciduous tree to 24.5m or so in the *Leguminosae* from N. America. It has PINNATE leaves composed of 11–23 ovate, 25–30mm long leaflets. The 19mm long, fragrant white pea-flowers are borne in short pendulous RACEMES. It was once much planted in Britain as a potential timber tree but proved unsatisfactory.

FALSE OAT GRASS *(Arrhenatherum elatius)*. A RHIZOMATOUS perennial in the *Gramineae* from Europe to W. Asia and N. Africa. It is naturalised elsewhere, and has stems to 1.2m tall with flat, scabrid leaves and 100–200mm nodding PANICLES of 8.0mm long spikelets. The extensively creeping rhizomes make this as bad a pest as couch.

▲ **FAMILY** The next unit of classification above genus and distinguished by the ending *-ae* or *-eae*., e.g. the lily family is *Liliaceae* and the cabbage family, *Cruciferae*.

FASCIATION Curiously flattened or crested growths appearing as if several stems had been squashed together side by side. Leafy and flowering stems can be affected.

There appear to be several causes, but very little scientific research into them. The COCK'S COMB plant is a mutant of this form which comes true from seed. Some of the others may be due to fungal, bacterial or mite damage to the growing tip of a shoot. *See also* PELORIA.

FASCICLE [FASCICULAT|US-, -A, -UM] Having a columnar habit with the branches and twigs erect.

FASTUOS|US, -A, -UM Appearing haughty or proud.

FEN An area of wet, peaty soils, often under water for at least part of the year, but differing from a BOG by being rich in mineral salts from drainage water, and often alkaline in reaction. The dominant tree is alder, with smaller, plants such as great fen sedge *(Cladium mariscus)*, common reed *(Phragmites communis)*, tussock sedge *(Carex paniculata)*, royal fern *(Osmunda regalis)*, marsh fern *(Thelypteris palustris)* milk parsley *(Peucedanum palustre)* and marsh orchids *(Dactylorhiza* spp.*)*

FENNEL *(Foeniculum vulgare)*. An erect perennial in the *Umbelliferae* from Europe. It has polished stems 0.9–1.5m tall and leaves much divided into thread-like segments 13–50mm long. The tiny yellow flowers are carried in 38–75mm wide UMBELS. The foliage is used to flavour sauces, soups and salads. Florence or finnochio fennel *(F.v. dulce)* has a bulb-like stem base which is cooked and eaten. *See also* GIANT FENNEL.

FERNS Spore-bearing plants known as *Filices* or *Filicopsida*, which with the horsetails and club mosses make up the *Pteriodophyta*. They are found throughout the world, mainly in the warmer, wetter climates and often in shady places. Some 10,000 species are at present recognised and many more are being found. They may be clump-forming of one to several rosettes (e.g. male fern), or spreading by RHIZOMES and producing single fronds (e.g. bracken). Frequently the leaves (usually known as fronds) are deeply divided, either TRI-, BI- or PINNATE or PINNATISECT. They unfold when young like a crozier. The visible fern plant is the SPOROPHYTE and bears spores in SPORANGIA which are in turn gathered into SORI on the undersides of the fronds. Spores give rise to tiny sexual plants or gametophytes (*see* ALTERNATION OF GENERATIONS). Several ferns have entire, undivided fronds, notably the hart's tongue and bird's nest ferns. Filmy ferns *(Hymenophyllum)* are a distinctive group of plants, mainly small, with creeping rhizomes. Their fronds are pinnate or bi-pinnate and semi-translucent, being only one cell thick except at the few veins; many species are EPIPHYTES in rain forests, two occurring on trees in Britain. Some ferns produce a stout, erect rhizome, thickly woven with aerial roots for support and with a head of leaves at the top. These are known as tree ferns and can attain 12m or more in height. Most of them are tropical, but species of *Dicksonia, Cyathea* and *Hemitelia* come from the cooler parts of Australia.

FEROX Fiercely spiny.

FERRUGINE|US, -A, -UM Rust-coloured, e.g. hairs on leaves.

FERTIL|IS, -E Producing abundant seeds; fertile.

FERTILISATION After pollination takes place, each pollen grain germinates, sending a tube down the STYLE. The generative NUCLEUS of the grain enters the tube and divides into two. When the pollen tube reaches the EMBRYO sac within the ovule, it liberates the two nuclei, one of which fuses with the egg cell, the other with the central fusion nucleus. Thus fertilisation is affected. Pollen from the same plant causes self-fertilisation, from another plant of the same species, cross-fertilisation.

FESCUE A name applied in a general way to a number of grass species in several genera of the *Gramineae*. The main genus is *Festuca* with 100 perennial species found throughout the temperate world. Some have narrow, rolled leaves, others flat, some are tufted, others creeping: creeping or red fescue *(F. rubra)* is stoloniferous, having bristle-like leaves, 150–600mm stems and reddish or purplish spikelets in 38–150mm PANICLES; sheep's fescue *(F. ovina)* is similar but of tufted habit, usually with greenish or bronzy-purple spikelets; blue fescue *(F. glauca)* is densely tufted with bright glaucous-blue leaves and is often cultivated as an ornamental.

FEVERFEW *(Chrysanthemum parthenium* syn. *Matricaria eximia)*. A 300–600mm perennial in the *Compositae* from S.E. Europe and naturalised elsewhere. It is a bushy, erect plant with OVATE, PINNATIFID leaves and large terminal CORYMBS of 13mm wide, white daisy flowers. Formerly grown as a medicinal herb.

FIBRES Strands or threads of hard, elastic tissue that act as strengthening in plant stems. If is formed of slender, tapered, dead cells thickened with lignin, the substance which gives wood its strength. Extracted from the softer tissues, fibres are much used commercially, e.g. flax, hemp.

FIBROUS [FIBROS|US, -A, -UM] Composed of or containing fibres.

FIBRILLOSE Covered with fibres or fibre-like strands.

FIG Many of the 800 species of *Ficus*, deciduous and evergreen trees, shrubs and climbers, mainly from the tropics and sub-tropics. They vary much in shape and size but are characterised by their curious 'inside-out' inflorescences. These are enlarged and hollowed out stem tips, globular to pear-shaped, with a tiny hole at the top. Tiny petal-less flowers are borne within these structures and are pollinated by minute gall-wasps. After fertilisation, the individual carpels and the hollow inflorescence swell up and become fleshy in some species, forming a syncomium or fig-fruit. Flowers and fruits may be borne on young shoots directly from the trunk, or on special underground stolons (geocarpic). Aerial roots are also a feature of many fig species; in the banyan they form prop roots which thicken to become secondary trunks. The so-called strangler figs usually start as EPIPHYTES high up in a tree. Later they send down aerial roots that reach the ground and form a network around the host trunk. Eventually the roots form a sheath around the tree trunk and strangle it to death. The fig branches eventually swamp and take over the host branches so that the one tree is replaced by another. The leaves of most figs are entire, LANCEOLATE to OVATE, OBLONG to OBOVATE; a few are PALMATE and lobed: indiarubber tree or rubber plant *(F. elastica)* grows to 30m with glossy, oblong-elliptic,

150–450mm leaves. It is often a strangler in the wild, and when young is much used as a house plant; climbing fig *(F. pumila)* is a small plant that climbs like ivy, with OVATE-CORDATE leaves. When it reaches the top of its support it produces fruiting branches with large, elliptic leaves; common fig *(F. carica)* is a deciduous shrub or tree to 9.1m with PALMATE 3–5 lobed leaves and the familiar fruit of commerce which may be ovoid to onion-shaped greenish to dark-purple. *See* BANYAN; BO TREE.

FIGWORT *(Scrophularia nodosa)*. A perennial plant in the *Scrophulariaceae*, from temperate Europe and Asia. It has erect, 450–900mm stems which are square in cross section, and opposite pairs of 50–125mm coarsely SERRATE leaves. The 6mm long green and red-brown flowers are pouch-like. Water figwort or water betony *(S. aquatica)* is similar, but the angles of the stem are winged and the narrower leaves are CRENATE.

FILAMENT The stalk which attaches the anthers to the base of the flower.

FILBERT
see HAZEL

FILIFORM [FILIFORMIS-E] Thread-like, e.g. stems.

FIMBRIATE [FIMBRIAT|US, -A, -UM] Bearing a fringe.

FIR Used indiscriminately for several coniferous trees, but strictly applying to *Abies*, a genus of pyramidal evergreen trees in the *Pinaceae* from the mountains and colder parts of the north temperate zone. The LINEAR, round or notched tipped leaves densely clothe the regular WHORLS of horizontal branches. The ovoid cones are erect, falling to pieces when ripe. The seedlings have 5–7 COTYLEDONS. Several species are important timber trees and many are grown as ornamentals. Giant, silver or grand fir *(A. grandis)* occasionally reaches 90m in its native N. America, the 25–50mm long, deep green leaves being blue-white beneath, the cylindrical cones are 75–100mm long. *See also* CANADA BALSAM.

FIRE THORN *(Pyracantha coccinea)*. A somewhat thorny evergreen shrub or small tree in the *Rosaceae* from S. Europe and Turkey. It has narrowly OBOVATE, toothed leaves and small clusters of 5-petalled 8mm wide white flowers. The profusely borne, berry-like fruits are bright red. *P.c.* 'Lalandii' has orange-red fruits and is commonly seen in gardens.

FISH POISON PLANTS Several tropical plants in the *Leguminosae* have the property of stupefying fish when crushed and thrown into the water. Of particular interest are several species of *Derris*, woody climbers from S.E. Asia, extracts of which kill both fish and insects. *D. elliptica* grows to 15m or more with PINNATE leaves of 11–15 OVATE leaflets each 25–125mm long. The pink and white pea-flowers are carried in AXILLARY RACEMES.

FISSION The simple dividing into two of unicellular plants such as algae and bacteria. Rapid multiplication can take place by this method.

FISTULOS|US, -A, -UM Hollow and pipe-like.

▲

■

FLABELLAT|US, -A, -UM Like a wide-opened fan, e.g. leaves.

FLACCID|US, -A, -UM Weak and floppy.

FLAGELLA
see CILIA

FLAGELLAR|IS, -E Having long, whip-like stems.

FLAMBOYANT TREE *(Poinciana regia)*. Also known as flame tree, an almost evergreen tree to 12.2m in the *Leguminosae* from Malagasy, but much grown throughout the tropics and subtropics. It has bipinnate leaves to 600mm long and terminal RACEMES of bright scarlet flowers, each with 5 long-stalked rounded petals.

▲ **FLAMINGO FLOWER** *(Anthurium scherzerianum)*. A tufted, evergreen plant in the *Araceae* from Costa Rica. It has long-stalked, leathery, oblong-lanceolate leaves 150–200mm long, and ovate, palette-like scarlet spathes with 50–100mm spirally twisted SPADICES.

FLANNEL WEED
see ALGA

FLAVESCENS Pale yellow.

FLAVID|US, -A, -UM Yellowish.

FLAV|US, -A, -UM Yellow, but paler than LUTEUS.

■ **FLAX** Several species of *Linum*, annuals and perennials in the *Linaceae* from throughout the temperate and subtropical world. They have LINEAR to OVATE leaves and terminal branched RACEMES of funnel-shaped or rotate flowers in shades of blue, purple, red, pink, yellow and white. Common flax *(L. usitatissimum)* provides the flax fibre of commerce. It is an erect plant to 600mm with linear-lanceolate leaves and blue rotate flowers to 32mm wide. The seeds yield linseed oil and the residue (cake) is fed to cattle.

FLEABANE Applied in a general way to several species in the genera *Conyza*, *Inula*, *Erigeron* and *Pulicaria*, all members of the *Compositae*: common fleabane *(Pulicaria dysenterica)* is an erect, 250–600mm perennial from Europe, Turkey and N. Africa, with OBLONG to LANCEOLATE, densely softly hairy leaves and 13–45mm wide, rich yellow flower heads; Canadian fleabane *(Conyza canadensis)* is a stiff, hairy annual to 900mm from N. America, but naturalised in many countries, with LINEAR-LANCEOLATE leaves and very small, white or lavender tinted flower heads in large, loose PANICLES. It is a weed of waste and cultivated land; alpine fleabane *(E. borealis)* is a low-growing perennial from European mountains, Scandinavia and Iceland.

FLEUR-DE-LYS
see IRIS

FLEXIL|IS, -E Flexible, capable of being bent easily.

FLEXUOS|US, -A, -UM Zig-zag; stems that bend alternately left and right from each node.

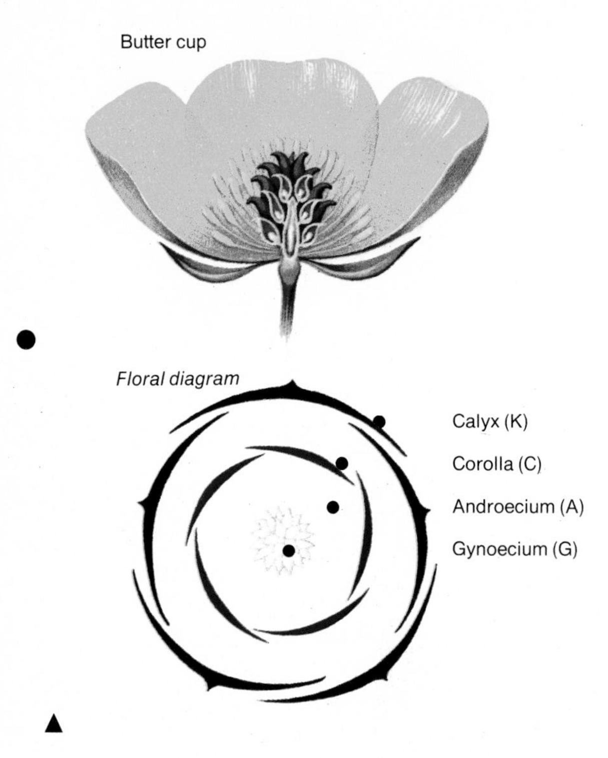

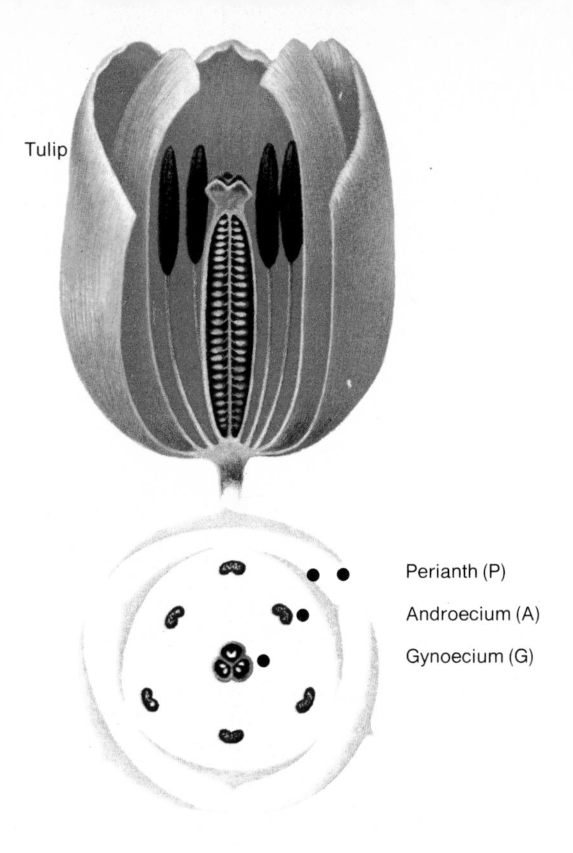

Floral formula (See ABBREVIATIONS)

$+ K_5\, C_5\, A_{\infty}\, \underline{G}_{\infty}$

$+ P_{3+3}\, A_{3+3}\, G_{(3)}$

FLOCCOS|US, -A, -UM Bearing tufts of woolly hairs.

FLORA All the plants of a particular region or country; a book describing them.

FLORAL Of or belonging to flowers; floral axis
see RECEPTACLE

● **FLORAL DIAGRAM** In effect a ground plan of a flower bud just before it opens; a cross section, showing the positioning of sepals, petals, stamens and ovaries, their size (width) and linkage.

FLORAL ENVELOPE The perianth, i.e. sepals or calyx, petals or corolla.

▲ **FLORAL FORMULA** An easy way of showing the main features of a flower by means of abbreviations. For example, the formula 'K(5) C(5) A5 $\overline{G}$(5)' is explained as follows: K=calyx of 5 united sepals (the brackets signifying joined, C=corolla of 5 united petals, A=androecium of 5 stamens and $\overline{G}$ 5=inferior gynoecium, an ovary of 5 joined carpels (a line above the letter indicating inferior, beneath it, superior).

FLORAL MECHANISMS The various ways that the parts of a flower are arranged or timed to ensure, as much as possible, cross-fertilisation, e.g. the ripening of stigmas and pollen at different times; loose, buoyant pollen easily carried away by the wind; sticky pollen for transportation by insects; sensitive stamens or stigmas that move when visited by insects.

FLORE PLENO Bearing double flowers.

FLORETS Usually very small flowers, aggregated together to form larger ones, e.g. the daisy flower and other composites.

FLORIBUND|US, -A, -UM Flowering profusely.

FLORISTIC REGIONS The division of the earth's surface into areas typified by the presence of particular plant genera and species, which in turn relate to climate, e.g. the Mediterranean region is typified by mild moist winters and long hot dry summers and is found also in California, Chile, S. Africa and W. Australia.

FLOS-JOVIS Jove's flower.

FLOTE GRASS *(Glyceria fluitans)*. Also called floating sweet grass, this is an aquatic plant in the *Gramineae* from the north temperate zone. It grows in wet ground or shallow water, the flat leaves often floating. The ascending 300–600mm stems bear loose, 100–450mm long PANICLES of 13–25mm long cylindrical spikelets.

■ **FLOWER** The specialised organ of the seed-bearing plants *(Spermatophyta)* concerned with reproduction. The essential parts are the pistils and stamens surrounded by petals, usually coloured for attracting pollinating insects, the whole enclosed and protected by the sepals. All these flower parts are specialised modified leaves.

FLOWERING FERN
see ROYAL FERN

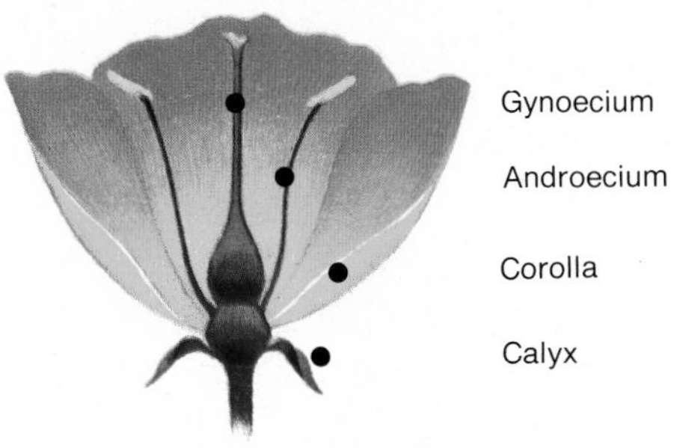

Whorls of a typical flower

Structure of a typical flower

■

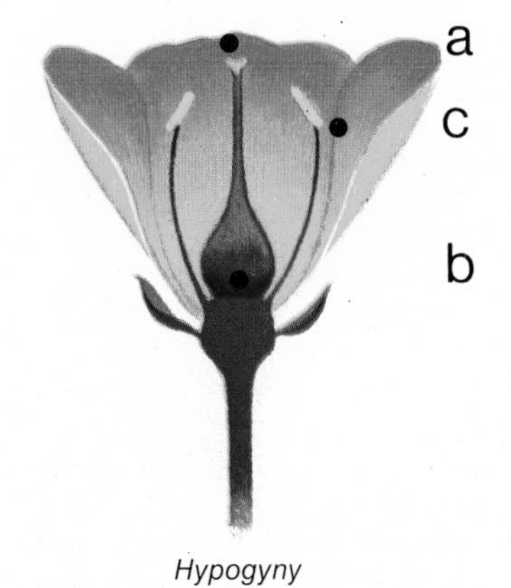

Hypogyny

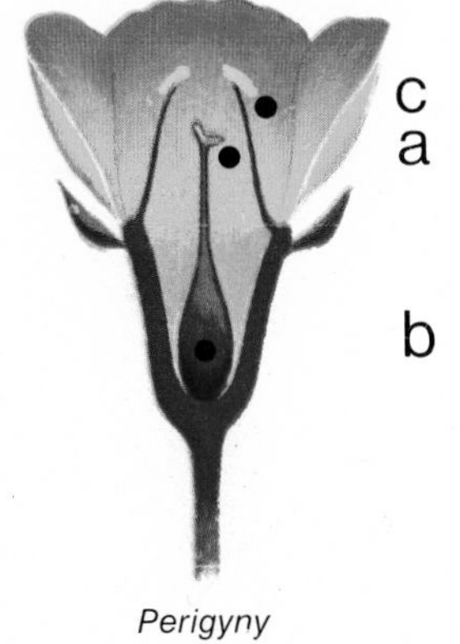

Perigyny

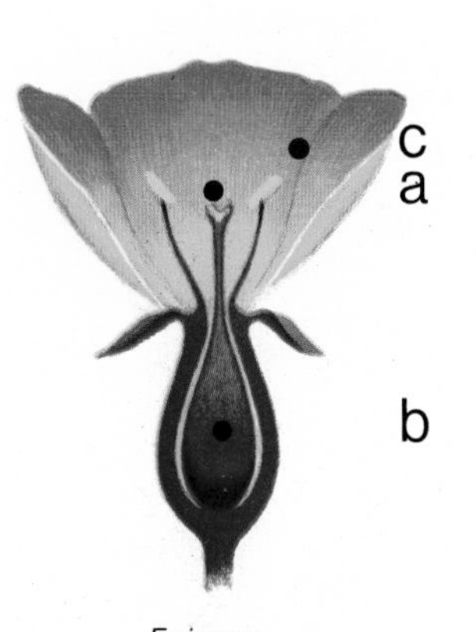

Epigyny

a Stigm
b Ovary
c Anthe

FLOWERING MAPLE Several species of *Abutilon*, shrubby plants in the *Malvaceae* from the tropics and subtropics, a few elsewhere. They have PALMATE to OVATE-CORDATE leaves and usually bell-shaped, nodding flowers, borne singly or in small groups from the leaf AXILS. Several are cultivated for their ornamental flowers and foliage: *A. megapotamicum* has ovate, cordate leaves and striking 38–50mm long, red and yellow flowers; *A. vitifolium* is a large shrub or small tree from Chile with hairy, palmate leaves and rotate, 50–75mm wide, blue-purple flowers.

● **FLOWERING RUSH** *(Butomus umbellatus)*. A RHIZOMATOUS rush-like perennial in the *Butomaceae* from Europe and Asia, naturalised in America. It has slender, triangular leaves to 1.5m and stems of the same height bearing 25–32mm pink flowers with 3 petals.

FLOWER OF A DAY
see DAY LILY

FLUITANS Floating in water.

FLUVIATIL|IS, -E Growing in streams or rivers.

FLY AGARIC
see AMANITA

▲ **FLY ORCHID** *(Ophrys insectifera)*. A slender, erect plant in the *Orchidaceae* from Europe. It has oblong to elliptic leaves 38–125mm long and 150–600mm stems. The flowers have 3 oblong, yellow-green sepals, 2 filiform velvety purple-brown petals and an oblong purple-brown LABELLUM bearing a shining bluish patch.

FLY TRAP
see VENUS'S FLY TRAP

FOEMINA Female.

FOLIACEOUS [FOLIAT|US, -A, -UM] Leafy or leaf-like.

FOLLICLE A dry, DEHISCENT fruit of one carpel opening along one side only, e.g. columbine, larkspur.

FONTAN|US, -A, -UM Growing in or by a spring.

FOOD OF THE GODS
see GIANT FENNEL

FOOL'S PARSLEY *(Aethusa cynapium)*. A 50–100mm tall, branched annual in the *Umbelliferae* from Europe and N. Africa. It has TERNATELY, bipinnate triangular leaves and tiny white flowers in 25–38mm wide UMBELS. A weed of cultivated soil.

FOOL'S WATERCRESS *(Apium nodiflorum)*. An aquatic perennial in the *Umbelliferae* from Europe and Asia. A procumbent or ascending plant with rooting stems 300–900mm long and bright green, PINNATE leaves. The leaflets are 13–32mm long, LANCEOLATE or OVATE, often slightly lobed. The tiny white flowers are borne in small UMBELS. Sometimes mistaken for watercress.

FOREST A term now equated with WOODLAND, but formerly any area outside common law, reserved for the King's hunting and including large areas without trees.

▲ 

■ 

FORGET-ME-NOT Several species of *Myosotis* a genus of 50 annual and perennial plants from temperate regions. They have lanceolate or oblanceolate to oblong, ovate leaves and SCORPIOID CYMES of small, rotate flowers with five lobes. They come in shades of blue, purple, yellow and white; common forget-me-not *(M. arvensis)* is an erect annual to 150mm or more with obovate-oblong, hairy leaves and bright blue flowers opening from pink buds; wood forget-me-not *(M. sylvatica)* is perennial and larger with flowers to 9mm wide. It is much cultivated; water forget-me-not *(M. scorpioides)* is a RHIZOMATOUS perennial 150–220mm tall which grows in wet places.

■ **FORREST, GEORGE** (1873–1932) A Scottish collector who became interested in plants at an early age. After a short period as an apprentice to a pharmaceutical chemist, he went to Australia. Fortune did not however favour him there and he returned to Scotland in 1902, later obtaining a position in the herbarium at the Edinburgh Royal Botanic Garden. In 1904, Mr A. K. Bulley (of the Bees Seeds Company) asked for a person to collect for him in Western China and Forrest was recommended.

Western China at that time was almost untouched by the horticultural collector and presented a vast area with perhaps the richest flora in the world. However, it was by no means a politically stable region, and on one occasion Forrest narrowly escaped with his life. For the next 28 years, during which he led 7 expeditions all to the Yünnan area, a vast number of plants new to science or to cultivation were introduced to Britain. Forrest's success were a direct result of his organising ability and his use of a large team of trained native collectors, sent out to cover as wide an area as possible. He died suddenly at the end of his seventh expedition just before he was due to set out for home. Among the many plants he introduced were 300 *Rhododendron* species and 154 distinct kind of *Primula*, plus magnolias, camellias, lilies and species of *Meconopsis*.

FORTUNE, ROBERT (1812–80) Born in Berwickshire, Scotland, he trained as a gardener, working at the Royal Botanic Garden, Edinburgh from 1839–1842, then at the Horticultural Society's garden at Chiswick. The ending of the Opium War gave opportunities for plant collecting in China, and he was sent out by the Horticultural Society in 1843 with instructions as to the plants to be collected.

He faced many problems on his arrival; the population at large were far from friendly to Westerners, and his movements were restricted to within 40 miles of the treaty ports, but he managed to travel further inland, usually in Chinese costume. He sought plants from Chinese gardens which were not in Western cultivation and sent home winter flowering jasmine *(Jasminum nudiflorum)*, *Anemone japonica*, bleeding heart *(Dicentra spectabilis)*, and many tree peonies. He came back to England in 1846, becoming curator of the Chelsea Physic garden, but returned to China 1848–51 on behalf of the East India Company. This time he concentrated on collecting tea plants for the Company's new plantations in India. Another visit (1853–56) produced *Rhododendron fortunei*, the parent of many modern hybrids, and from 1860 to 1862 he was in Japan bringing back umbrella pine *(Sciadopitys verticillata)* and *Primula japonica*. Chief books: *Three years wanderings in the Northern Provinces of China* (1847), *A Journey to the Tea Countries of China* (1852).

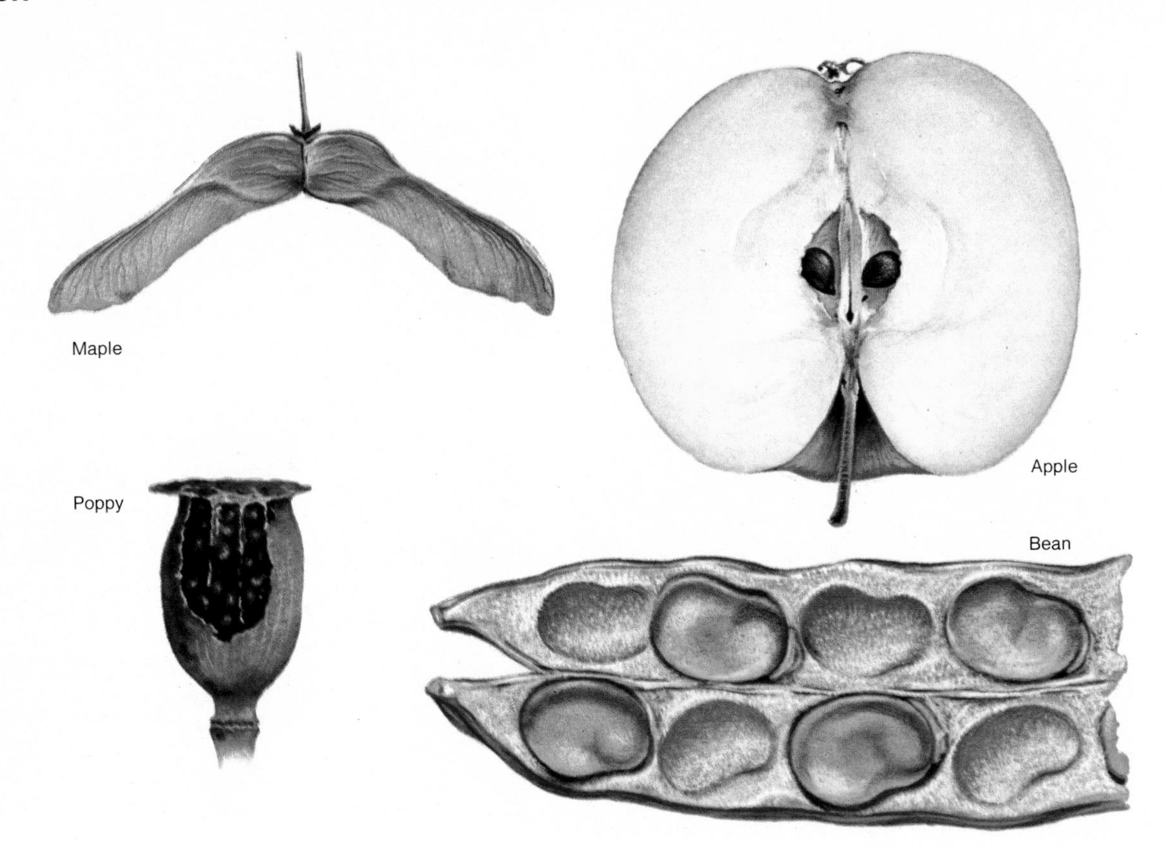

FOXGLOVE *(Digitalis purpurea)*. A biennial or short-lived perennial plant in the *Scrophulariaceae* from W. Europe and Morocco. It has 150–300mm long OVATE, CRENATE, softly pubescent leaves and unbranched stems to 1.5m with pendent, 25–38mm flattened bell-flowers, rose-purple bearing dark purple spots within.

FOXTAIL GRASS Several perennial species of *Alopecurus*, tufted or creeping plants in the *Gramineae* from Europe, Asia and N. America. They are characterised by having cylindrical, tail-like PANICLES, each spikelet having a fine AWN. Meadow foxtail *(A. pratensis)* is a robust grass to 900mm with 38–74mm flowering panicles, opening earlier than most grasses.

FRAGILIS, -E Brittle, e.g. of stems.

FRAGRANT ORCHID *(Gymnadenia conopsea)*. A tuberous-rooted, ground-dwelling plant in the *Orchidaceae* from Europe and Asia. It has somewhat keeled, narrowly oblong, 63–150mm long leaves and stems to 300mm or more carrying dense spikes of fragrant, reddish-lilac flowers. These have deeply 3-lobed flared LABELLUMS.

FRANGIPANI *(Plumeria rubra)*. A large deciduous shrub or small tree in the *Apocynaceae* from Central America, but grown in the tropics and subtropics. It has LANCEOLATE leaves to 300mm long and stiff clusters of 5-petalled, rather fleshy, tubular, fragrant flowers in shades of white, pink and yellow.

FRANKINCENSE The fragrant gum resin used in incense obtained by wounding the stems of *Boswellia carteri*. This is a large evergreen shrub or small tree to 6m, in the *Burseraceae* from Somalia and Arabia. It has PINNATE leaves with up to 21, OVATE, 38mm long, undulate leaflets. The small white flowers are borne in AXILLARY RACEMES.

FREE Separate; of plant organs such as petals and sepals.

FREESIA A genus of 20, bulbous-rooted perennials in the *Iridaceae* from S. Africa. They have long, sword-shaped leaves in flat fans which rise from an ovoid corm. The funnel-shaped, fragrant 25–50mm long, cream flowers are borne on 450–756mm stems. Many hybrid cultivars are grown; these have flowers in shades of yellow, cream, pink, red and purple.

FRENCH MARIGOLD *(Tagetes patula)*. A 300–450mm annual in the *Compositae* from Mexico. It has PINNATISECT leaves with LINEAR toothed lobes and erect, terminal flower heads. The latter are daisy-shaped, with a tubular INVOLUCRE and broad, brownish-yellow to crimson-brown RAY florets. Many cultivars derived from this, also hybrids with African marigolds, are grown.

FRITILLARY Species of *Fritillaria*, a genus of 85 bulbous-rooted perennials in the *Liliaceae* from the north temperate zone, with unbranched, erect stems and linear to ovate leaves. The bell-shaped flowers are pendent or carried horizontally: crown imperial *(F. imperialis)* is the largest, with robust stems to 1.2m, narrowly LANCEOLATE leaves and a terminal UMBEL of nodding, 50–75mm long red or yellow bells; snake's head fritillary *(F. meleagris)*

	Type	Example
Dry indehiscent	Achene	Strawberry
	Nut	Oak
	Samara	Maple
	Schizocarp	Geranium
Dehiscent	Capsule	Poppy
	Follicle	Larkspur
	Legume	Bean
	Siliqua	Mustard
Succulent	Berry	Tomato
	Drupe	Cherry
	Pepo	Cucumber
	Pome	Apple

●

▲

is slender, the stems 200–380mm tall having linear leaves and 1–2, nodding purple 32–50mm chequered bells, sometimes white. These and others are grown in gardens.

FROG BIT *(Hydrocharis morsus-ranae)*. A floating aquatic in the *Hydrocharitaceae* from Europe and Asia. It is stoloniferous, with orbicular, floating leaves and tufts of roots which hang in the water from each node. In autumn, large resting buds called turions sink to the bottom of the pond or ditch, rising again the following spring and breaking into growth. The 19mm wide 3-petalled flowers are white.

FROND The usually dissected leaves of palms and ferns.

FRONDOSE [FRONDOS|US, -A, -UM] Leafy.

FRUCTIFICATION The fruits and any attendent stalks and bracts.

● **FRUIT** In the strict sense the fruit is directly derived from the ovary of the flower which enlarges after fertilisation to make room for the maturing seeds. It may be dry like a capsule or nut, or fleshy like a berry. The dry fruits may liberate the seeds by a slit or pore or explode like a gorse pod. In the broad sense a fruit includes the above and those attendent structures which look like and act as true fruits. For example, the strawberry is a swollen receptacle which bears true fruits upon its surface, the achenes or 'seeds'. The apple and pear are also receptacles, but in this case they surround the true fruit, the core.

FRUTESCENS [FRUTICANS] Becoming shrubby.

FRUTICOS|US, -A, -UM [FRUTEX] Shrubby or bushy.

▲ **FUCHSIA** A genus of deciduous and evergreen shrubs and trees in the *Onagraceae* from Central and S. America and New Zealand. Named for Leonhard Fuchs (1501–66), a German physician and herbalist. The pendulous flowers of fuchsia have a tubular perianth from which are usually borne 4 spreading, narrow sepals and 4, broad petals which form a bell. The leaves are LANCEOLATE to OVATE, often toothed, either in opposite pairs or WHORLS of 3 or more. Common or hardy fuchsia *(F. magellanica)* grows to 1.8m or more with lanceolate to ovate leaves in groups of 2 or 3. The 25–50mm flowers have scarlet tubes and sepals and purple-blue petals. It has given rise to hundreds of cultivars.

FUGACIOUS Of fleeting or short duration.

FULGENS [FULGID|US, -A, -UM] Bright or shining, usually of red flowers.

FULV|US, -A, -UM Yellow-brown or tawny.

FUMITORY Several species of *Fumaria*, a genus of 55 mainly annual plants in the *Fumariaceae* from Europe, Asia and the mountains of Africa. They are bushy or sprawling plants, sometimes climbing by their leaflet stalks. The leaves are PINNATISECT, 2, 3 or 4 times, and the slender ZYGOMORPHIC flowers are spurred. Common fumitory *(F. officinalis)* may be sub-erect or semi-climbing, with bipinnate leaves having deeply pinnatisect leaflets and AXILLARY RACEMES of 6–9mm long pink flowers.

● **FUNGI** A unique group of plants classified between the *Algae* and *Bryophyta*. They lack CHLOROPHYLL and live as SAPROPHYTES or PARASITES. The actual body of a fungus consists of a mass of white threads known as HYPHAE which actively feed and grow. They reproduce by various kinds of spores: minute, often air-borne structures. The more primitive fungi bear their spores directly on the tips of the hyphae, e.g. potato blight. The more specialised kinds build the hyphae into elaborate fruiting bodies on which the spores are carried, e.g. the familiar toadstools and mushrooms.

FUNICLE The stalk which attaches the ovule to the placenta in the ovary.

FURCAT|US, -A, -UM Forked.

FURZE
see GORSE

FUSCOUS [FUSC|US, -A, -UM] Dark or dark brown.

FUSIFORM|IS, -E Spindle-shaped, tapering to each end.

▲ **GALL** Variously shaped structures of plant tissue, resulting from the irritation set up by the presence of parasitic bacteria, fungi, eelworms, mites or insects. Some of the most obvious galls are caused by insects such as gall midges and gall wasps. The latter are responsible for the familiar oak apples.

GALLANT SOLDIER *(Galinsoga parviflora)*. An erect annual in the *Compositae* from S. America but now a cosmopolitan weed. It grows to 750mm tall, with stalked, OVATE-ACUMINATE leaves in opposite pairs and dichasial CYMES of tiny, 3–4mm wide daisy flower heads.

GAMA GRASS
see BOUTELOUA

GAMETANGIUM [GAMETANGIA] The specialised cells in the filamentous ALGA *Oedogonium* which produces GAMETES.

GAMETS The male and female reproductive cells. In some primitive plants, e.g. the filamentous algae *Chlamydomonas*, they are alike in size and appearance, but in the majority of plants the female cell (the egg or ovum) is larger.

GAMETOPHYTE The GAMETE-bearing plant. In the life cycles of many plant groups, the sex cells are borne on individuals quite unlike the familiar plant. *See* ALTERNATION OF GENERATIONS.

GAMOPETALOUS [GAMOSEPALOUS] With petals or sepals united to form a tube, cup or bell shape.

GARDEN CRESS *(Lepidium sativum)*. A 200–450mm unbranched annual in the *Cruciferae*, probably from W. Asia or Egypt, but widely grown and naturalised elsewhere. The basal leaves are LYRATE with OBOVATE lobes the stem ones PINNATE or bipinnate. The small, fragrant white or pinkish flowers are borne in RACEMES. The seedlings of this plant form the familiar cress in salads.

GARDENER'S GARTERS The popular name for the variegated form of reed grass.

GARDENIA A genus of 250 species in the *Rubiaceae* from tropical Africa and Asia. One species is familiar in its double form as a fragrant buttonhole or corsage flower. This is *G. jasminoides*, an erect evergreen shrub to 1.8m from China with elliptic glossy leaves and salviform, solitary white flowers.

GARLIC *(Allium sativum)*. A bulbous perennial in the *Alliaceae* from Asia, but long cultivated in Mediterranean countries and elsewhere. The familiar garlic bulb is in fact a cluster of bulblets known as cloves. They are pungently onion-flavoured. The leaves are flattened, LINEAR and the 300–600mm stems bear greenish-white to purple, starry flowers in UMBELS.

GARLIC MUSTARD *(Alliaria petiolata* syn. *A. officinalis)*. Also called Jack-by-the-hedge, this is a biennial or short-lived perennial in the *Cruciferae* from Europe, Asia and N. Africa. It has erect, simple or shortly branched stems and long-stalked, RENIFORM, CORDATE leaves. The small, white 4-petalled flowers are carried in terminal racemes. The whole plant smells of garlic.

GARRIGUE A type of plant community that covers large areas around the Mediterranean, usually on hot, dry rocky ground. The main constituents are low-growing shrubs (mainly under 600mm), often spiny and aromatic, such as rosemary, sage, thyme, hyssop, spiny burnet and spiny spurge. In between the shrubs are a scattering of smaller plants, many of them bulbous, such as crocus, tulip, iris and fritillary. The plant species vary in different areas, but the overall effect is the same.

GASTEROMYCETES A sub-order of the *Basidiomycetes*, fungi which produce spores within a globular or flask-shaped fruiting body, e.g. puffballs and earth stars.

GEAN
see CHERRY

GELID|US, -A, -UM From cold or icy regions.

GEMMAE Small bud-like organs which are readily detached and soon grow into new plants, particularly of liverworts and mosses.

GENE [GENES] Units of inheritance which are carried on the CHROMOSOMES. Each one may be looked upon as a set of instructions for duplicating a particular character or part of one. All the characters that make up an individual species, e.g. height, flower shape and colour, leaf shape, speed of maturity etc. are controlled by genes, though climate and soil may modify their expression.

GENERATIVE CELL The main one of the two nuclei which form in the pollen grain before fertilisation.

GENERIC NAME
see GENUS

GENETICS The study of heredity relative to the gene theory of transmitted characters.

▲

●

GENICULATE [GENICULAT|US, -A, -UM] Sharply bent, like a knee.

GENOTYPE The full genetic potential of a plant, as distinct from how that plant may develop in less than ideal conditions under a certain set of environmental factors (the phenotype); e.g. strong winds may cause a tree to be lop-sided and stunted whereas in shelter the coding of its genes would ensure a tall and shapely specimen.

GENTIAN All the 400 species of *Gentiana* in the *Gentianaceae*, mainly alpine plants from throughout the world except Africa. Most are of low, tufted growth with entire, opposite pairs of leaves, but a few are tall or climbing. The tubular, funnel or bell-shaped flowers may be blue, purple, yellow, red or white. Many species are grown in gardens: trumpet gentians, *(G. clusii* and *G. kochiana* syn. *G. acaulis)* are common in the mountains of Europe, forming hummocks of elliptic, sometimes undulate leaves and deep rich blue 50mm long flowers, spring gentian *(G. verna)* is smaller, the flowers tubular with brilliant blue petal lobes; great yellow gentian *(G. lutea)* is a robust, erect plant 0.6–1.8m tall with spikes of yellow flowers.

GENUS A category of plant classification which groups together all the species with characters in common, e.g. all the different sorts of buttercup are grouped in *Ranunculus* which is the genus or generic name. *See* NOMENCLATURE.

GEOCARPIC Fruiting under ground, e.g. peanut, where the young fruits are pushed into the soil to mature. *See also* FIG.

GEOPHYTE The fourth of a seven-category system of classifying life forms: it covers herbaceous plants with winter resting buds beneath the soil. Most of the plants with bulbs, corms, tubers and rhizomes are included.

GEOTROPISM The response of a plant to gravity. Roots are attracted downwards towards it and are said to be positively geotropic. Stems grown upwards away from it and are said to be negatively geotropic. Organs such as RHIZOMES and STOLONS which grow horizontally are diageotropic. Leaves and some stems which normally grow at an oblique angle to the ground are plagiotropic.

GERMINATION The breaking into growth of the embryo within the seed. Before this can happen, the seed must be moist and have sufficient oxygen and warmth. The emergence of the young root is regarded as germination, though gardeners use this term for shoots as they break the soil. The growth of spores is also germination.

GHERKIN A small-fruited cucumber, *see* CUCURBITS.

GHOST TREE
see DOVE TREE

GIANT FENNEL Several species of *Ferula*, giant perennials to 4.5m in the *Umbelliferae* from the Mediterranean region to C. Asia. But for their large size and several tiny botanical characters, they resemble FENNEL. They yield several kinds of medicinal gum when the roots are damaged: gum ammoniac from *F. communis*, gum asafoetida (food of the gods) from *F. asafoetida* and *F. narthex*, and gum galbanum from *F. galbaniflua*.

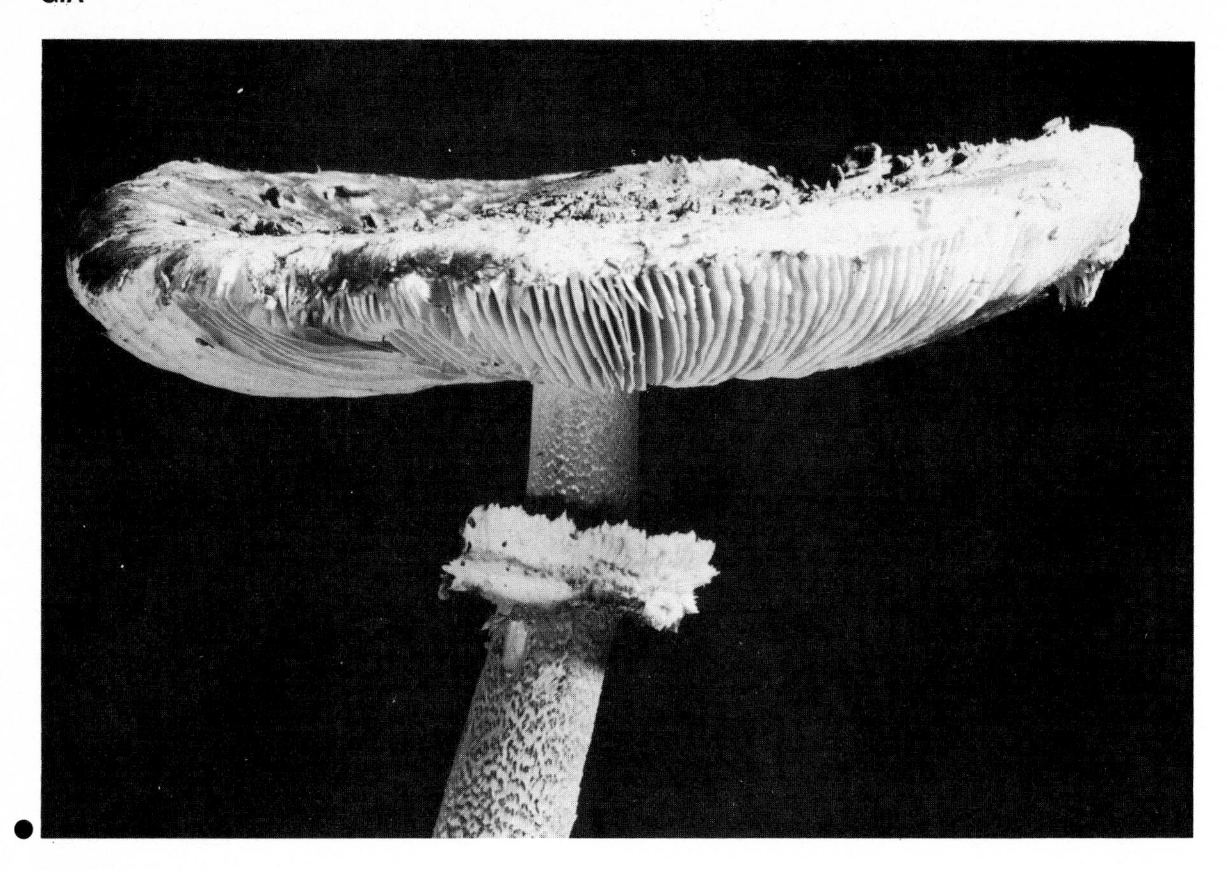

●

GIANT GROUNDSEL Several species of *Senecio* from the mountain tops of tropical East Africa. They have thick woody trunks bearing rosettes of large, somewhat cabbage-like leaves 600mm or more long in some species, densely covered with silvery hairs. The centrally borne inflorescence to 1.5m carries numerous flower heads, some species having RAY florets, others being without. *S. brassica* from M. Kenya has a short trunk and silvery white leaves.

GIBBOUS [GIBB|US, -A, -UM] Humped or swollen, often one-sidedly.

● **GILLS** The plates of tissue on the underside of a mushroom cap, which bears the spores.

GIN BERRY
see JUNIPER

GINGER *(Zingiber officinale)*. A clump-forming perennial in the *Zingiberaceae* from S.E. Asia. It has irregularly branched, fleshy aromatic RHIZOMES which yield ginger. The erect 900mm stems bear two ranks of LINEAR-LANCEOLATE leaves. The yellow, purple spotted and striped, orchid-like flowers are borne in 250mm spikes from ground level.

GINSENG The tuberous roots of *Panax quinquefolius*, a perennial plant in the *Araliaceae* from Asia and N. America. It has 300mm stems and DIGITATE leaves with 5, OBOVATE, toothed leaflets. The small yellow-green flowers are borne in UMBELS. The root is said to have medicinal properties and is popular as an aphrodisiac.

GIPSYWORT *(Lycopus europaeus)*. An ascending perennial in the *Labiatae* from Europe and Asia, naturalised in N. America. It has hairy, OVATE-LANCEOLATE, PINNATELY lobed leaves to 100mm long. The tiny, white, 5-lobed, tubular flowers are carried in dense WHORLS in the upper leaf AXILS.

GLABROUS [GLAB|ER, -RA, -RUM; GLABRAT|US, -A, UM] Smooth; without hairs.

GLACIAL|IS, -E From frozen, icy places.

GLADDON
see IRIS

GLADIAT|US, -A, -UM Sword-shaped, usually of leaves.

GLADIOLUS A genus of 300 perennials in the *Iridaceae* from the Canaries eastwards to Asia and Africa. They grow from corms, often stoloniferous, producing fans of sword-shaped leaves and spikes of ZYGOMORPHIC, tubular flowers with 6, spreading, or sometimes hooded TEPAL lobes. *G. primulinus* grows to 450mm or more, with 50–75mm long yellow and red flowers, the upper central tepal hooded. The many hundreds of cultivars are derived from this and several other African species.

GLAND A tiny secretory organ, often simply a pore which secretes a volatile oil, water or sugar. *See* CHALK GLANDS; NECTARY.

GLANDULAR HAIRS Short or long hairs that have glands (usually oil) at their tips.

▲ 

■ 

GLANDULOS|US, -A, -UM Bearing glands.

GLAREAL [GLAREOS|US, -A, -UM] Of dry open, gravelly places.

GLASSWORT Sometimes called marsh samphire, these are several annual and perennial species of *Salicornia* from the salt marshes of Europe. They have small, fused pairs of fleshy leaves which entirely envelope the stems and tiny, barely visible, petal-less flowers. Common glasswort *(S. europaea)* is a branched, erect annual, 150–300mm tall, dark to yellow-green sometimes flushed red. It is eaten as a vegetable or pickled.

GLASTONBURY THORN A form of common hawthorn (*Crataegus monogyna* 'Biflora') which produces some of its flowers in winter.

• **GLAUCESCENT** Somewhat blue or grey-green.

GLAUCOUS [GLAUC|US, -A, -UM] Blue or grey-green.

GLOBE ARTICHOKE
see ARTICHOKE

▲ **GLOBEFLOWER** Several species of *Trollius*, clump-forming perennials in the *Ranunculaceae* from the north temperate and arctic zones. They have long-stalked, PALMATE lobed leaves and erect stems bearing large globular to cup-shaped yellow to orange-yellow flowers of 5 or more petaloid sepals. Common globeflower *(T. europaeus)* has stems to 600mm.

GLOBE THISTLE Several species of *Echinops*, clump-forming perennials in the *Compositae* from Europe, Asia and Africa. They are distinguished by having flowering INVOLUCRES with one floret only, but many of these in dense, globose heads. The leaves are PINNATISECT and sometimes spiny. Several species are grown in gardens. *E. sphaerocephalus* has leaves white-tomentose beneath and 0.6–2.1m stems bearing 38–63mm wide heads of pale bluish tubular florets.

GLOBOS|US, -A, -UM Rounded or spherical.

GLOCHID A barbed bristle found on the areoles of some cacti.

GLOMERAT|US, -A, -UM Gathered into dense clusters.

GLOMERULE A head-like condensed CYME.

GLORY OF THE SNOW Several species of *Chionodoxa*, small bulbous-rooted plants in the *Liliaceae* from E. Mediterranean and Turkey. They have LINEAR to chanelled leaves and racemes of 6-petalled starry, blue flowers: *C. sardensis* has 13–19mm wide, rich blue flowers on 75–125mm stems: *C. luciliae* grows to 150mm with starry, 25mm wide, white-eyed, blue flowers.

■ **GLORY PEA** *(Clianthus formosus)*. An evergreen trailing plant in the *Leguminosae* from Australia. It has white hairy PINNATE leaves of 11–21 OVATE-OBOVATE leaflets. The brilliant red, beak-like flowers are 100mm long with an erect, standard petal and a long slender, pointed KEEL.

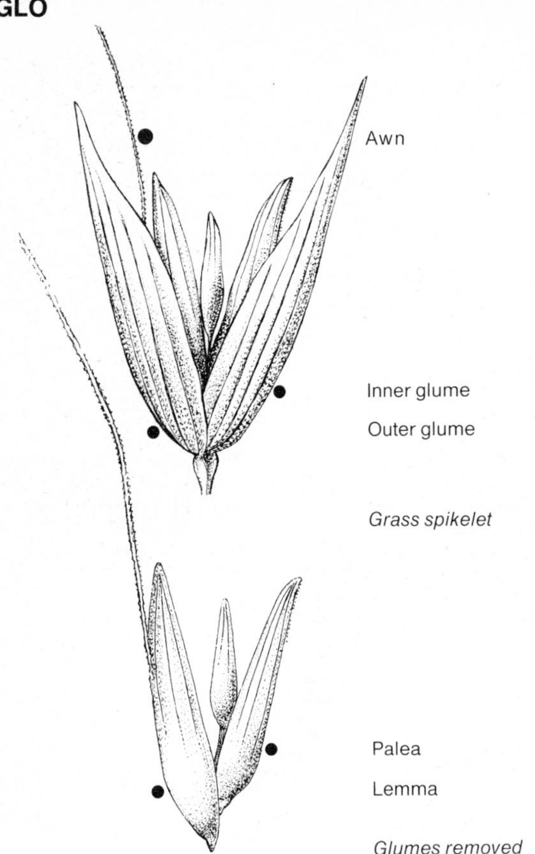

● *Grass spikelet*

Glumes removed

▲

GLOXINIA Several species now classified as *Sinningia*, tuberous-rooted plants in the *Gesneriaceae* from Brazil. *S. speciosa* and its hybrids are popular pot plants. They have oblong, velvety leaves to 200mm in length, borne in a stalked rosette and large waxy, velvety bell flowers to 75mm long, in shades of purple, red and white.

GLUMACE|US, -A, -UM Bearing glumes.

● **GLUMES** The two chaffy bracts at the base of a grass spikelet.

GLUTINOS|US, -A, -UM Gluey or sticky.

GOAT'S BEARD Two different plants go by this name. *Aruncus dioicus*, syn. *A. sylvestre*, a clump-forming 1.2m perennial in the *Rosaceae* from the northern hemisphere. It has tri-pinnate leaves with SERRATE oblong leaflets and large terminal PANICLES of tiny, white flowers. *Tragopogon pratensis* is an annual or perennial in the *Compositae* from Europe and Asia. It is 300–600mm tall with slender pointed leaves and terminal heads of long, involucral bracts and yellow LIGULATE florets.

GOAT'S RUE *(Galega officinalis)*. A 0.6–1.5m, clump-forming perennial in the *Leguminosae* from Europe and W. Asia. It has erect stems with PINNATE leaves of 11–17 OVATE, entire, leaflets. The 13mm long blue-purple, lilac or white pea-flowers are carried in stalked RACEMES.

GOATWEED
see GROUND ELDER

GOLDEN CHAIN [GOLDEN RAIN]
see LABURNUM

▲ **GOLDEN CLUB** *(Orontium aquaticum)*. An aquatic perennial in the *Araceae* from N. America. It has rosettes of oblong, elliptic, glaucous leaves to 300mm long, most of which are above water. The tiny, yellow flowers are borne in cylindrical club-like spikes. The seeds were formerly eaten by Amer-Indians.

GOLDEN ROD Most of the 100 species of *Solidago*, herbaceous perennials from N. America (one in Europe and Asia). They have erect, unbranched stems, lanceolate to obovate leaves and tiny flower heads in larger terminal panicles: common golden rod *(S. virgaurea)* reaches 300–750mm (sometimes much less on mountains) with OBLANCEOLATE leaves and 6mm flower heads; Canadian golden rod *(S. canadensis)* grows 0.6–1.5m with lanceolate leaves and 5mm flower heads. Hybrid cultivars of it are commonly grown.

■ **GOLDEN SAXIFRAGE** Several species of *Chrysosplenium*, small perennials in the *Saxifragaceae* from north temperate regions. They are mainly decumbent, with rooting stems or stolons and rounded, stalked leaves. The small petal-less flowers are borne in flattened CYMES surrounded by coloured bracts. Opposite-leaved golden saxifrage *(C. oppositifolium)* has opposite pairs of orbicular leaves and 50–150mm tall flowering stems with bright, green-yellow bracts.

GOOD KING HENRY
see GOOSEFOOT

GOOSEBERRY *(Ribes uva-crispa)*. A deciduous, bushy, spiny shrub to 900mm or so in the *Grossulariaceae* from Europe and N. Africa. The leaves are OVATE to OBOVATE with 3–5 broad lobes and the small, greenish, purple-tinged flowers are pendulous. The familiar ovoid gooseberry fruits are 13–38mm long, the many CULTIVARS ranging from smooth or hairy to yellow or reddish.

GOOSEFOOT Covers many of the 110 species of *Chenopodium*, annuals, perennials and shrubs in the *Chenopodiaceae* from temperate regions. The leaves are linear to ovate often lobed and toothed, and the tiny flowers are gathered in small clusters which in turn form PANICLES: many-seeded goosefoot *(C. polyspermum)* is an annual with angled, erect or decumbent stems to 900mm, having ovate-elliptic, 50mm leaves and lax inflorescences in separated small clusters; fat hen *(C. album)* is an erect, often red-stemmed annual plant with RHOMBOID to LANCEOLATE toothed leaves and crowded flower clusters; good King Henry *(C. bonus-henricus)* is a clump-forming perennial to 450mm with broadly HASTATE, 50–100mm leaves and the clustered flowers in a pyramidal inflorescence. It was formerly much grown as a vegetable.

GOOSEGRASS
see CLEAVERS

○ **GORSE** Several species of *Ulex*, green-stemmed, spiny shrubs in the *Leguminosae* from W. Europe and N.W. Africa, but introduced to many parts of the world, originally as stock feed. Seedling plants have TRIFOLIATE leaves which are replaced by small scales in the adult stage. The golden flowers are pea-shaped and the seed pods explode when ripe: common gorse *(U. europaeus)* is 0.6–1.8m tall with deeply furrowed spines to 25mm long, and flowers 13–19mm long; western gorse or dwarf furze *(U. gallii)* is usually shorter with faintly furrowed spines and 13mm flowers; dwarf gorse *(U. minor)* is smaller, often less than 600mm with flowers to 10mm.

GOURD
see CUCURBITS; CALABASH TREE

GOUTWEED
see GROUND ELDER

GRACIL|IS, -E Thin, slender, graceful.

GRAM
see CHICK PEA

GRAMA [GRAMMA] GRASS
see BOUTELOUA

GRAMINE|US, -A, -UM Grass-like.

GRAMINIFOLI|US, -A, -UM With grass-like leaves.

GRANADILLA
see PASSION FLOWER

GRANNY'S BONNET
see COLUMBINE

GRANULAT|US, -A, -UM Bearing small knobs or tubercles, e.g. bulbils or buds.

GRAPEFRUIT
see CITRUS

GRAPE HYACINTH Most of the 60 species of *Muscari*, bulbous-rooted perennials from Europe, Mediterranean and W. Asia. They have linear, channelled leaves and mainly dense RACEMES of bell or urn-shaped flowers. In some species the upper flowers are sterile but brightly coloured. *Muscari armeniacum* is commonly seen in gardens, with 150–300mm stems and purple-blue flowers; feather or tassel hyacinth *(M. comosum)* has 200–450mm stems and loose racemes of olive-purple, fertile flowers and blue sterile ones; *M.c.* 'Monstrosum' has all the flowers transformed into purple-blue FILAMENTS: *M. latifolium* is similar to *M. armeniacum* but has only one, broad leaf per bulb.

● **GRAPE VINE** *(Vitis vinifera)*. A deciduous, woody climber in the *Vitidaceae* probably from the Caucasus area but cultivated in most temperate countries. It grows to 18.2m with PALMATE, deeply 3–5 lobed leaves and pendent PANICLES of tiny green flowers, the petals falling as the anthers open. The familiar fruits may be black-purple or amber yellow.

GRAPPLE PLANT *(Harpagophytum procumbens)*. A PROCUMBENT perennial in the *Pedaliaceae* from S. Africa. It has PALMATE, lobed leaves, the lobes deeply cut, and blue and purple tubular flowers in the leaf AXILS. The unique capsules have several woody, grapple-like 25mm long hooks which catch in the feet or hair of animals and are thus distributed.

GRASS All members of the *Gramineae*, the largest family of MONOCOTYLEDON plants with about 600 genera and ten thousand species dispersed throughout the world. They may be tufted, stoloniferous or RHIZOMATOUS, prostrate or erect, and range in height from two centimetres or so to the giant bamboos of 30m. The narrow, parallel-sided leaf-blades are attached to a semi-cylindrical stalk known as a sheath and which usually fits closely to the stem. The flowers lack the usual sepals and petals of flowering plants, each consisting of an ovary with two plumy STIGMAS, three STAMENS and two delicate scale-like organs analogous to petals celled lodicules. These are enclosed within two bracts, the narrow inner one known as a palea, and a broader, larger outer one called the lemma. This forms a floret, one or more of which are enclosed in lemma-like bracts (GLUMES). The whole forms the spikelet. The seed-like fruit is a CARYOPSIS. *See* individual grasses under separate headings.

GRASS NUT [TIGER NUT, CHUFA] *(Cyperus esculentus)*. A grassy perennial in the *Cyperaceae* from S. Europe and Asia. It has LINEAR, arching leaves to 300mm or so, in tufts, and tiny greenish flowers in spikes. The underground stolons bear 13–25mm ovoid, nut-like tubers with crisp, edible flesh.

GRASS OF PARNASSUS *(Parnassia palustris)*. A tufted, herbaceous perennial in the *Parnassiaceae* from wet ground in Europe, Asia and N. Africa. It has OVATE-CORDATE, stalked basal leaves and erect stems, bearing solitary, 5-petalled, white flowers to 25mm across.

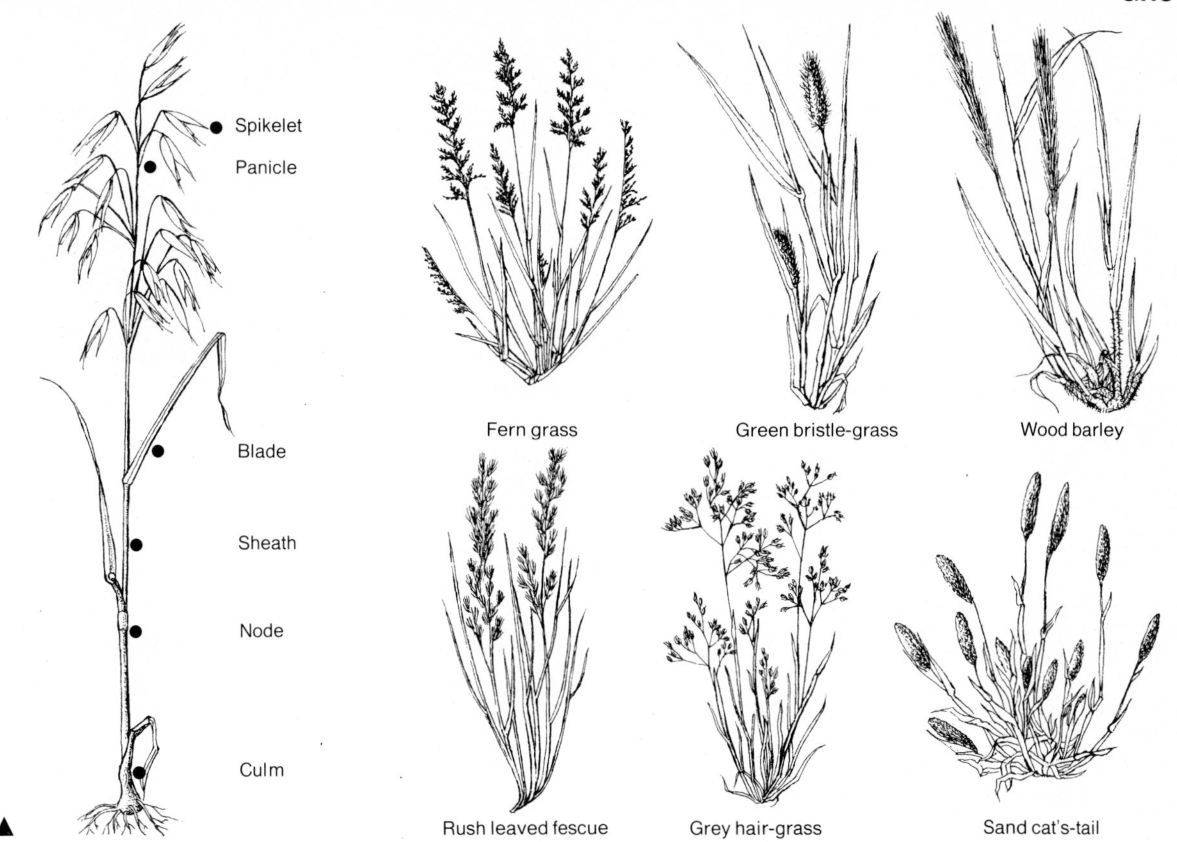

GRASS TREE [BLACKBOY] *(Xanthorrhoea preissii)*. A curious woody-stemmed perennial in the *Xanthorrhaceae* (syn. *Liliaceae*) from W. Australia. It forms a trunk-like stem to 4.5m topped by a dense tuft of arching, linear 0.6–1.2m leaves. Small white flowers are carried in 300–900mm long dense, cylindrical spikes.

GRAVEOLENS Strong or pungently smelling.

GREAT BURNET
see SALAD BURNET

GREATER CELANDINE *(Chelidonium majus)*. An erect perennial in the *Papaveraceae* from Europe and Asia. The leaves are almost PINNATE with 5–7 OVATE-OBLONG leaflets, somewhat CRENATE toothed. The 4-petalled, bright yellow flowers are 19–25mm wide on branched stems. The orange, latex-like sap was formerly used for warts and eye ailments.

GREAT VALERIAN
see JACOB'S LADDER

GREENGAGE
see PLUM

GREENHEART The timber of *Nectandra rodiaei*, an evergreen tree to 18m in the *Lauraceae* from Guyana. It has oval, prominently veined leaves and AXILLARY PANICLES of bell-shaped flowers. The smooth grey bark (bibisi bark) has medicinal properties, being used for fevers.

GREENWEED
see DYERS' GREENWEED

GROUND ELDER *(Aegopodium podagraria)*. Also called goutweed and bishops weed, this is a RHIZOMATOUS perennial in the *Umbelliferae* from Europe and Asia. Said to have been introduced into Britain as a vegetable in Roman times. It has BITERNATE leaves with ovate, irregularly SERRATE leaflets and erect, branched stems to 900mm with UMBELS of tiny white flowers. A serious weed in some gardens owing to its fast-growing rhizomes.

GROUND IVY *(Glechoma hederacea* syn. *Nepeta hederacea)*. A prostrate perennial in the *Labiatae* from Europe and Asia. It has opposite pairs of stalked, RENIFORM to OVATE-CORDATE, 13–25mm wide leaves and ascending flower stems to 150mm or more. The tubular, 2-lipped 13mm long, lavender-purple flowers are borne in AXILLARY WHORLS.

GROUND NUT [PEANUT, MONKEY NUT] *(Arachis hypogaea)*. A spreading annual in the *Leguminosae* from Brazil, but much cultivated in the tropics. It has PINNATE leaves of 4, ELLIPTIC to OBOVATE leaflets and small, yellow pea-flowers close to the soil. After fertilisation, the young pods are forced into the soil where they mature and ripen to the familiar ribbed, woody peanut pod, containing one to three ovoid seeds.

GROUNDSEL *(Senecio vulgaris)*. An erect, overwintering annual in the *Compositae* from Europe, Asia and N. Africa and naturalised elsewhere. The obovate, pinnatifid leaves are borne on 75–450mm stems which are topped by a

▲

CORRYMBOSE cluster of narrow flower heads formed of yellow, tubular florets. The less common *S. v. radiatus* has flower heads with short, yellow ray florets.

GROWING POINT
see MERISTEM

GROWTH Increase in size by cell division and expansion rendered permanent by the thickening of cell walls and the formation of supporting tissues, such as fibre and wood. *See also* CELL DIVISION, GERMINATION, PHOTOSYNTHESIS, VASCULAR BUNDLE.

GUARD CELLS
see STOMA

● **GUAVA** Two species of *Psidium*, evergreen shrubs to small trees in the *Myrtaceae* from S. America, but grown elsewhere in the tropics: purple or strawberry guava *(P. cattleianum)* grows to 6.0m with 50–75mm obovate, leather leaves and white flowers with a mass of small stamens. The deep red-purple, 38mm fruits are nearly globular; common guava *(P. guajava)* is a suckering shrub or tree to 4.5m with angular stems and oblong, elliptic leaves. White flowers are followed by 50–100mm long, globose to pear-shaped fruits.

▲ **GUELDER ROSE** *(Viburnum opulus)*. A deciduous shrub to 3.0m or more in the *Caprifoliaceae* from Europe and W. Asia. It has PALMATE, 3–5 lobed, toothed leaves and flattened UMBELS of florets, the inner small and fertile, the outer sterile with broad, white petals. The sub-globose juicy fruits are translucent red.

GUERNSEY LILY *(Nerine sarniensis)*. A bulbous-rooted perennial in the *Amaryllidaceae* from S. Africa with strap-shaped leaves and erect stems 450–750mm tall, bearing umbels of 6-petalled, red or salmon lily-like flowers after the leaves have died down.

GULF WEED
see OAR WEED

GUM ARABIC The gum resins from several species of *Acacia*, notably *A. arabica*, a tree in the *Leguminosae* from N. Africa. It has hard, sharp 50–75mm long spines and bipinnate leaves of tiny, OBLONG-LINEAR leaflets. Clusters of 13mm wide, globular heads bear many small, white florets.

GUM TRAGACANTH The gum resin from several species of *Astragalus*, mainly *A. gummifer*, a 300mm, thorny shrub from S.E. Europe, Turkey and Iran. It has PINNATE leaves of 8–14 elliptic leaflets and clusters of elongated, yellow pea-flowers. The gum is used in medicines, pills and ointments.

■ **GUM TREE** Many of the 500 species of *Eucalyptus*, evergreen trees in the *Myrtaceae* mainly from Australia. The juvenile leaves of some species are very different from the adult, being in opposite pairs and AMPLEXICAUL or PERFOLIATE. Later they become alternate and stalked, usually narrowly LANCEOLATE to OVATE, sometimes FALCATE. The petals fall as a cap when the flowers open, the showy part being the dense mass of stamen filaments usually white, but sometimes red. Many species are used as timber and for the extraction of eucalyptus oil. Several are grown

as ornamentals: blue gum *(E. globulus)* can reach 60m with blue-white, ovate-cordate juvenile foliage and dark green, lanceolate falcate leaves 150–300mm long; cider gum *(E. gunnii)* grows to 30m with ORBICULAR, GLAUCOUS juvenile leaves and 100mm lanceolate adult ones; urn-fruited gum *(E. urnigera)* is similar, to 15.2m with very glaucous, orbicular to ovate juvenile leaves.

GUNPOWDER PLANT
see ARTILLERY PLANT

GUTTATION The exudation of droplets of water from certain STOMATA which never close, usually situated on leaf margins at the ends of the fine veins. The droplets of water hanging from the tips of grass leaves (often confused with dew) are exuded from small gaps in the epidermis. *See also* CHALK GLANDS.

GYMNOSPERMAE [GYMNOSPERMS] One of the two divisions of the seed-bearing plants, characterised by having the OVULES sitting naked on flattened, scale-like carpels which are arranged in spikes called strobili in conifers and cycads.

GYNODIOECIOUS Having separate female and hermaphrodite flowers on different individual plants, e.g. members of the dead nettle family.

GYNOECIUM The female part of a flower, primarily the CARPELS but covering the whole PISTIL.

GYNOMONOECIOUS Having separate female and hermaphrodite flowers on the same plant, e.g. members of the daisy family *(Compositae)*.

GYNOPHORE An elongation of the RECEPTACLE of a flower bearing the PISTIL only, e.g. spider flower *(Cleome)*.

GYRANS Circling or revolving
see TELEGRAPH PLANT

HABIT The general or overall appearance of a plant relating to its manner of growth, e.g. erect, straggling, mat-forming, bushy, compact etc.

HAEMATODES Blood-red.

HAIR GRASS Several species of annual and perennial grasses from four genera in the *Gramineae* from temperate regions in both hemispheres: (grey hair grass, *Corynephorus canescens*, from Europe only); crested hair grass *(Koeleria cristata)* is a compact, tufted perennial species with rolled, bristle-like or narrowly LINEAR leaves and 150–600mm talls stems bearing spike-like PANICLES TO 100mm long, usually silvery-green or purplish; early hair grass *(Aira praecox)* is a slender annual, 25–150mm tall from acid, sandy soils having spike-like panicles of tiny silvery or purplish spikelets; tufted hair grass *(Deschampsia caespitosa)* is a robust perennial forming large tussocks of broad, arching leaves and stems to 1.8m.

HAIR MOSS Several species of *Polytrichum*, particularly *P. commune*, an erect ericoid species in the *Bryophyta*. It is the tallest moss in Britain, growing to 300mm in bogs, with

long-stalked spore capsules covered by pointed, yellow, fringed caps until ripe; juniper hair-moss *(P. juniperinum)* is similar but much smaller, forming large patches on the drier parts of moors and heaths.

HALOPHIL|US, -A, -UM Salt-loving.

HALOPHYTE [HALOPHYTIC] Plants which normally grow in salty soils, e.g. dunes and marshes by the sea.

HAPLOID A single set of CHROMOSOMES, as in the sex cell nuclei just before fertilisation, i.e. half the normal diploid number. *See* CELL DIVISION.

HAPTOTROPISM A sensitivity response to contact, e.g. when a twining stem or tendril comes into contact with a support, the side farthest from the support grows faster causing a curvature round it.

HARD FERN *(Blechnum spicant)*. An evergreen, RHIZOMATOUS fern in the *Blechnaceae* from the northern hemisphere with tufts of LANCEOLATE, PINNATE 100–450mm long fronds, the numerous PINNAE are SESSILE, LINEAR-OBLONG. The SPORANGIA are carried on special, taller erect fronds with narrower pinnae.

HAREBELL *(Campanula rotundifolia)*. (Bluebell in Scotland). A slender perennial with underground stolons in the *Campanulaceae* from north temperate regions. It has long-stalked rounded-crenate 13mm wide basal leaves and linear ones on the wiry, 150–300mm stems. The 13–19mm long purple-blue bell-flowers are pendent.

HARE'S TAIL GRASS *(Lagurus ovatus)*. An annual plant in the *Gramineae* from the Mediterranean region and naturalised elsewhere. It has a tuft of erect stems to 450mm or so with shortish, flat, softly hairy leaves and 25–50mm dense, ovoid, white-hairy PANICLES of spikelets.

HARICOT BEAN
see RUNNER BEAN

● **HART'S TONGUE FERN** *(Asplenium scolopendrium,* syn. *Phyllitis scolopendrium)*. An evergreen species in the *Aspleniaceae* from Europe, Asia and N. Africa. It forms tufts of strap-shaped CORDATE-based fronds 300–600mm long, the backs of the fertile ones bearing short bands of SPORANGIA at right angles to the mid-rib.

HASHISH
see HEMP

HASTATE [HASTAT|US, -A, UM] Shaped like a spear head: i.e. triangular-OVATE with two basal, barb-like lobes.

HAULM The stems of beans, peas and potatoes.

HAUSTORI|UM, -A Food-absorbing, sucker-like structures produced by parasitic fungi and flowering plants. Once in position, the haustorium sends out piercing branches which link up with the vascular system of the host.

HAWKBIT Three species of *Leontodon*, rosette-forming perennials in the *Compositae* from Europe and Asia. The

▲

commonest and most widespread is autumnal hawkbit *(L. autumnale)* with smooth OBLANCEOLATE leaves varying from sinuate-toothed to deeply PINNATIFID. The 13–25mm wide, yellow flower heads are composed of LIGULATE florets the outer ones streaked red on the backs.

HAWKSBEARD Several sorts of *Crepis*, annual, biennial and perennial plants in the *Compositae*, mainly from Europe. Beaked hawksbeard *(C. versicaria* ssp. *taraxacifolia)* is a biennial with stalked, oblanceolate, LYRATE or RUNCINATE-PINNATIFID basal leaves. The erect, 300–600mm stems, often purplish at the bases, branch above and bear many 19–25mm wide yellow flower heads of LIGULATE florets, the outer streaked brown below.

HAWKWEED Most of the 1000 species of *Hieracium*, a very variable genus of perennials in the *Compositae*, by some authorities divided into 5–20,000 MICROSPECIES most of them from the northern hemisphere but a few from the south. They have stalked or SESSILE, LINEAR-LANCEOLATE to broadly OVATE usually distantly toothed leaves, and terminal PANICLES of yellow flower heads of LIGULATE florets.

HAWTHORN [MAY] Two species of *Crataegus*, large deciduous shrubs or small trees in the *Rosaceae* from Europe. They have thorny stems and obovate, 25–50mm long, 3–7 lobed leaves. The white or pink, 5-petalled flowers have pink or purple anthers and are arranged in flattened CORYMBS: common hawthorn *(C. monogyna)* has deeply lobed leaves and flowers with one STYLE; midland hawthorn *(C. laevigata* syn. *C. oxyacanthoides)* has broad, shallowly lobed leaves and two stigmas to each flower (sometimes one or three). Rarer in Great Britain than *C. monogyna* and mainly in wooded areas. Pink and red and double-flowered forms are cultivated.

▲ **HAZEL** *(Corylus avellana)*. A large deciduous shrub or small tree to 6.0m in the *Corylaceae* from Europe to Turkey. It has 50–100mm long, sub-orbicular leaves, often shallowly lobed and doubly SERRATE. The male flowers are borne in the familiar pendulous, yellow catkins. The female flowers remain in the bud, only a few red STYLES protruding. The globose or ovoid cob or hazel nuts are borne in small clusters, each surrounded by a lobed, leafy INVOLUCRE. Filbert *(C. maxima)* is similar, but larger in every way with the nuts almost enclosed by a long tubular involucre. Its purple-leaved form is often grown as an ornamental.

HEAD A group of small flowers forming a larger one, e.g. daisy, globe thistle, teasel. Also used of larger, dense clusters of bloom, e.g. hydrangea.

HEART'S EASE
see PANSY

HEART WOOD
see DURAMEN

HEATH Many of the 500 species of *Erica*, also known as bell heather, in the *Ericaceae* from Europe, the Mediterranean region and S. Africa. They are bushy, wiry shrubs or small trees with linear or needle-like leaves and small,

●

bell or urn-shaped flowers often profusely borne: European cross-leaved or bog heath *(E. tetralix)* has erect stems to 450mm, hairy leaves in cross-shaped WHORLS and UMBELS of pink, urn-shaped flowers; bell heather *(E. cinerea)* also European, is a spreading shrub to 300mm or so in height with 3 leaves to each whorl and dense RACEMES of red-purple bells.

HEATHER [LING] *(Calluna vulgaris)*. An evergreen, very variable bushy shrub in the *Ericaceae* from Europe and Morocco. It is similar to the heath, but the leaves are scale-like and the showy part of each flower is the 4-lobed, purple CALYX which covers the tiny bell-like COROLLA.

● **HEATHLAND [HEATHS]** Areas of sharply-drained sandy soils, overlain with a thin covering of peat and largely dominated by heaths and heathers, bilberry, mat grass etc. Moorland is very similar but is usually moister and more peaty, particularly that at higher elevations. Associated plants are cotton grasses and purple moor grass.

HEDERACE|US, -A, -UM Ivy-like, usually of leaves.

HEDGE MUSTARD *(Sisymbrium officinale)*. An annual or biennial plant in the *Cruciferae* from Europe, W. Asia and N. Africa, and naturalised in many temperate countries. It has deeply PINNATIFID basal leaves with large, terminal lobes and erect, wiry stems bearing short, lateral branches. The tiny 4-petalled flowers are carried in RACEMES.

HEDGE PARSLEY Three species of *Torilis*, annuals in the *Umbelliferae* from Europe, Asia and N. Africa. They have bipinnate leaves and branching stems bearing small UMBELS of tiny white or pinkish flowers. *See also* COW PARSLEY.

▲ **HEDGEROW** A continuous line of shrubs and trees, usually originally planted from native species and cut back at regular intervals. Sometimes they originate as very thin strips of woodland left at field edges and by roads. Spiny shrubs such as HAWTHORN and BLACKTHORN were commonly planted; but many other species come in as bird-voided or wind-blown seeds, e.g. wild roses, field maple, BUCKTHORN, BLACKBERRY, WILLOW, HAZEL, DOGWOOD, GUELDER ROSE. At the base of the hedge, mainly perennial plants become established, e.g. climbers such as black and white bryony, bindweed and vetches. Free standing plants such as FOXGLOVE, NETTLES, GARLIC MUSTARD, COW PARSLEY and PRIMROSE also occur. Hedges form a valuable refuge for all sorts of plant and animal life, particularly those normally found in woodland.

HELIOTROPE *(Heliotropium peruvianum)*. An evergreen shrub in the *Boraginaceae* from Peru. It grows to 1.8m in the wild (600–900mm in gardens) with slightly BULLATE, oblong-lanceolate leaves and terminal clusters of small, very fragrant salver-shaped violet or lilac flowers. Most of the garden plants are of hybrid origin.

HELLEBORE *(Helleborus)* A genus of 20 perennials from Europe to the Caucasus with one species in China. They have long-stalked, PALMATE, PEDATE or TRIFOLIATE leaves and cup-shaped green, white or red-purple flowers of 5 or 6 PETALOID SEPALS, the petals being transformed

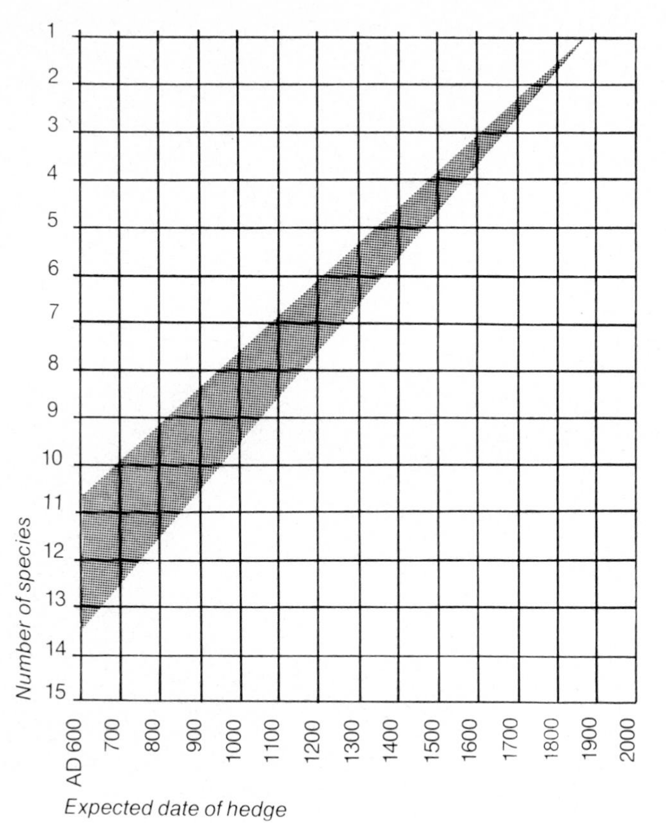

The graph shows the expected date of hedges in S.E. England, based on the average number of species of woody plants in a 30 metre stretch:

For example:

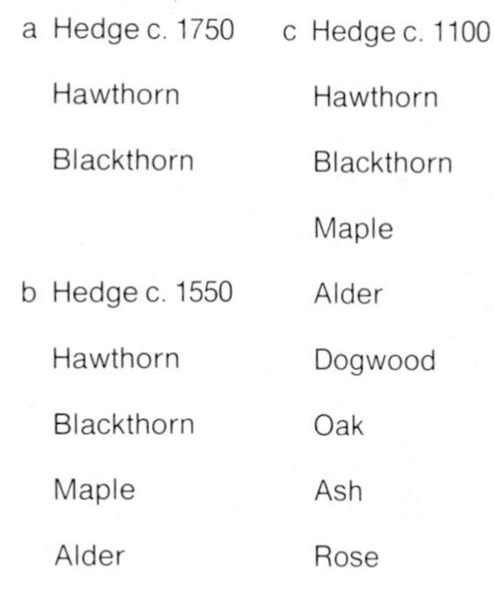

into NECTARIES. The fruits are FOLLICLES: Christmas rose *(H. niger)* has 7–9-lobed leaves and large white flowers in winter; stinking hellebore, bear's foot or setterwort *(H. foetidus)* has overwintering leafy stems and pedate leaves with up to 9 narrow lobes. The green 13–25mm flowers are carried in terminal clusters.

HELLEBORINE Several different kinds of orchids in the genera *Epipactis* and *Cephalanthera*, ground dwelling members of the *Orchidaceae*. One of the commonest and most widespread is broad-leaved helleborine *(E. helleborine)* from the woods of Europe, Asia and N. Africa and naturalised in N. America. It has 300–750mm stems with spirally arranged leaves, the lowest OVATE-ELLIPTIC, the highest LANCEOLATE. The nodding dull purple to greenish flowers are borne in RACEMES: marsh helleborine *(E. palustris)* is common in fens and dune slacks with 150–450mm stems and oblong-ovate to lanceolate leaves. The drooping to horizontal flowers have purple-green sepals, and purple-veined, whitish petals and labellum (lip).

HELODOXA Glory of the marsh.

HELOPHYTES Marsh plants; the fifth of a 7-category system of classifying life forms.

HEMICRYPTOPHYTES Herbaceous plants with winter buds at soil level, e.g. GOLDEN ROD, DOCK, NETTLE (common). It is the third of a 7-category system of classifying life forms.

HEMIPARASITE
see PARASITE

HEMLOCK *(Conium maculatum)*. An erect biennial to 1.8m in the *Umbelliferae* from Europe and Asia, naturalised elsewhere. It has furrowed, GLAUCOUS, purple-spotted stems and OVOID to DELTOID bi- or tripinnate leaves. The tiny white flowers are borne in numerous small UMBELS. A very poisonous plant.

HEMLOCK SPRUCE *(Tsuga)*. A genus of 15 species of evergreen conifers in the *Pinaceae* from the Himalaya, E. Asia, and N. America. They have flattened sprays of small LINEAR leaves and small, ovoid, pendulous cones: eastern hemlock *(T. canadensis)* grows to 30m with an irregularly pyramidal outline. The dark green leaves have two silvery bands beneath and the cones are 13-25mm long; western hemlock *(T. heterophylla)* grows to 60m or more, with a shapely pyramidal outline, but is otherwise similar.

HEMP *(Cannabis sativa)*. A DIOECIOUS, annual plant to 2.4m in the *Cannabidaceae* from Asia, but much grown elsewhere. It has DIGITATE leaves with 5–7 slender, toothed, tapered leaflets and terminal PANICLES of tiny, greenish petal-less flowers. The fibres it produces are used for rope-making and dried leaves form hashish or bhang, a reputedly poisonous, intoxicating drug. The seeds were once much used in birdseed mixtures: manilla hemp is the fibre from *Musa textilis*, a kind of banana; sunn or Madras hemp comes from *Crotalaria juncea*, a 1.2m annual in the *Leguminosae* from tropical Asia, with elliptic leaves and showy golden-pea flowers.

HEMP AGRIMONY *(Eupatorium cannabinum)*. An erect, woody-based perennial in the *Compositae* from Europe to Asia and N. Africa. It has downy stems to 1.2m

●

▲

with OVATE to LANCEOLATE, toothed leaves in opposite pairs. The tiny red-mauve or paler flower heads are gathered into dense terminal CORYMBS. Joe-Pye weed *(E. purpureum)* is similar but taller with magenta flowers.

HEMP NETTLE Several species of *Galeopsis*, bristly, hairy annuals in the *Labiatae* from Europe and Asia. They have nettle-like leaves in opposite pairs and dense WHORLS of tubular, 2-lipped flowers in the upper leaf AXILS: common hemp nettle *(G. tetrahit)* grows from 150–900mm with 13–19mm long pink, purplish or white flowers; large-flowered hemp nettle *(G. speciosa)* is similar, but the flowers are yellow and violet-purple, 22–34mm long.

HEN AND CHICKENS Used for a curious mutation of pot marigold and daisy flowers when small flower heads arise from the top of the normal ones.

● **HENBANE** *(Hyoscyamus niger)*. A sticky-hairy, strong-smelling annual in the *Solanaceae* from Europe, W. Asia and N. Africa. It grows to 750mm, with oblong-ovate leaves, the basal ones to 200mm long, sometimes with a few, large teeth. The tubular flowers expand to 25mm across, the yellow ground colour being netted with purple. The CALYX becomes larger and urn-shaped after flowering with a round capsule at the bottom. A poisonous, narcotic plant.

HENBIT
see DEAD NETTLE

HENNA *(Lawsonia inermis)*. An evergreen, 3.0m shrub in the *Lythraceae* from Africa, India and S.W. Asia. It has opposite pairs of ovate to lanceolate leaves and panicles of small, pink, red or white flowers. The dried, ground leaves produce the red dye, henna.

HEPATICS [HEPATICAE]
see LIVERWORT

HERB A plant with non-woody stems that die back after flowering and seeding. Annual and biennial plants die completely, but perennials (herbaceous perennials) grow again from the root each year.

HERBARIUM A collection of dried, pressed plants, usually arranged systematically and kept for reference purposes.

HERB BENNET
see AVENS

HERB CHRISTOPHER
see BANEBERRY

▲ **HERB PARIS** *(Paris quadrifolia)*. A woodland, herbaceous perennial in the *Trilliaceae* from Europe to Siberia. It has erect, unbranched stems with a single whorl of 4, OBOVATE, 63–113mm long leaves. The solitary green flower arises from the centre of the leaves and has 4, LANCEOLATE, 25mm sepals and 4 thread-like petals. The black, berry-like fruit splits when ripe.

HERB ROBERT
see CRANESBILL

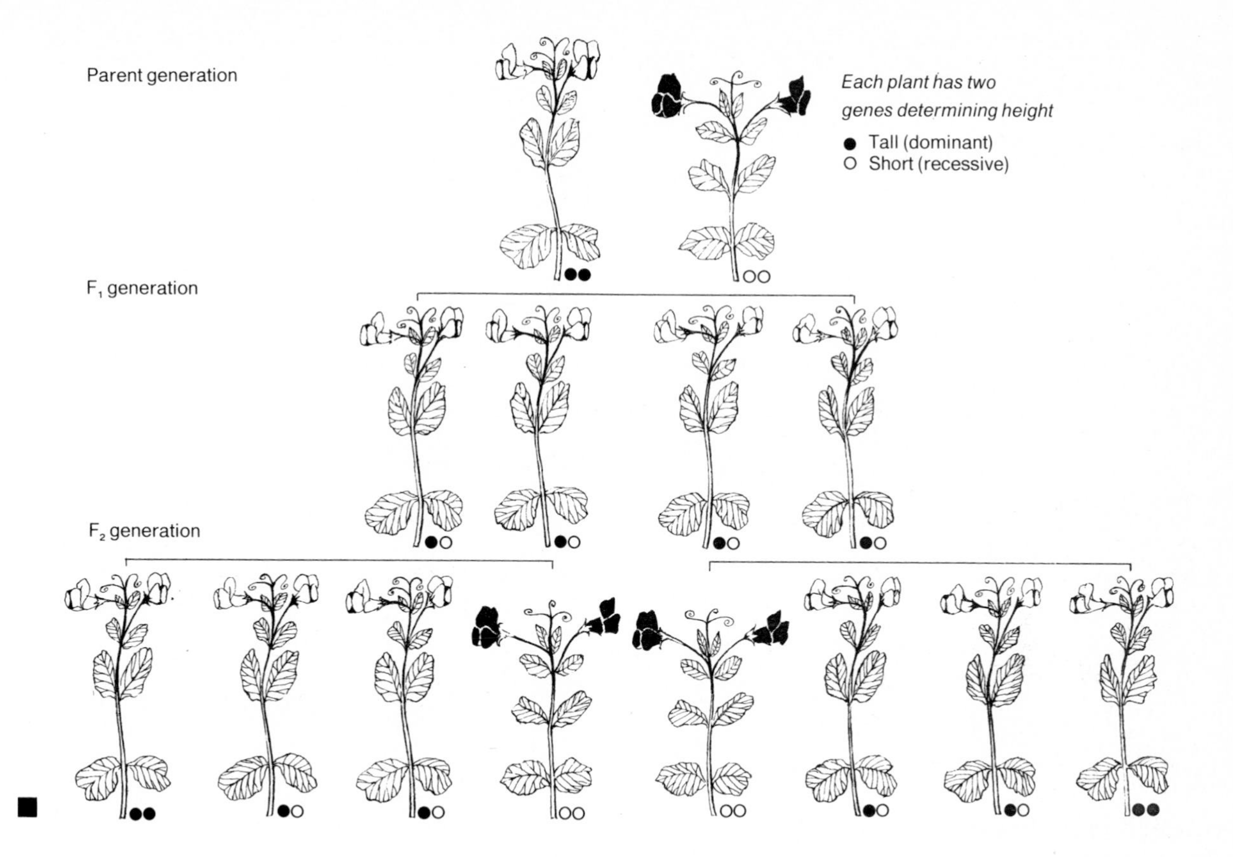

■ **HEREDITY** The passing on of characters from a plant to its progeny, so maintaining the identity of a species. The basic principles of heredity were worked out in the 1850s and 1860s by Gregor Mendel, a monk and science teacher at Brno, Czechoslovakia. CHROMOSOMES and GENES and the way they behave were not known then, but his experiments clearly showed that the characters of any given plant were transmitted by pollen and egg cells from one generation to the next. He showed that though some characters dominate others when combined (dominant and recessive), the recessive ones were not lost and will show up again under the right circumstances. One of Mendel's best known experiments was with tall and short forms of garden peas. These he crossed together and saved all the resulting seeds. The plants produced by these seeds (the First Filial or F_1 generation) were all intermediate between the parent plants, the dominant tall and short factors seeming to have reached a compromise. When these F_1 plants were self-pollinated and a further generation (F_2) grown, the resultant plants were a mixture of tall and short to a ratio of 3 tall to 1 short. The seeds from these plants, again self-pollinated, were then grown to provide an F_3 generation. Seeds from the short plants produced another crop of short plants only, but those from the tall plants gave both tall and short seedlings to the same ratio of 1:3. This experiment proved both that the character controlling units (GENES) can be carried through several generations unchanged, and also that one character can have dominance over another.

HERMAPHRODITE Having functional STAMENS and PISTIL in the same flower.

HETEROCARPOUS Producing fruits of different kinds on the same plant, e.g. the 'seeds' of pot marigold.

HETEROCHLAMYDEOUS Of flowers which have a distinct COROLLA (petals) and CALYX (sepals), e.g. primrose. The opposite is monochlamydeous where the PERIANTH is uniform, e.g. tulip which has 3 coloured sepals almost identical to the petals.

HETEROECIOUS Of rust fungi which have different generations on totally different host plants.

HETEROPHYLL|US, -A, -UM Of plants with differently shaped leaves on the same individual, e.g. common ivy.

HETEROSPOROUS Of plants producing both large and small spores (megaspores and microspores) which give rise to distinct male and female GAMETOPHYTES respectively, e.g. some clubmosses *(Lycopodium)* and the GYMNOSPERMS.

HETEROSTYLY Of flowers with styles of different lengths to ensure cross-pollination, e.g. in primrose, some flowers have a short style and long stamens (thrum-eyed) and others are the reverse (pin-eyed). Only pollen from a thrum-eyed plant will fertilize a pin-eyed one effectively.

HETEROTROPHIC Of plants which depend upon others, or their dead remains, for part or all of their nutrition: e.g. total parasites such as broomrape; partial parasites such as mistletoe; and saprophytes such as bird's nest orchid.

HETEROZYGOUS Of hybrid plants which carry characters they do not show, e.g. the F_1 generation of tall and short pea hybrids described under HEREDITY.

HEXAGON|US, -A, -UM With 6 angles.

HEXANDRIUS, -A, -UM With 6 stamens.

HIANS Gaping, e.g. capsules.

HIBERN|US, -A, -UM Flowering or growing in winter.

HIBISCUS A varied genus of 300 trees, shrubs, perennials and annuals in the *Malvaceae* from the tropics and subtropics. The leaves range from OVATE to PALMATE, often deeply divided, and the showy flowers have 5 petals. Among the annual species are: okra *(H. esculentus)*, a 0.9–1.8m plant with 3–5 lobed leaves and crimson centred yellow flowers followed by cylindrical fruits which are eaten when young; flower of the hour *(H. trionum)* a 300–750mm tall plant having 3-lobed leaves and purple-brown centred yellow flowers 50mm or more wide which may only stay open for an hour or two or up to half a day; *H. rosa-sinensis* is a perennial shrub or small tree commonly grown in the tropics, with ovate leaves and flowers to 125mm across in a wide range of colours; *H. schizopetalus* has drooping branches and pendulous 50–75mm flowers, the orange-red petals deeply cut and REFLEXED in a unique way.

HICKORY Several species of *Carya*, deciduous trees in the *Juglandaceae* from E. Asia and eastern USA. They have PINNATE leaves and petal-less, unisexual flowers, the males in catkins, the females in terminal spikes or clusters: shagbark hickory *(C. ovata* syn. *C. alba)* grows to 36m with fairly small branches and 100–150mm long greenish catkins. The green to brown spherical fruits each contain one subglobose edible nut: pecan *(C. pecan* syn. *C. olivaeformis)* grows to 45m with 38–50mm oblong, red-brown, highly edible nuts.

HIEMAL|IS, -E [HYEMAL|IS, -E] Of winter, e.g. plants which bloom then.

HILUM The scar on the seed where it was attached to the fruit by the FUNICLE. Broad bean seeds show it well.

HIP [HEP] The fruit of the rose.

HIRSUT|US, -A, -UM [HIRT|US, -A, -UM] Hairy.

HIRTELL|US, -A, -UM Somewhat hairy.

HISPID|US, -A, -UM Harshly or bristly hairy.

HOGWEED Several species of *Heracleum*, mainly biennials or perennials in the *Umbelliferae* from north temperate regions and tropical African mountains: common hogweed, cow parsnip or keck *(H. sphondylium)* is a biennial, 75–180cm high with PINNATE, bristly leaves to 600mm and UMBELS of small whitish or pinkish flowers; giant hogweed *(H. mantegazzianum)* can reach 3.0–3.6m with deeply PINNATISECT basal leaves and large wheel-like UMBELS of white flowers.

HOLDFAST The sucker-like root attachment of many seaweeds.

HOLLY *(Ilex aquifolium)*. A spiny evergreen tree in the *Aquifoliaceae* from Europe. It is mainly DIOECIOUS, with AXILLARY clusters of small, white waxy flowers followed by globose red, yellow or orange fruits. There are several variegated leaved CULTIVARS. The similar but duller-leaved American holly is *I. opaca*.

HOLLY FERN *(Polystichum lonchitis)*. An evergreen mountain fern in the *Aspidiaceae* from the north temperate zone. Has PINNATE fronds 150–600mm long, the PINNAE being LINEAR-LANCEOLATE. The SORI are carried along either side of the mid-rib of the pinnae.

HOLLYHOCK *(Althaea rosea)*. A 1.8–3m perennial in the *Malvaceae* with long-stalked rounded, CORDATE leaves, somewhat 5–7 lobed or angled. The 50–75mm widely funnel-shaped flowers are white, yellow and red in terminal RACEMES. Most of the garden cultivars are of hybrid origin with the allied *A. filicifolia*, distinguished by its deeply-lobed leaves.

HOLOPHYTIC Of plants which manufacture their own food by aid of chlorophyll and the sun's energy.

HOMINY
see MAIZE

HOMOSPOROUS Of plants producing spores of one kind which give rise to a uniform GAMETOPHYTE generation producing both male and female cells, e.g. most ferns.

HONESTY *(Lunaria annua* syn. *L. biennis)*. A chiefly biennial plant in the *Cruciferae* from S.E. Europe, but much grown and often escaping. It is an erect plant with stalked OVATE-CORDATE ACUMINATE leaves and RACEMES of 4-petalled, 25mm wide red-purple flowers. After the 32–45mm flattened pods have shed their seeds the silvery septa remain and are used for winter decoration.

HONEY DEW The sticky exudations of greenfly (aphids) particularly noticeable on certain kinds of lime *(Tilia)*. The sweet, sticky layer soon becomes blackish due to the presence of a minute sooty mould fungus.

HONEY [BOOTLACE] FUNGUS *(Armillaria mellea)*. Also known as Honey agaric, it is a *Basidiomycete* fungus from temperate regions. The yellow-brown, 50–100mm wide toadstools are a common feature at the base of dead trees or stumps, often appearing in dense clusters. The fungus spreads through the soil from tree to tree by blackish, bootlace-like structures of HYPHAE called rhizomorphs. It is a serious pest of woodland, orchards and gardens.

HONEY LOCUST *(Gleditsia*, syn. *Gleditschia, triacanthos)*. A thorny tree to 30m or more in the *Leguminosae* from E.N. America. It has PINNATE or partly pinnate and bipinnate leaves to 200mm long and tiny, green 3–5 petalled flowers in 50mm long RACEMES. The dull red 300–450mm long pods are lined with a sweet pulp.

HONEYSUCKLE Several, often fragrant flowered, plants are given this name, but the true honeysuckles are species of *Lonicera*, evergreen and deciduous climbers and shrubs in the *Caprifoliaceae* from Europe. They have opposite pairs of OVATE leaves and tubular, 2-lipped flowers followed by fleshy berries; common honeysuckle *(L. periclymenum)* can climb to 6m or trail through hedges, and has 38mm long, cream flowers which darken after pollination.

● **HOOKER, JOSEPH DALTON** (1817–1911). Son of Sir William Jackson Hooker, he was born in Suffolk and became interested in botany at an early age. In 1839 he took a medical degree at Glasgow and in the same year joined Sir James Ross's expedition in the *Erebus* to the south temperate and sub-Antarctic regions. On his return he wrote the superbly illustrated *Flora Antarctica* and, during the years ahead, separate floras of New Zealand and Tasmania. In 1848 he led an expedition to the Central and E. Himalaya, then un-mapped and untravelled by Europeans (he produced the first accurate map of the region). Enormous numbers of plants were collected, among them many unsuspectedly gorgeous rhododendrons which helped to create a new facet of gardening, built on later by another generation of plant hunters in China. During subsequent years Joseph Hooker travelled in Lebanon, Morocco, Russia and N. America as well as on shorter journeys in Europe.

In 1855 he was appointed Assistant Director to his father at Kew and in 1865 Director. During the next 20 years he made Kew undisputedly the finest botanic garden of all. Between 1862 and 1883, Sir Joseph and George Bentham collaborated on the great task of classifying and describing all the flowering plants then known. Entitled *Genera Plantarum*, it for long remained a standard work of reference and all new plants were classified according to the Bentham and Hooker system. In 1873 he was elected President of the Royal Society, and in 1877 was knighted.

On retiring from Kew in 1885, he devoted the remainder of his long life to writing. During this period he also extensively revised his late friend George Bentham's *Handbook of the British Flora*, which then became the standard flora and for generations of botany students was known simply as 'Bentham and Hooker'.

▲ **HOOKER, WILLIAM JACKSON** (1785–1865). British botanist and naturalist born at Norwich; a talented artist and a diligent and highly competent writer. Once he had decided upon botany as his prior interest, original monographs of many plant groups flowed from his pen, beautifully illustrated by his own line drawings. Most of his books became collectors' items within a very short time, in particular those on ferns and mosses. In 1819, he was elected Professor of Botany at Glasgow, and despite lack of experience in lecturing soon became the most popular professor in the university, inspiring the students with his enthusiasm and dedication. William Hooker was knighted in 1836, and 5 years later after much scheming – in the nicest possible way – became the first Director of the newly created Royal Botanic Garden at Kew. This position brought out his administrative talents to the full, and it is he who created the garden we know today, taking over the old royal gardens of Kew Palace and filling them with unusual plants from all over the world. He also had the first of Kew's vast tropical greenhouses built and founded the great library. Edited the Journal of Botany and the

Botanical Magazine. Main books: *British Jungermanniae* (1812–16); *Flora Londinensis* (1818); *Muscologia Britannica* (1818); *Musci Exotica* (1820); *Flora Exotica* (1822).

HOP *(Humulus lupulus)*. A DIOECIOUS, herbaceous perennial twiner in the *Cannabidaceae* from Europe and W. Asia. It grows 3–6m each season, with opposite pairs of PALMATELY lobed, coarsely toothed leaves. The tiny greenish male flowers are borne in leafy PANICLES, the female in cone-like spikes, the latter becoming the hops of commerce, used to flavour beer.

HOREHOUND Two perennial members of the *Labiatae* from Europe, W. Asia, and N. Africa: black horehound *(Ballota nigra)* is a hairy plant to 900mm with an unpleasant smell and pairs of coarsely CRENATE, OVATE to ORBICULAR, 19–50mm leaves. The purple, tubular 2-lipped 13–19mm flowers are borne in dense AXILLARY clusters; white horehound *(Marrubium vulgare)* is white TOMENTOSE all over, the stems 300–600mm tall, bearing 13–38mm leaves and 13mm white flowers.

HORMONES Organic substances which promote the growth of plant organs, e.g. root, flower and fruit production.

HORNBEAM *(Carpinus betulus)*. A deciduous tree to 27m in the *Corylaceae* from Europe to Turkey. It has a fluted trunk, smooth grey bark and ascending branches. The leaves are ovate, 25–75mm long, prominently veined and doubly toothed. The MONOECIOUS flowers are borne in greenish catkins, the females about 19mm, the males 25–50mm. The fruit is a small ovoid ribbed nutlet, attached to a 3-lobed bract which acts as a wing for dispersal.

HORNED POPPY *(Glaucium flavum)*. A biennial or perennial in the *Papaveraceae* often by the sea in Europe, around the Mediterranean and W. Asia. It has rosettes of GLAUCOUS, LANCEOLATE, PINNATELY lobed or PINNATIFID leaves and 300–600mm branched stems bearing 4-petalled poppy-like yellow flowers. The horn-like capsule is 150–300mm long.

HORN OF PLENTY *(Craterellus cornucopioides)* a curious, funnel-shaped, blackish brown, wavy-margined *Basidiomycete* fungus from temperate woods, particularly beech. It is edible and pleasant when cooked.

HORNWORT *(Ceratophyllum demersum* and *C. submersum)*. Rootless aquatic perennials in the *Ceratophyllaceae* from Europe, Asia and N. Africa. The branched stems can reach 900mm, bearing regular WHORLS of firm-textured leaves divided into forked, LINEAR segments. The tiny, petal-less, green flowers are unisexual, one at each node.

HORRID|US, -A, -UM Very bristly or prickly.

● **HORSE-CHESTNUT** *(Aesculus hippocastanum)*. A deciduous tree to 24m in the *Hippocastanaceae* from Albania and Greece, but much planted in temperate areas. It has grey-brown bark which flakes off when mature, and PALMATE leaves with 5–7 LANCE-OBOVATE irregularly toothed leaflets, 75–200mm long. The erect, pyramidal PANICLES bear 4–5 petalled flowers with basal yellow spots which age red. The green prickly globose fruits contain one or two polished seeds, the familiar conkers. The red flowered horse-chestnut *(A. X carnea)* is a hybrid between this and the American red buckeye *(A. pavia)*.

HORSE RADISH *(Armoracia rusticana)*. A perennial plant in the *Cruciferae* from S.E. Europe and W. Asia, but often cultivated and naturalised elsewhere. It has a long whitish root which is used for sauces and long-stalked OVATE-OBLONG leaves 300–450mm long. The small, 4-petalled white flowers are carried in much branched, CORYMBOSE PANICLES.

HORSETAIL Several species of *Equisetum*, primitive plants in the *Equisetaceae* and placed in the same division as the ferns *(Pteridophyta)*. They are found throughout most of the world, but not in Australasia, often in moist or wet soil. They have deep, creeping RHIZOMES and hollow stems with WHORLS of green, segmented slender branches. The spores are carried in SPORANGIOPHORES which are attached to PELTATE scales, forming a terminal cone-like structure: field horsetail *(E. arvense)* has two kinds of stems, green, sterile ones 200–750mm tall and brown, unbranched fertile ones 100–250mm tall bearing the sporing cones; it can be a serious pest in gardens; great horsetail *(E. telmateia)* is only found in permanently wet soils. It has sterile stems 0.9–1.8m high and fertile ones 200–400mm in height.

HORTENS|IS, -E Relating to a garden; plants raised there. Also a synonym of HYDRANGEA.

HORTUS SICCUS
see HERBARIUM

HOST The plant upon which a parasite lives.

HOSTA (syn. *Funkia*). A genus of 40 species of herbaceous perennials in the *Liliaceae* from China and Japan. They are clump-forming with LANCEOLATE to broadly OVATE spreading leaves which may be less than 75mm to over 450mm long. The white to lilac, lily-like flowers are tubular with 6, spreading TEPALS and are borne in terminal RACEMES. Several species and variegated-leaved cultivars are frequently met with in gardens.

HOTTENTOT FIG *(Carpobrotus edulis* syn. *Mesembryanthemum edule)*. A prostrate SUCCULENT in the *Aizoaceae* from S. Africa but naturalised in many frost-free countries including parts of Great Britain. It has pairs of fleshy leaves 63–100mm long, triangular in cross section, and daisy-like, rose-purple or yellow 50mm wide flowers. The fleshy capsule is edible.

HOT WATER PLANTS A general name for all 50 species of *Achimenes*, perennials in the *Gesneriaceae* from tropical America. They have small, tuber-like RHIZOMES covered with fleshy scales and erect or lax, hairy stems bearing ovate, toothed leaves in pairs or WHORLS. The tubular flowers expand to 5 broad lobes and range from white and yellow to red and purple. Many hybrid CULTIVARS are grown as pot plants. The vernacular name seems to be a corrupted form of the recommendation to water them with tepid or just warm water.

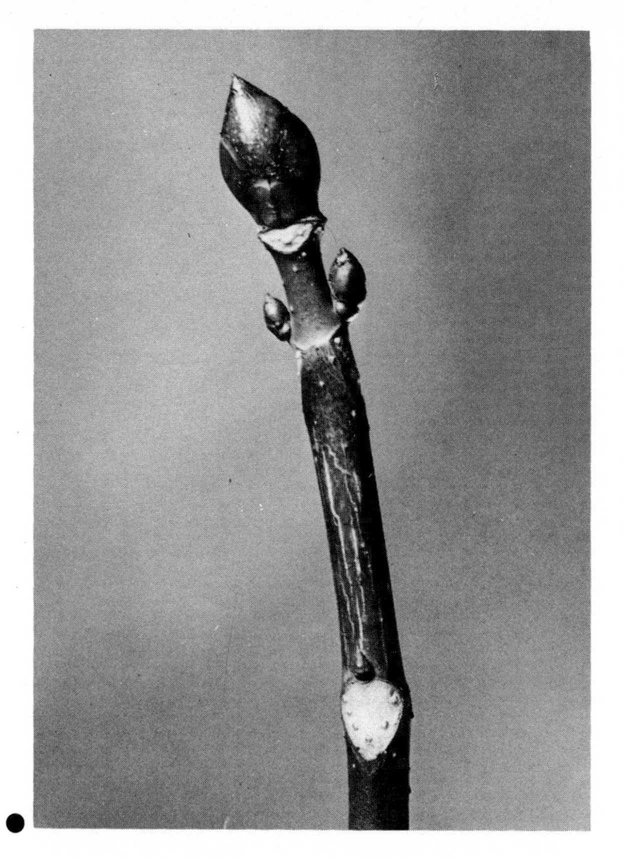

HOUND'S TONGUE *(Cynoglossum officinale)*. A softly pubescent biennial in the *Boraginaceae* from Europe and Asia, introduced into N. America. It has basal LANCEOLATE leaves to 300mm and erect, leafy stems bearing branched CYMES of dull-red-purple 10mm long funnel-shaped flowers. The fruit comprises 4, OVATE, flattened 6mm long nutlets covered with short barbed spines.

HOUSELEEK All the 25 species of *Sempervivum*, SUCCULENT plants in the *Crassulaceae* from the mountains of Europe to the Caucasus and N. Africa. They form dense rosettes of SPATHULATE, fleshy leaves which in turn form hummocks or clumps. Some have smooth leaves, others are hairy, the cob-web houseleek having web-like hairs all over the rosette. The flowers have 6–18, LINEAR petals which may spread out in daisy-fashion or remain joined in a tube. The latter are mainly yellow-flowered species known as *Jovibarba*. The largest species, *S. tectorum*, was formerly planted on roofs to seal leaks and supposedly to ward off lightning.

HUCKLEBERRY *(Gaylussacia)*. A genus of 49 deciduous and evergreen shrubs in the *Ericaceae* from N. and S. America. They vary in habit from mat-forming to 1.8m shrubs, with small, often ovate leaves and bell or urn-shaped flowers: black huckleberry *(G. baccata)* is a usually erect shrub to 900mm with 25–50mm long OBOVATE deciduous leaves and drooping RACEMES of dull red, URCEOLATE flowers. The lustrous black 15mm wide fruits are edible.

HUMIFUS|US, -A, -UM Spreading out flat on the ground.

HUMIL|IS, -E Low-growing, dwarf.

HUMUS Organic matter resulting from the decomposition of plant and animal remains in the soil. It gives soil its dark colour and improves its ability to hold water and dissolved minerals so important for plant growth.

HYACINTH Several species of *Hyacinthus*, bulbous plants in the *Liliaceae* from the Mediterranean and Africa. They vary from small and slender to robust plants with linear, often channelled leaves and erect racemes of horizontal or pendent bell-flowers. Common hyacinth *(H. orientalis)* is robust to 300mm tall with the narrow flowers having 6, spreading lobes. They are fragrant in shades of white, yellow, purple, red and blue.

HYACINTH BEAN *(Dolichos lablab)*. Also called lablab, this is a perennial or annual climber in the *Leguminosae* probably from Asia, but now grown widely in the tropics and subtropics. It has TRIFOLIATE leaves and purple and white pea-flowers in the leaf AXILS. The 75mm long pods may be purple or yellow and contain oval beans, white, red, black or mottled depending on the cultivar.

HYALINE [HYALIN|US, -A, -UM] Transparent, e.g. leaf or BRACT margins.

HYBRID A plant resulting from crossing two species. It is indicated in writing by placing a cross before the name under which it is described; or, if the names of the two parent species are used, then the cross is placed between them, e.g. the winter heath is *Erica X darleyensis* or *E. carnea X mediterranea*.

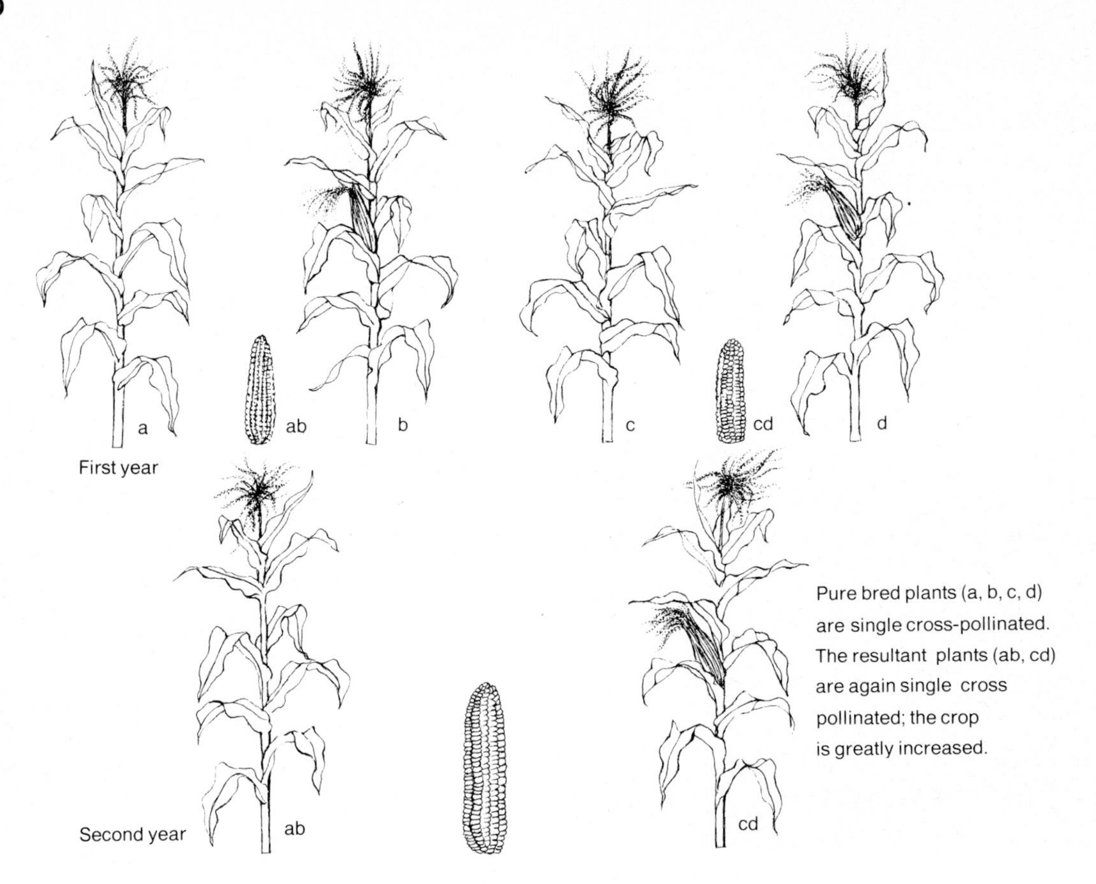

Pure bred plants (a, b, c, d) are single cross-pollinated. The resultant plants (ab, cd) are again single cross pollinated; the crop is greatly increased.

● **HYBRID VIGOUR** The phenomenon of increased vigour, uniformity and yield when two pure-bred plants of the same species or CULTIVAR are crossed together. Nurserymen and seedsmen now concentrate on this aspect of plant breeding to produce superior flowers, vegetables and fruits.

HYDRANGEA A genus of 80 evergreen and deciduous shrubs in the *Hydrangeaceae* from N. and S. America, and Asia from the Himalaya to Japan and Java. They have opposite pairs of OVATE to LANCEOLATE leaves and small flowers in flattened or rounded CORYMBS, often with larger, showy sterile florets around the outside. *H. macrophylla* is the commonest hydrangea in gardens, a deciduous shrub to 2.4m or more, with broadly ovate leaves. Two forms are grown: Hortensia, having mop-like heads of almost entirely sterile flowers, and Lacecaps, with fertile flowers in the middle of each cluster surrounded by a ring of sterile florets. Shades of pink, red, purple and blue occur, but the blue flowers are only produced on acid soil. Climbing hydrangea *(H. petiolaris)* climbs by aerial roots like those of ivy, to 18m or more with broadly OVATE leaves and lace-cap heads of white flowers.

HYDROME [LEPTOME] Special cells in the stems of the larger mosses (e.g. common hair moss), which carry water and dissolved foods to and from the roots and leaves. They are the primitive equivalents of XYLEM and PHLOEM.

HYDROPHYTE The 6th of a seven-category system of classifying life forms: water plants, either submerged, floating or rooted in mud with most of the stems and leaves above water.

HYPERBOREAN [HYPERBORE|US, -A, -UM] From the north or northern temperate regions.

HYPHAE
see FUNGI

HYPOCOTYL The stem below the seed leaves of an EPIGEAL seedling.

HYPOCRATERIFORM Salver-shaped; a tubular flower which spreads out flat at the mouth, e.g. primrose.

HYPOGEAL
see COTYLEDON

▲ **HYPOGYNOUS** A flower in which the PISTILS (GYNOECIUM) is at the summit of the RECEPTACLE, the stamens, petals and sepals all arising lower down; a superior flower.

HYSSOP *(Hyssopus officinalis)*. An aromatic, woody based perennial in the *Labiatae* from S. Europe, W. Asia and Morocco. It has erect stems 200–600mm tall with opposite pairs of LINEAR to LANCEOLATE leaves and WHORLS of violet-blue, tubular, 2-lipped flowers to 13mm long.

HYSTRIX Porcupine-like, bristly.

ICELAND MOSS *(Cetraria islandica)*. A lichen from mountains and moorlands in north temperate and Arctic areas. The brown or reddish-brown fronds are ribbon-like and branching, their margins bearing stiff, blunt bristles. An edible jelly can be made from it by boiling.

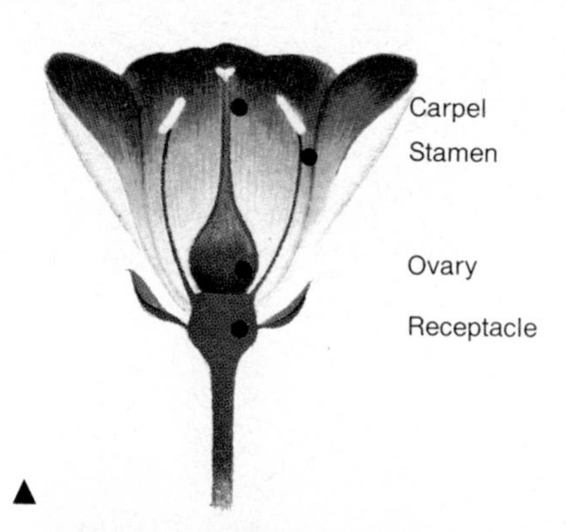

▲

■

○

ICE PLANT A somewhat ambiguous name for several SUCCULENT plants, mainly *Cryophytum* (syn. *Mesembryanthemum*) *crystallinum*, a prostrate annual in the *Aizoaceae* from S. Africa and naturalised in the Mediterranean region and California. It has OVATE, 50–75mm long leaves covered with glistening papillae, like ice crystals, and 13mm wide, white daisy flowers. In British gardens, the perennial *Sedum spectabile* is often called ice plant. It has bright, GLAUCOUS, OBOVATE leaves on erect, 300–400mm stems and flat heads of small, starry pink flowers much beloved by butterflies.

IGNE|US, -A, -UM Fiery, red.

IMBIBITION The absorption of water by the seed, mainly through the MICROPYLE pore, but also via the seed coat (testa).

IMBRICATE [IMBRICAT|US, -A, -UM] Overlapping, like tiles on a roof.

○ **IMPARIPINNATE** A pinnate leaf with an odd, terminal leaflet.

■ **IMPATIENS** A genus of almost six hundred annuals, perennials and sub-shrubs in the *Balsaminaceae* from tropical and temperate regions. They have lanceolate to ovate, often toothed leaves and pendulous red, purple, yellow or white flowers with large funnel-shaped spurs with the tips usually curving under. The cylindrical or ovoid fruits have elastic valves which, when ripe, fall inwards, violently discharging the seeds to a distance of up to several feet. *I. wallerana* and its forms *I. w. holstii* and *I. w. sultanii* are the popular Busy Lizzies grown as house plants.

IMPEDIT|US, -A, -UM Tangled or hindering; e.g. the low thickets of *Rhododendron impeditum*.

IMPERIALIS, -E Noble or showy.

IMPRESS|US, -A, -UM Impressed or sunken, e.g. leaf veins.

IMPUDIC|US, -A, -UM Immodest or lewd.

INCAN|US, -A, -UM Grey or hoary.

INCARNAT|US, -A, -UM Flesh-coloured.

INCENSE CEDAR *(Calocedrus decurrens* syn. *Libocedrus decurrens)*. A pyramidal to columnar, 30–45m evergreen coniferous tree in the *Cupressaceae* from Oregon and California. It has flattened sprays of pointed, laterally compressed scale leaves and 19mm cylindrical cones. The close-grained red-brown timber has a fragrance of incense and is resistant to decay. It is used for pencils and for exposed woodwork.

INCERTAE SEDIS Of uncertain position, i.e. plants inadequately described or difficult to classify.

INCISED [INCIS|US, -A, -UM] Sharply or deeply cut; of leaf margins.

INCOMPLETE Lacking an expected part or organ, e.g. flowers without petals.

▲

INDEHISCENT Of fruits which do not open to liberate the seeds, e.g. fleshy berries.

INDIAN BEAN *(Catalpa bignonioides)*. A deciduous tree to 12.2m in the *Bignoniaceae* from east N. America. Has broadly OVATE, CORDATE leaves 150–250mm long and erect PANICLES of flattened, bell-shaped white flowers having yellow and purple markings. The slender bean-shaped pods are 200–380mm long.

INDIAN CRESS
see NASTURTIUM

INDIAN FIG
see CACTUS *(Opuntia)*

INDIAN LIQUORICE
see CRAB'S EYE VINE

INDIAN PAINTBRUSH *(Castilleja)*. A genus of 200 species of partial root parasites in the *Scrophulariaceae* from N. and S. America (two in Arctic Eurasia). They have LINEAR to OBOVATE leaves, entire or incised and erect stems bearing terminal spikes of tubular, 2-lipped flowers with large, often brightly coloured BRACTS.

INDIAN SHOT *(Canna indica)*. A perennial plant with fleshy RHIZOMES in the *Cannaceae* from C. and S. America. It has OBLONG-OVATE basal leaves to 450mm, and erect, 0.9–1.5m stems with smaller ones. The 50mm long, red or pink flowers have 3 petals and 3 PETALOID stamens. The extremely hard, spherical seeds are said to have been used as a substitute for shot.

INDIGENOUS Plants native to a particular area or country.

INDIGO *(Indigofera tinctoria)*. A 1.2–1.8m shrubby perennial in the *Leguminosae* from S.E. Asia. It has pinnate leaves with 9–15 obovate leaflets. The small red pea-flowers ar borne in short AXILLARY RACEMES. The blue dye is obtained by macerating the plant in water and then aerating the liquid.

INDUSIUM The membranous covering over the SPORANGIA in ferns and which gives each SORUS its particular shape.

INERM|IS, -E Unarmed; without spines or stinging hairs.

INFERIOR OVARY
see EPIGYNOUS

INFLAT|US, -A, -UM Swollen or bladder-like.

INFLEXED [INFLEX|US, -A, -UM] Bent inwards.

● **INFLORESCENCE** The arrangement of flowers on the stem; *see also* SPIKE, RACEME, PANICLE, CYME (dichasial and monochasial), CORYMB, CAPITULUM, UMBEL (single and compound).

INFUNDIBULIFORM|IS, -E Funnel-shaped.

▲ **INK CAP** Several species of *Coprinus*, fungi in the *Basidiomycetes* from temperate climates. They are so-

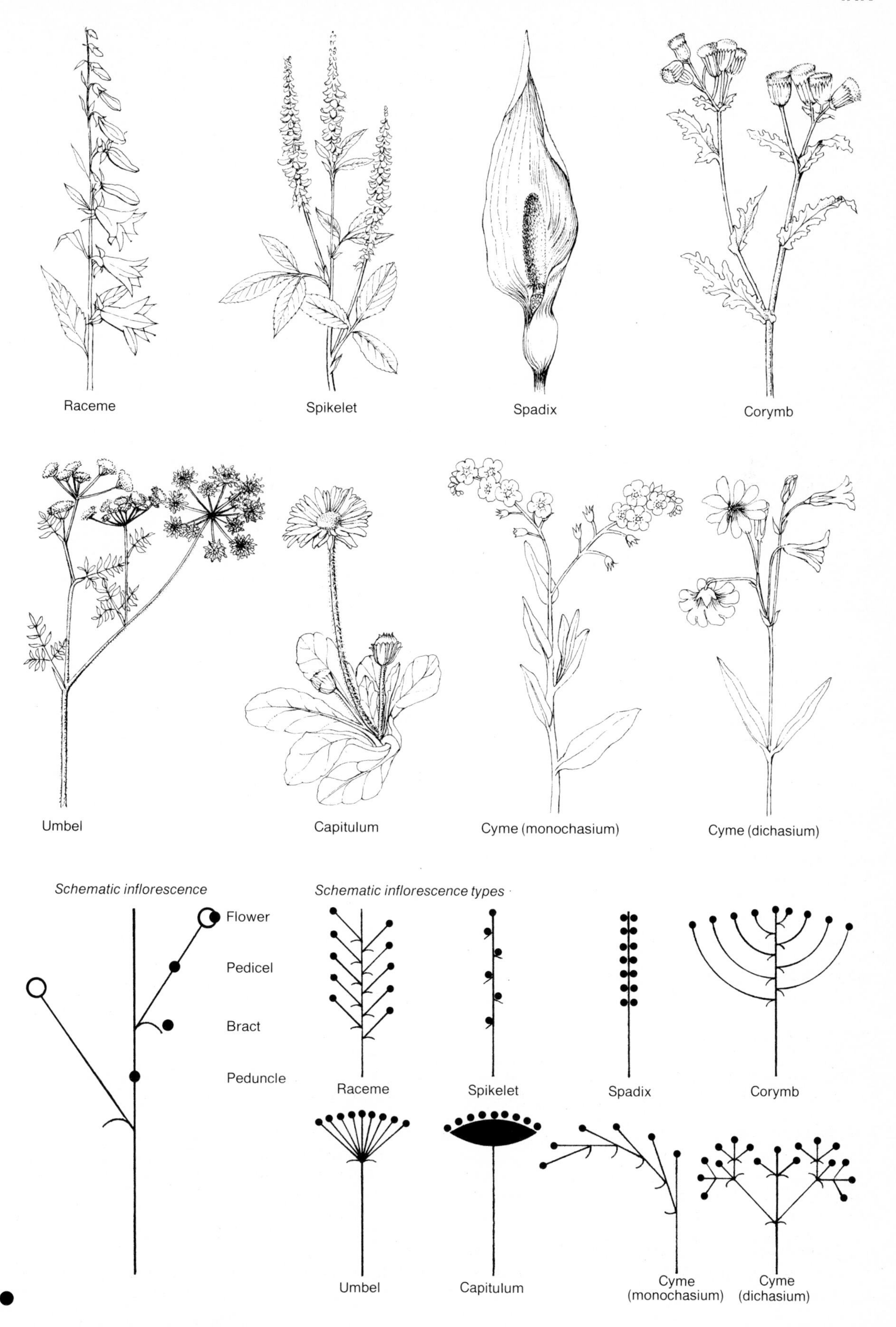
Raceme
Spikelet
Spadix
Corymb
Umbel
Capitulum
Cyme (monochasium)
Cyme (dichasium)
Schematic inflorescence
Flower
Pedicel
Bract
Peduncle
Schematic inflorescence types
Raceme
Spikelet
Spadix
Corymb
Umbel
Capitulum
Cyme (monochasium)
Cyme (dichasium)

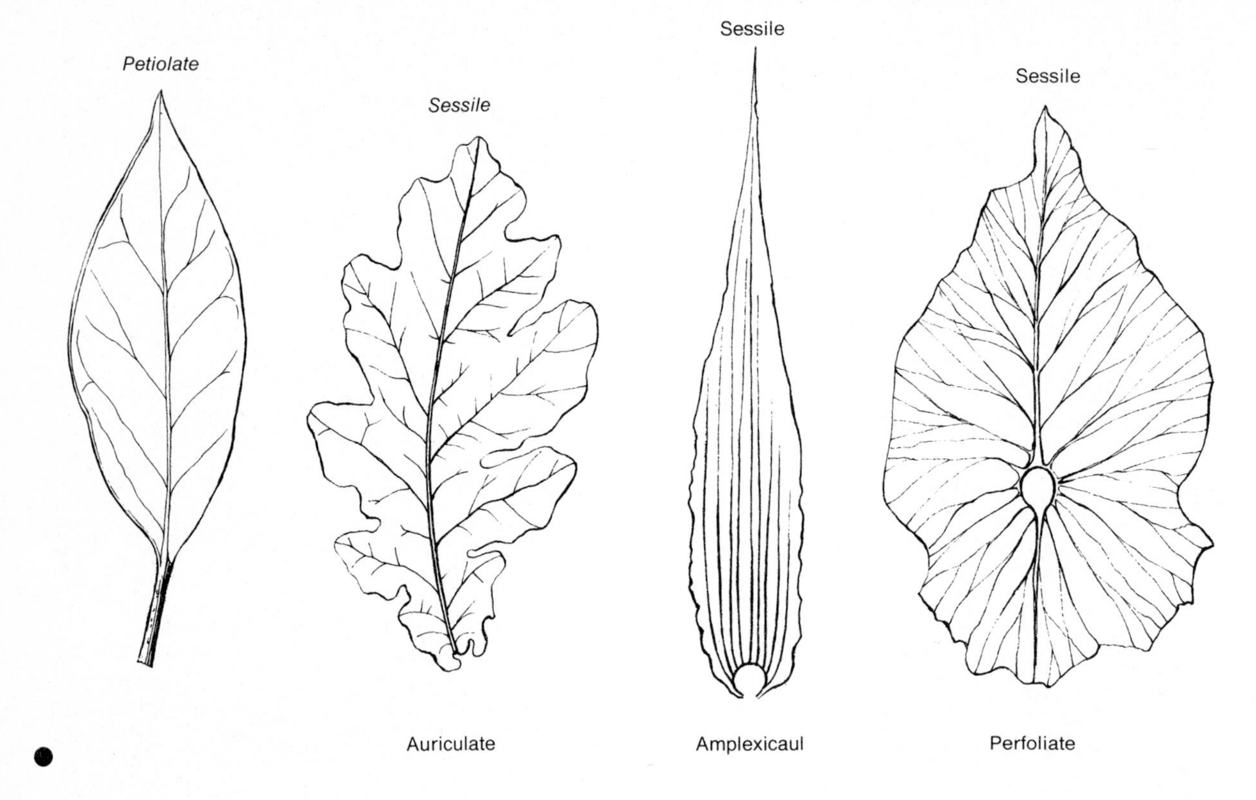

called because when the caps are mature, they dissolve from the bottom upwards into an inky liquid containing the spores. Shaggy ink cap *(C. comatus)* has cylindrical white caps with shaggy scales fancifully resembling a lawyer's wig, an alternative common name.

INNATE An anther joined to the FILAMENT by its base.

INODOR|US, -A, -UM Without scent or odour.

● **INSERTION** The way a leaf joins a stem, i.e. with a PETIOLE or stalk (petiolate), clasping (amplexicaul), forming a sheath (sheathing), etc.

INSIGN|IS, -E Outstanding, distinguished, remarkable.

INTEGERRIM|US, -A, -UM Entire, without lobes or teeth.

INTEGRIFOLI|US, -A, -UM With entire leaves.

INTEGUMENTS Protective layers around the NUCELLUS and EMBRYO SAC within the OVULE.

INTERCELLULAR SPACES Air spaces between the cells in leaf, stem or root tissue.

INTERNODE The length of stem between two nodes, the points on a stem where leaves and AXILLARY buds arise.

INTERRUPTEDLY PINNATE Pinnate leaves made up of large and small leaflets, e.g. potato, silverweed, agrimony.

INTINE The inner layer of the double coat of a pollen grain.

INTRORSE ANTHER LOBES that face the centre of a flower and shed pollen inwards.

INTRUDED Projecting forwards.

INVOLUCRAT|US, -A, -UM Having an INVOLUCRE.

INVOLUCRE A WHORL of BRACTS surrounding a cluster of flowers or florets (as in the daisy family) or a single flower (as with winter aconite or anemone).

INVOLUTE Rolled inwards; as in the margins of leaves in a bud.

IPECACUANHA *(Cephaelis ipecacuanha)*. A perennial plant in the *Rubiaceae* from S. America, with curiously thickened, beaded roots. It has OVATE leaves and small white flowers in stalked heads with an INVOLUCRE. The root provides the well-known emetic medicine and extracts for coughs and dysentery.

▲ **IRIS** A genus of 300 species of perennials in the *Iridaceae* from the northern temperate zone. Some are RHIZOMATOUS, others BULBOUS. The rhizomatous species have folded, sword-shaped leaves (ensiform); the bulbous ones, LINEAR, channeled or cylindrical. The flowers have a coloured PERIANTH with 3 large outer segments often arching or pendulous, and 3 smaller, inner ones, usually erect. Between the inner lobes and almost resting on the outer ones are 3, arching, PETALOID STYLES: winter iris *(I.*

▲

■

unguicularis syn. *stylosa)* is an evergreen, tufted plant with long-tubed, 75mm, bright lilac purple flowers; nameless iris *(I. innominata)* from N.W. America has grassy leaves and slender 100–150mm stems with yellow, net-veined flowers; Spanish iris *(I. xiphium)* is bulbous-rooted with linear, channeled leaves and 450–600mm stems bearing white, yellow or blue, 100mm wide flowers with orbicular falls; yellow iris or flag *(I. pseudacorus)* grows in wet ground or in water with sword-shaped leaves and stems to 25cm with 75–100mm yellow flowers; stinking iris or gladdon *(I. foetidissima)* is evergreen, with 300–600mm stems carrying small, purplish flowers followed by large ovoid-oblong capsules which split open to show bright red seeds; the many CULTIVARS of bearded iris grown in gardens are hybrids derived from *I. pallida* and other species, in a wide range of colours; orris root *(I. germanica florentina)* resembles a white *I. pallida* and is grown for its rhizomes which are used in perfumery and also have medicinal properties.

IRISH MOSS
see CARRAGHEEN

IRON WOOD Several hard-wooded trees, but mainly *Ostrya virginiana*, a deciduous species in the *Betulaceae* from east N. America. It resembles HORNBEAM.

ISOBILATERAL Leaves with both surfaces alike.

ISOGAMETES Male and female sex cells alike in shape and size.

■ **IVY** *(Hedera helix)*. An evergreen, woody climber in the *Araliaceae* from Europe and Iran. It climbs by means of AERIAL ROOTS to 30m in tall trees, but also forms vast mats on the ground, where supports are lacking. The leaves of creeping and climbing stems are PALMATELY 3–5 lobed, those on the non-climbing flowering stems LANCEOLATE to OVATE. Flowering only takes place when ivy has reached the top of its support and non-climbing stems form. The 5-petalled, yellow-green flowers are borne in rounded UMBELS and followed by black fruits. There are many CULTIVARS with variously shaped and variegated leaves.

IVY-LEAVED BELLFLOWER *(Wahlenbergia hederacea)*. A slender, creeping perennial in the *Campanulaceae* from Europe. It has almost orbicular, somewhat angled or lobed stalked leaves to 9mm wide and nodding, pale blue bell-flowers 6–9mm long.

IVY-LEAVED TOADFLAX *(Cymbalaria muralis)*. A creeping, evergreen perennial in the *Scrophulariaceae* from S. Europe, but naturalised elsewhere. It has stalked, rounded, 5-lobed leaves to 25mm long and lilac tubular, 2-lipped flowers each with a curved spur.

JACARANDA *(Jacaranda mimosifolia* syn. *J. ovalifolia)*. A small tree to 6.0m in the *Bignoniaceae* from Argentina. It has 450mm long, bipinnate leaves of oval to oblong leaflets and pyramidal PANICLES of pendent blue trumpet flowers. Commonly planted in the tropics and sub-tropics and confused with the allied Brazilian *J. acutifolia.*

JACK [JAK] FRUIT *(Artocarpus integrifolius)*. An evergreen tree to 9.1m in the *Moraceae* from S.E. Asia. It has 100–158mm long OBOVATE to OBLONG leaves and

MONOECIOUS clusters of green flowers, the males in catkins, the females in heads, borne directly on the trunk (cauliflory). The fruits fuse together to form oblong or lobed, somewhat spiky, green, multiple fruits to 18kg in weight.

JACK-BY-THE-HEDGE
see GARLIC MUSTARD

JACK-IN-THE-PULPIT In Britain, LORDS AND LADIES. In USA, used for *Arisaema triphyllum*. *See* ARISAEMA.

JACOBEAN LILY *(Sprekelia formosissima)*. A bulbous-rooted perennial in the *Amaryllidaceae* from Mexico. It has several arching, LINEAR leaves which die down when mature and are followed by solitary crimson, 75–100mm wide, ZYGOMORPHIC flowers on 300–450mm stems. The 6 narrow petals are arranged in an orchid-like fashion, the 3 upper spreading, the 3 lower rolled together at the base to form a false LABELLUM.

JACOB'S LADDER *(Polemonium caeruleum)*. A clump-forming herbaceous perennial in the *Polemoniaceae* from the north temperate zone. Has 100–300mm PINNATE leaves with 13–25 OVATE-LANCEOLATE leaflets and erect stems to 900mm with CORYMBOSE clusters of drooping, 19–32mm wide, 5-petalled blue flowers.

● **JADE VINE** *(Strongylodon macrobotrys)*. A vigorous, evergeen climber in the *Leguminosae* from the Philippines. It has TRIFOLIATE leaves with ovate leaflets and dense pendulous, 0.6–1.5m RACEMES of curved, pointed, 75–100mm jade-green pea-flowers. The large ovoid pods are 150mm or more long.

JAGGERY
see PALM (sugar palm)

JALAP Two different plants go under this name; true jalap *(Ipomoea purga* syn. *Exogonium purga)* is a tuberous-rooted climber in the *Convolvulaceae* from Mexico. It has CORDATE, triangular OVATE leaves and rose-purple, widely funnel-shaped flowers. Jalap is obtained from the sliced ▲ and dried roots; false jalap or marvel of Peru *(Mirabilis jalapa)* is a tuberous-rooted, erect perennial in the *Nyctaginaceae* from tropical America. It has ovate, LANCEOLATE leaves and terminal clusters of long-tubed, trumpet-like flowers in shades of red, yellow and white.

JAMAICA HONEYSUCKLE
see PASSION FLOWER

JAMESTOWN WEED
see THORN-APPLE

JAPANESE ARALIA *(Fatsia japonica* syn. *Aralia sieboldii)*. An evergreen shrub in the *Araliaceae* from Japan. It bears 150–380mm wide glossy, leather, PALMATE leaves having 7–9 lobes, and PANICLES of globose UMBELS. The small, 5-petalled, milky-white flowers are followed by black berries. This shrub is much grown as a pot plant when young.

JAPANESE CEDAR *(Cryptomeria japonica)*. An evergreen coniferous tree to 45m in the *Taxodiaceae* from Japan. It has reddish fibrous bark and much branched stems bearing 6–13mm long awl-shaped leaves. The brownish, globular cones are about 19mm across.

▲

JAPANESE CHERRIES A general term for many different sorts of cherry grown for ornamental purposes. They resemble the fruiting CHERRY, but flower colours range from cream shot with greenish yellow, through pink to red-purple. In autumn the foliage turns red and yellow.

JAPANESE LAUREL *(Aucuba japonica)*. An evergreen shrub to 3m in the *Cornaceae* from Japan. It has glossy, leathery, OVATE to LANCEOLATE leaves, toothed at the tips. The small, purple-green 4-petalled flowers are borne in terminal clusters, male and female on separate plants. The ovoid, fleshy fruits are 13–19mm long. Cultivated forms often have a variegation of yellow spots caused by a VIRUS.

JAPANESE QUINCE ['JAPONICA']
see QUINCE

JASMINE *(Jasminum)*. A genus of 300 species of evergreen and deciduous shrubs and climbers in the *Oleaceae* from Europe and Asia. The climbers may twine or scramble, and the shrubs are often straggling. The leaves are usually PINNATE, sometimes TRIFOLIATE or simple, and the tubular flowers expand to 5 or more flat lobes. The fruit is a berry: winter jasmine *(J. nudiflorum)* is a scrambling or almost prostrate, deciduous shrub with trifoliate leaves and bright yellow, 19–25mm long flowers on leafless twigs; common jasmine *(J. officinale)* is a deciduous twiner to 12.2m with pinnate leaves and 25mm wide, white, very fragrant blooms in terminal clusters.

■ **JELLY FUNGI** A class of *Basidiomycete* fungi known as *Tremellales* which produces fruiting bodies having a jelly-like consistency. They may be cup or ear-shaped, take the form of pustules or stags' horns, or appear like convoluted ribbons of jelly; many are coloured.

JERUSALEM SAGE *(Phlomis fruticosa)*. A 0.6–1.2m evergreen shrub in the *Labiatae* from the Mediterranean region. The opposite pairs of OVATE to OBLONG leaves are finely corrugated and white hairy. The tubular yellow, 2-lipped flowers, the upper lip large and hooded, are borne in terminal WHORLS.

JESSAMINE
see JASMINE

JIMSON WEED
see THORN-APPLE

JOB'S TEARS *(Coix lacryma-jobi)*. An annual grass in the *Gramineae* from the tropics and subtropics. It grows 0.6–1.5m tall, with narrowly LANCEOLATE, 150–600mm long leaves and terminal, nodding RACEMES of MONOECIOUS spikelets. The female spikelets are ovoid and harden later into tear-shaped, glossy bead-like fruits in shades of white, grey, brown or black.

JOE-PYE WEED
see HEMP AGRIMONY

JOHN-GO-TO-BED-AT-NOON
see GOAT'S BEARD

JONQUIL
see NARCISSUS

JUDAS' BAG
see BAOBAB

JUDAS' TREE *(Cercis siliquastrum)*. A deciduous shrub or tree to 12m in the *Leguminosae* from S. Europe to Turkey. It has almost rounded, deeply CORDATE, GLAUCOUS-green leaves and crowded clusters of bright rose-purple pea-flowers. The latter are 12–19mm long, often borne on older stems and branches and are followed by flat, wing-like pods distributed by the wind.

JUJUBE *(Zizyphus jujuba)*. A deciduous shrub or tree to 9m in the *Rhamnaceae* from the Mediterranean region. It has zigzag stems armed with pairs of spines, one straight, the other shorter and hooked. The leaves are ovate to lanceolate and the tiny, yellowish flowers followed by 12–32mm long black, edible fruits.

JUNCE|US, -A, -UM Rush-like.

JUNEBERRY
see AMELANCHIER

JUNGLE Tropical forest in wet, low-lying areas, or regions of high rainfall. There is usually a framework of well-spaced tall trees with shorter ones fitting in underneath, and often a third layer of shade-tolerant trees beneath that. The heads of the trees are often laced together by high growing climbers (lianes) and many EPIPHYTES grow on the branches. On the ground there may be tangled thickets of shrubs and other plants, but this is usually a feature of jungles that have been exploited for timber and then allowed to grow up again.

● **JUNIPER** *(Juniperus)*. A genus of 60 species of evergreen conifers in the *Cupressaceae* from the northern hemisphere. Most characteristic feature is the berry-like cone, formed of 1–4 WHORLS of scales, only one of which is fertile. Later the scales fuse and become fleshy. The leaves are either scale-like and closely pressed to the stem, or longer and awl-shaped, standing out from the stem. Junipers vary much in habit, from trees 30m tall to completely prostrate mat-formers; common juniper *(J. communis)* is usually a large shrub, but can be completely prostrate or a tree to 9m or more. It has awl-shaped leaves up to 16mm long. The cones are globose, blue-black to 13mm across; pencil cedar or juniper *(J. virginiana)* is a tree to 30m with a pyramidal head and tiny, overlapping scale leaves often mixed with 6mm long awl-shaped juvenile ones. The 6mm long cones are sub-globose and GLAUCOUS. The wood is much used for pencil-cases; a fragrant oil is distilled from the shavings and used to scent soap, etc.

JUTE *(Corchorus capsularis)*. An annual plant to 3.6m or so in the *Tiliaceae* from S. Asia. It has slender-pointed, OVATE leaves and small, 5-petalled yellow flowers in the upper leaf AXILS. The fibres are very woody and after RETTING are treated with fish-oil or kerosene as a softener. Jute is used for sacking, hessian, linoleum, etc.

KAFFIR LILY A name given to two quite different plants. (1) *(Clivia miniata)*. A robust, evergreen perennial in the *Amaryllidaceae* from S. Africa. It has broad, arching, leathery strap-shaped leaves to 600mm long and UMBELS of 50–75mm funnel-shaped bright orange-red flowers.
▲(2) *(Schizostylis coccinea)*. A slender RHIZOMATOUS

perennial in the *Iridaceae* from S. Africa. It has narrow, sword-shaped foliage and stems to 900mm bearing RACEMES of red or pink, 6-TEPALLED, crocus-like blooms.

KALE [BORECOLE] *(Brassica oleracea acephala)*. Originally MUTANT forms of wild cabbage in the *Cruciferae*, which proved to be very hardy in inland gardens. They produce dense heads of arching leaves which are variously crisped or curled along the margins. There are green, purple and variegated leaved forms. Russian and hearting kales tend to form small, cabbage-like hearts but are seldom grown in Britain.

KANGAROO APPLE *(Solanum aviculare* and *S. laciniatum)*. Evergreen, rather straggling shrubs in the *Solanaceae* from Australia and New Zealand. They have LANCEOLATE leaves to 250mm long on adult plants, deeply 3-lobed on younger ones. The 25–32mm wide rotate purple flowers of *S. aviculare* have 5 prominent lobes; the 32–50mm flowers of *S. laciniatum* have short, broad lobes. The yellowish berries are ovoid, to 25mm long.

KANGAROO PAW Several species of *Anigozanthus*, evergreen perennials in the *Haemodoraceae* from S.W. Australia. They have narrow, sword-like foliage and stems to 900mm or more with PANICLES of tubular flowers. The latter have the COROLLA tubes divided into 6, pointed lobes which fan out like the claws of a foot: *A. manglesii* has green 75mm flowers; *A. flavidus* has 39–50mm blooms, yellow-green flushed red.

KANGAROO THORN
see ACACIA *(armata)*

KANGAROO VINE *(Cissus antarctica)*. An evergreen climber to 6m or more in the *Vitidaceae* from Australia. It has OVATE, glossy, leathery leaves 75–125mm long with SINUATE margins and well-spaced sharp teeth. The tiny greenish flowers are followed by globular, dark purple fruits. Much grown as a house plant in Britain and very tolerant.

KAPA CLOTH
see PAPER MULBERRY

KAPOK The cotton-like floss found in the seed capsules of several members of the *Bombacaceae*, tall trees from the tropics. The two main ones are: silk cotton tree *(Ceiba pentandra)* with a prickly trunk when young, PALMATE leaves of 5–7 leaflets, and large fragrant, silky-hairy yellowish flowers; cotton tree *(Bombax malabaricum)*.

■ **KAURI PINE** *(Agathis australis)*. A massive, evergreen coniferous tree in the *Araucariaceae* from New Zealand. It forms a scarcely tapering, smooth grey trunk branching high up at an ascending angle. The 25–75mm leaves are leathery, narrowly lanceolate to oblong and the 50–75mm wide cones are almost globose. A valuable timber tree in New Zealand where most of the best specimens are now gone. In Waipoua forest there are still trees 51m in height and 13.7m in girth. The dried resin of kauri (copal) is used for paints, varnishes and linoleum. Much of it was dug from the soil on the site of former forests.

KEEL The two lower petals of a pea-type flower *(Leguminosae)* which are pressed together around the stamens and pistil.

KELP
see SEAWEED

KENTUCKY BLUE GRASS
see MEADOW GRASS

KENTUCKY COFFEE TREE *(Gymnocladus dioicus)*. A deciduous tree to 18.2m in the *Leguminosae* from N. America. It has large, bipinnate leaves to 900mm long and 600mm wide and small, DIOECIOUS, 5-petalled greenish-white flowers in PANICLES. The large, thickened pods can be 175–225mm long.

KERRIA *(Kerria japonica)*. A slender, erect, deciduous shrub in the *Rosaceae* from China and Japan, sometimes called Jew's mallow. It has sharply and doubly toothed ovate, lanceolate leaves and terminal clusters of orange-yellow, 5-petalled flowers like small single roses. In gardens, the double-flowered form is commonest (*K.j.* 'Flore-pleno').

KEYS
see SAMARA

KIDNEY BEAN
see RUNNER BEAN

KIDNEY VETCH [LADIES' FINGERS] *(Anthyllis vulneraria)*. An erect or spreading perennial in the *Leguminosae* from Europe and N. Africa. The leaves are PINNATE, the terminal leaflet being much larger than the rest. Each stem terminates in a close head of white, woolly CALYCES from which emerge the 13mm long, yellow or red flushed pea flowers.

KING ALFRED'S CAKES
see CRAMP BALLS

KINGCUP
see MARSH MARIGOLD

● **KINGDON WARD, F. (1885–1958)** Son of the Cambridge botanist, H. M. Ward. He longed to visit the tropics and in 1906 took up a teaching post in Shanghai. He remained there for only two years, leaving in 1909 to join M. Anderson, an American zoologist on a visit to the interior. On the strength of plants he collected on this trip, Kingdon Ward was invited in 1911 to make a further journey into Yünnan by A. K. Bulley of Bees Seeds.

From that moment he spent the rest of his life on expeditions, largely into the remote and little explored areas of Burma, Assam, Tibet and China where the great rivers of S.E. Asia carve their way through the mountain barrier. In all he made 22 expeditions, many without any companions, and not only collected vast numbers of plants and seeds – his collecting numbers reached 23,000 on his last trip – but also mapped many areas previously unsurveyed. On his last six journeys he was accompanied by his second wife. He also kept extensive journals and wrote twenty-three books, many about his travels, among them: *The Land of the Blue Poppy* (1913), *The Romance of Plant Hunting* (1924), *Plant Hunting on the Edge of the World* (1930), and *Burma's Icy Mountains* (1939).

No fewer than 45 plants bear his name, a fitting tribute to his many discoveries. Of his finds, the first successful introduction of the legendary Himalayan blue poppy *(Meconopsis betonicifolia)* and the giant cowslip *(Primula florindae)* are perhaps the best known. He was renowned for his toughness and refusal to let physical discomforts stand in the way of exploration.

KLINOSTAT A clockwork apparatus used in the laboratory to revolve a plant continuously in a horizontal position, so that the roots or shoots are never still long enough to respond to gravity and display GEOTROPISM.

KNAPWEED [HARDHEAD] Several species of *Centaurea*, erect, perennial plants in the *Compositae* from Europe and W. Asia. They are characterised by the INVOLUCRES of the flower heads being hard and rounded, composed of tightly, overlapping bracts. The florets are all tubular with 6 slender lobes; greater knapweed *(C. scabiosa)* has stems to 900mm and 100–250mm, deeply PINNATIFID, OBLANCEOLATE leaves. The 32–50mm wide flowers heads are red-purple; lesser knapweed *(C. nigra)* grows to 600mm or so with oblanceolate leaves, sometimes irregularly toothed and PINNATISECT at the bases. The 19–38mm wide flower heads are pale red-purple.

KNOTGRASS Several species of *Polygonum*, but mainly *P. aviculare*, a wiry, prostrate annual in the *Polygonaceae* of cosmopolitan distribution. It has stems 300–900mm long, radiating outwards and elliptic, SPATHULATE, LANCEOLATE or LINEAR leaves 25–50mm long. Small pinkish or whitish flowers appear in the upper leaf AXILS. Several names have been given to various forms of this variable and common weed.

KNOTWEED Several species of *Polygonum*, perennials in the *Polygonaceae* from Asia. The most commonly seen naturalised by roadsides in Britain is Japanese knotweed *(P. cuspidatum)*, a RHIZOMATOUS plant with erect, bamboo-like stems to 2.4m or more with OVATE-oblong leaves and AXILLARY PANICLES of tiny white flowers.

KOHL RABI *(Brassica oleracea caulorapa)*. A curious vegetable coming halfway between a cabbage and a turnip. The stems swell out like an aerial turnip and may be green or purple. The somewhat cabbage-like leaves arise directly on the swollen area and must be removed before cooking.

KOWHAI Two closely related species of *Sophora*, sometimes called New Zealand laburnum. They are deciduous shrubs or small trees in the *Leguminosae* and provide the national flower of New Zealand. They have zigzag stems and PINNATE leaves, and 25–38mm long yellow pea-flowers of somewhat tubular form. The slender, 4-winged pods are constricted between the seeds, creating a chain of winged beads; *Sophora tetraptera* has LINEAR-OBLONG leaflets; *S. microphylla* is smaller with shortly-oblong ones.

KUDZU *(Pueraria phaseoloides)*. A perennial climbing or trailing plant in the *Leguminosae* from Malaysia. It is much planted in the tropics to prevent erosion, as a green manure dug into the soil when young, or as stock feed. The plant has TRIFOLIATE leaves of broadly OVATE 50–113mm long leaflets and AXILLARY RACEMES of mauve, pink, blue or white pea flowers.

KUMQUAT *(Fortunella)*. A genus of 6 evergreen, thorny shrubs in the *Rutaceae* from the Malay peninsula. It is very closely allied to CITRUS, but the fruits have 3–5 sections only, each with two seeds. The small fruits can be eaten whole and are often candied.

▲ **LABELLUM** The lowest of three petals in an orchid flower, specially modified to aid pollination by insects, in various ways. It may provide an alighting platform, or a trap so arranged that insects can only get out by either taking pollen or placing it on the STIGMA. An extreme example of adaptation is seen in those flowers which mimic insects (SPIDER, FLY ORCHIDS). Here the labellum is shaped like the body of a spider or fly and invites copulation.

LABIAT|US, -A, -UM Tubular flowers divided at the mouth into two lip-like structures; most are members of the *Labiatae* (dead nettle family).

LABLAB
see HYACINTH BEAN

LABRADOR TEA *(Ledum groenlandicum)*. A 900mm evergreen shrub in the *Ericaceae* from Greenland and N. America. It has 13–50mm long dark green, oblong-ovate leaves, densely rusty hairy beneath and terminal UMBELS of 13mm wide, white, 5-petalled flowers.

LABURNUM Also called golden rain or chain. A genus of 4 species of shrubs or small trees in the *Leguminosae* from Europe, W. Asia and N. Africa. They have TRIFOLIATE leaves and pendulous or erect RACEMES of yellow pea-flowers; common laburnum *(L. anagyroides)* is a tree to 6m with 25–75mm long grey-green leaflets, 100–200mm pendulous flowering stems and narrow seed pods; Scots or alpine laburnum *(L. alpinum)* is similar with flowering RACEMES to 300mm or more and flattened pods. Specimens in gardens are often hybrids between the two species *(L. X watereri)*.

LACE BARK *(Hoheria)*. 5 species of deciduous or evergreen shrubs and small trees in the *Malvaceae* from New Zealand. They have OVATE to LANCEOLATE, prominently toothed leaves and a profusion of wide open, 5-petalled, white flowers. Common lacebark or houhere *(H. populnea)* is a tree to 10.6m with ovate, doubly SERRATE leaves and 25mm wide, flowers.

LACERAT|US, -A, -UM Appearing torn or cut into fringe-like segments.

LACINIAT|US, -A, -UM Cut or slashed into narrow lobes.

LACQUER TREE *(Rhus verniciflua)*. A deciduous tree to 9m or more in the *Anacardiaceae* from China and Japan. It has 300–600mm long PINNATE leaves with 7–13 ovate leaflets. The tiny yellow-white flowers are carried in drooping PANICLES. It provides the essential ingredient in Chinese and Japanese lacquer.

LACTE|US, -A, -UM The colour of milk.

LACUSTR|IS, -E An inhabitant of lakes or ponds.

●

▲

LADY FERN *(Athyrium filix-femina)*. An elegant, herbaceous fern in the *Athyriaceae* from the north temperate zone, and the mountains of India and S. America. It has bipinnate, lanceolate fronds arranged in vase-like rosettes 600–900mm tall. The tiny SORI form a row on either side of the PINNULE.

LADY ORCHID *(Orchis purpurea)*. A tuberous-rooted ground dwelling species in the *Orchidaceae* from Europe to the Caucasus and Turkey. It has a cluster of ovate-oblong leaves and a stem to 375mm or so bearing a dense spike of 19mm long flowers. Each bloom has a large white or rose-pink skirt-like LABELLUM and the 3 green and purple sepals arch forwards to form a sunshade.

● **LADY'S MANTLE** Several species of *Alchemilla*, low growing perennials in the *Rosaceae* from the north temperate zone. They have woody RHIZOMES, tufts of PALMATE, toothed, often silky-hairy leaves and panicles of tiny greenish or yellowish flowers. The commonest and largest species in gardens is *A. mollis* with softly hairy leaves to 150mm across bearing 6–9 shallow lobes.

▲ **LADY'S SLIPPER** *(Cypripedium calceolus)*. A perennial ground-dwelling species in the *Orchidaceae* from the north temperate zone. A RHIZOMATOUS plant with stems to 300mm or more, bearing strongly ribbed, ovate leaves on the lower half. The solitary or paired flowers have slender, pointed, spreading maroon sepals and a pale yellow, slipper-like LABELLUM.

LADY'S SMOCK
see BITTER CRESS

LADY'S TRESSES Several species of *Spiranthes*, slender, ground-dwelling members of the *Orchidaceae* from the north temperate zone and S. America, characterised by having very small, almost tubular white flowers arranged in a spiral up the stem; autumn lady's tresses *(S. spiralis)* is a plant of short turf with GLAUCOUS, ovate 25mm leaves and 75–200mm in stems.

LAEVIGAT|US, -A, -UM, [LAEVIS, -E] Smooth, polished, hairless.

LAMB KILL
see CALICO BUSH

LAMB'S LETTUCE *(Valerianella locusta* syn. *V. olitoria)*. An annual in the *Valerianaceae* from Europe, W. Asia and N. Africa, introduced elsewhere. It forms rosettes of SPATHULATE to OBLONG leaves to 63mm (more in gardens) and branched, ascending 75–250mm stems bearing tiny, funnel-shaped, pale lilac flowers in dense CYMES. Sometimes grown as a salad crop.

LAMINA The blade of a leaf, without the stalk.

LANAT|US, -A, -UM [LANIGER|US, -A, -UM] Covered with woolly hairs.

LANCEOLATE Leaves, bracts or petals shaped like the head of a lance. For standardised botanical use it denotes a shape broadest below the middle, and between 3 and 6 times as long as wide. Leaves wider than this are described as OVATE.

LANOS|US, -A, -UM Softly hairy or woolly.

LANUGINOS|US, -A, -UM Woolly, downy, cottony.

LAPAGERIA ROSEA A wiry evergreen climber to 6m in the *Liliaceae* from Chile and providing that country's national flower. It has dark green ovate leaves and waxy, rose-crimson bell-flowers to 75mm long and more.

■ **LARCH** *(Larix)*. A genus of about 12 deciduous trees in the *Pinaceae* from the north temperate zone. They have needle-like leaves scattered along the vigorous leading shoots and borne in WHORLS on the short spurs or flowering shoots. The cones are ovoid and the strong, durable timber is in demand. Common larch *(L. decidua)* can reach 45m. It is pyramidal when young, broadening out when old. The female flowering cones are red and combine attractively with the bright green, spring foliage. Most of the larch plantations in Britain are of hybrid origin, being *L. decidua X L. kaempferi*, the Japanese larch.

LARKSPUR Several annual species of *Delphinium*, erect plants in the *Ranunculaceae* from Europe, Asia and N. Africa. Their leaves are PALMATELY cut into many LINEAR segments and the branched RACEMES of spurred flowers are formed of blue sepals. Eastern or oriental larkspur *(D. orientale)* is the most commonly met with, being frequently seen in gardens. It grows to 300–600mm in height and bears 25–38mm wide blue, purple, pink or white flowers.

LATERAL Side shoots arising on main or leading stems, usually referring to trees and shrubs.

LATEX A white, yellow or sometimes red milky fluid found in many plants, e.g. poppy, bellflower. The latex is contained in special cells or groups of cells that form a network through the plant. The exact purpose of the latex is not fully known; in some cases it is associated with nutritive substances, but in others it appears to be mainly waste material. *See also* RUBBER.

LATIFOLI|US, -A, -UM With broad leaves.

○ **LAUREL** A general name for several evergreen shrubs or trees, spread over a number of genera and species. *See* CHERRY LAUREL, JAPANESE LAUREL, BAY LAUREL, DAPHNE (spurge laurel).

LAURESTINUS *(Viburnum tinus)*. An evergreen shrub in the *Caprifoliaceae* from S.E. Europe. It has opposite pairs of 50–100mm OVATE leaves and 50–100mm flattened clusters of small pink-budded, white flowers. The small fruits are deep blue.

LAVENDER *(Lavandula)*. A genus of 28 species of low, evergreen, sweetly aromatic shrubs in the *Labiatae* from the Canary Islands, the Mediterranean region and India. They have linear, sometimes PINNATIFID, often grey-hairy leaves and dense spikes of blue or purple, 2-lipped tubular flowers: common or old English lavender *(L. spica)* is a 600–900mm shrub with ENTIRE 38–50mm grey-white hairy leaves and fragrant grey-blue flowers. There are a number of CULTIVARS, both tall and short, with pink, white and darker blue flowers; *L.s.* 'Vera' (Dutch lavender) is widely grown for the extraction of lavender oil and for drying for pot-pourri.

LAVENDER COTTON
see COTTON LAVENDER

LAVER Several species of *Porphyra*, red seaweeds in the *Bangiaceae* with a world-wide distribution. The main species is purple laver *(P. umbilicaulis)* which takes two forms, circular and LINEAR. The circular plants grow from a central HOLDFAST, forming waved, mantle-like fronds 150–200mm across. They are thin and silky, rose-purple to olive-brown. This marine alga is eaten in Wales and Ireland after boiling and frying with oatmeal in bacon fat. It has long been cultivated in China and Japan where it is eaten as a dry, crisp covering to rice, etc.

LAX|US, -A, -UM Of loose or lax growth.

LEADER The main or leading stem of a young tree or branch which is actively growing and extending year by year.

LEADWORT Several species of *Plumbago*, evergreen shrubs and climbers in the *Plumbaginaceae* from the tropics and sub-tropics. They have linear to oblong-ovate, entire leaves and spikes of slender, tubular flowers which open out to 5 broad petals. The best known is *P. capensis*, a sprawling or climbing shrub to 6m with terminal spikes of 19mm wide, sky-blue flowers. Leadwort is also sometimes used for *Ceratostigma*, closely allied, low deciduous shrubs with the same sort of blue flowers.

● **LEAF** A specially modified part of the stem with all the tissues spread out so that light and air can readily penetrate. Leaves are the factories of the plant, elaborating sugars and starch from carbon-dioxide and water with the energy of sunlight via the chlorophyll. They are also breathing organs, air entering the tissues via small pores (stomata). Leaves are extremely variable in shape and size. They may be millimetres long in heather and over 6m in coconut palms. They may be only one cell thick in mosses and filmy ferns and up to 25mm thick in some succulent plants. In shape they may be thread or strap-like to circular, with every grade in between. There are several off-beat shapes, such as LUNATE, FALCATE, FLABELLATE PANDURATE, RUNCINATE and LYRATE. Leaves may have smooth (entire), toothed or lobed margins or they may be so deeply dissected that several smaller leaflets or leaves are formed, e.g. the PINNATE leaves of an ash tree. The leaflets may again be so divided to form a bipinnate leaf, characteristic of many ferns. Leaves of one blade are called simple, those with several are known as compound.

LEAF COLOUR Most leaves are a shade of green, due to the presence of chlorophyll. The leaves of some plants take on shades of purple or red, e.g. copper and purple beech, beetroot etc., from the presence of colour pigments known as anthocyanins. Variegated leaves show patterns or spots of white or yellow as the result of loss of chlorophyll from certain cells due to a mutation or the presence of a virus. Other leaves take on a silvery sheen caused by the epidermis being raised up with a layer of air beneath. Leaves may be hoary or grey due to a coating of colourless hairs. During autumn in temperate climates, the leaves of deciduous trees turn bright colours before they fall. This is because of a breakdown of chlorophyll, leaving varying amounts of orange and yellow pigments to show through.

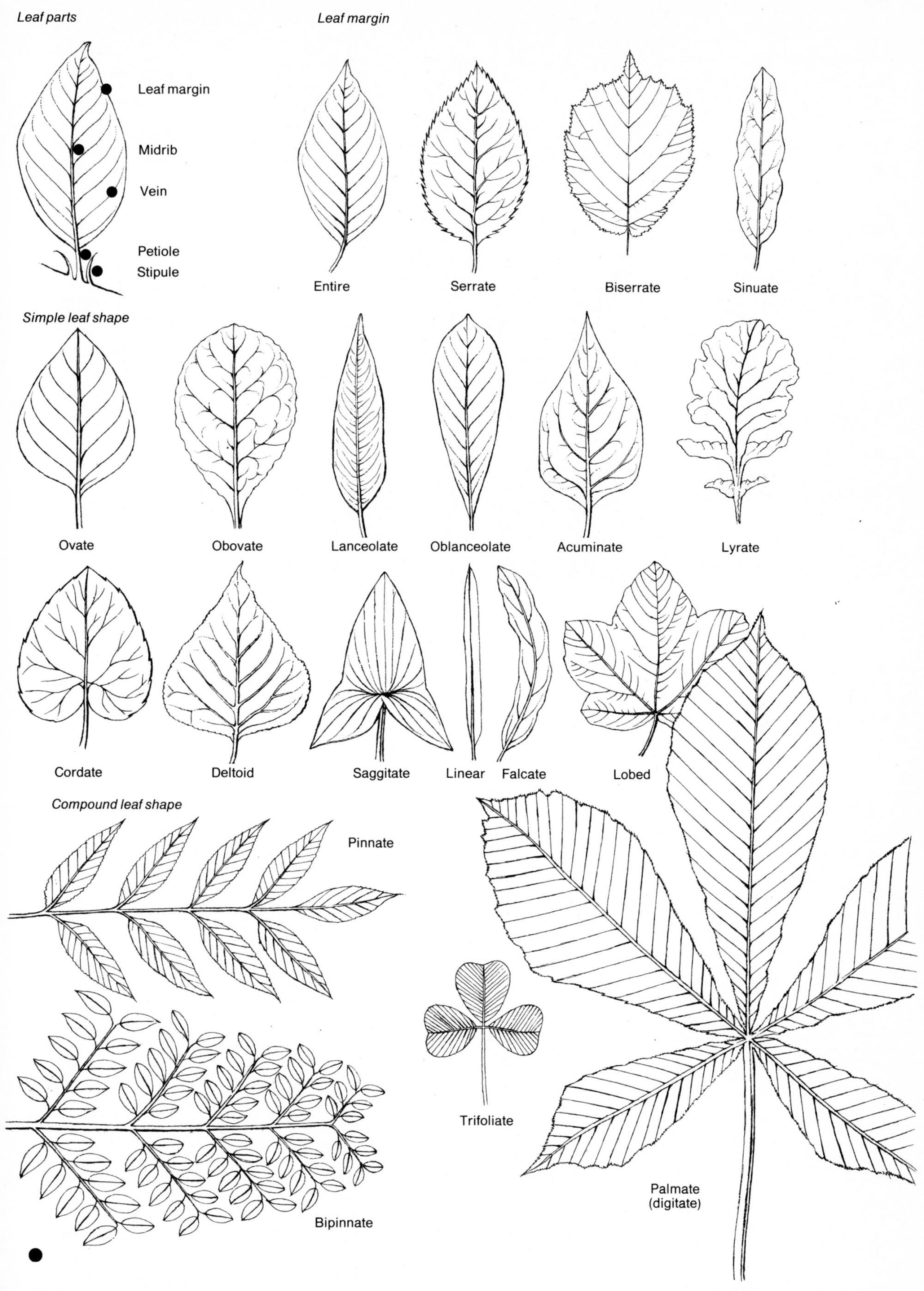
Leaf parts
Leaf margin
Midrib
Vein
Petiole
Stipule
Leaf margin
Entire
Serrate
Biserrate
Sinuate
Simple leaf shape
Ovate
Obovate
Lanceolate
Oblanceolate
Acuminate
Lyrate
Cordate
Deltoid
Saggitate
Linear
Falcate
Lobed
Compound leaf shape
Pinnate
Bipinnate
Trifoliate
Palmate
(digitate)

LEAF MOSAIC The natural arrangement of leaves on the stem to obtain the maximum amount of sunshine. *See also* PHYLLOTAXY.

LECANORA Several kinds of lichen that commonly encrust stone and slate. They form irregularly lobed to almost circular patches, usually grey with paler edges and a granular appearance. The spore-producing APOTHECIA erupt through near the middle in small, dark pustules.

● **LEEK** *(Allium porrum)*. A bulbous perennial in the *Alliaceae* probably derived by man from the wild sand leek, *A. ampeloprasum*. It has GLAUCOUS, KEELED leaves which arch outwards in two ranks and a globular UMBEL of whitish or purplish 6-petalled flowers. It is grown as an annual in the garden, planted deeply or earthed up to whiten (blanch) the leaf bases.

LEGUME The fruit or pod of most members of the pea family *(Leguminosae)* and derived from a single CARPEL. Also used as a general term for the crop plants themselves, particularly peas, beans and clovers.

LEMON
see CITRUS

LEMON-SCENTED VERBENA *(Lippia citriodora* syn. *Aloysia citriodora)*. A large deciduous shrub or small tree to 6m in the *Verbenaceae* from Chile. It has LANCEOLATE, 50–100mm long leaves, usually carried in WHORLS of three, and terminal PANICLES of tiny, 2-lipped, pale purple flowers. The foliage smells strongly of sweet lemons when bruised.

LENTICEL Special pores that form in woody stems, appearing as small raised spots brimming with red-brown, corky powder. They allow air to pass more readily into the inner tissues.

LENTICULAR Lens-shaped; i.e. a disk, thicker in the centre and tapering to the circumference.

LENTIL *(Lens culinaris* syn. *L. esculenta)*. An annual crop plant in the *Leguminosae*, probably from the Eastern Mediterranean, but cultivated for so long that this is no longer certain. It has erect or semi-erect stems to 450mm with PINNATE leaves of 10–12 oblong-lanceolate leaflets. The tiny white or bluish pea-flowers are borne in pairs and each small oblong pod which follows has two peas. The lentils are the cleaned peas split in two. They are high in protein and have long been used as a food by catholics during lent, hence the name. *Lens* has given its name to the familiar glass disk of similar shape to the seeds.

▲ **LEOPARD'S BANE** *(Doronicum pardalianches)*. An herbaceous perennial in the *Compositae* from Europe, formerly much used as a medicinal drug. It is STOLONIFEROUS with long-stalked, broadly OVATE, CORDATE leaves. The 450–900mm stems bear AMPLEXICAUL leaves and bright yellow daisy flowers 44–56mm across. Several similar species and hybrids are grown in gardens.

LEPIDOTE [LEPIDOT|US, -A, -UM] Covered with small scales, e.g. the leaves of some *Rhododendron* species.

LEPTOME
see HYDROME

LETTUCE *(Lactuca sativa)*. An annual or biennial member of the *Compositae* which arose as a mutant or hybrid between the wild prickly lettuce *(L. serriola)* and least lettuce *(L. saligna)*. This is deduced from recent evidence, the original change having happened many centuries ago, a long-headed lettuce being depicted on an Egyptian tomb dated 4,500 BC. Lettuce CULTIVARS can be placed in two groups, the long-headed cos and the round cabbage types. There are also hybrids between the two which combine their various characters (e.g. the popular 'Iceberg' and 'Great Lakes'). Cos lettuce have smooth, erect, oblong leaves of crisp texture; cabbage sorts have rounded, often heavily convoluted and soft textured leaves. When mature, a flowering stem 300–900mm tall appears carrying many small, cylindrical heads of yellow florets. *See also* LAMB'S LETTUCE, PRICKLY LETTUCE, SEA LETTUCE, WATER LETTUCE.

LEUCANTH|US, -A, -UM Having white flowers.

LEUCOPLAST
see CELLS

LEVERWOOD
see IRON WOOD

LIANE Giant woody climbing plants which loop through the high trees of jungles and rain forests.

■ **LICHENS [LICHENES]** A group of plants formed by the partnership of *Fungi* and *Algae*. The fungus forms the main outer body of the plant, while within is a mixture of fungal hyphae and algal cells. The two plants are dependent upon each other, a state known as symbiosis. The fungus obtains the products of PHOTOSYNTHESIS from the alga, while the alga obtains a water supply, mineral salts and perhaps some proteinaceous foods from the fungus. Lichens are very diverse in habit, from the erect, branched mini-tree form of reindeer moss to the flat, hard crusts of LECANORA. They grow on rocks near the wave splash zone, by the sea, on heaths, in woods – both on the ground and on the trees – and high on the mountains. Most of them can stand extreme desiccation for lengthy periods and they are very long-lived. Dog lichen *(Peltigera canina)* forms broad, leaf-like, branching lobes curled up at the edges and creeping over the ground. When moist it is brownish-green and flexible; when dry, papery and white-grey.

LIFE FORMS As an easy means of indicating how a plant passes the winter, the Danish botanist Raunkiaer has classified them into 7 categories according to the position of the resting buds in relation to soil level. The classification roughly follows the existing definitions of trees, shrubs, sub-shrubs, herbaceous perennials, bulbs (together with other forms of underground storage organs) plus annuals, marsh and water plants. *See* PHANEROPHYTES, CHAMAEPHYTES, GEOPHYTES, HELOPHYTES, HEMICRYPTOPHYTES, HYDROPHYTES, THEROPHYTES.

LIGHT Plants are primarily dependent upon light as the energy source for the working of PHOTOSYNTHESIS. Only total parasites and saprophytes can live without light. The actual intensity of light is critical for different plants, but whether they are sun or shade lovers, there is a level below which photosynthesis cannot take place sufficiently to keep them going and etiolation and death follow. The day

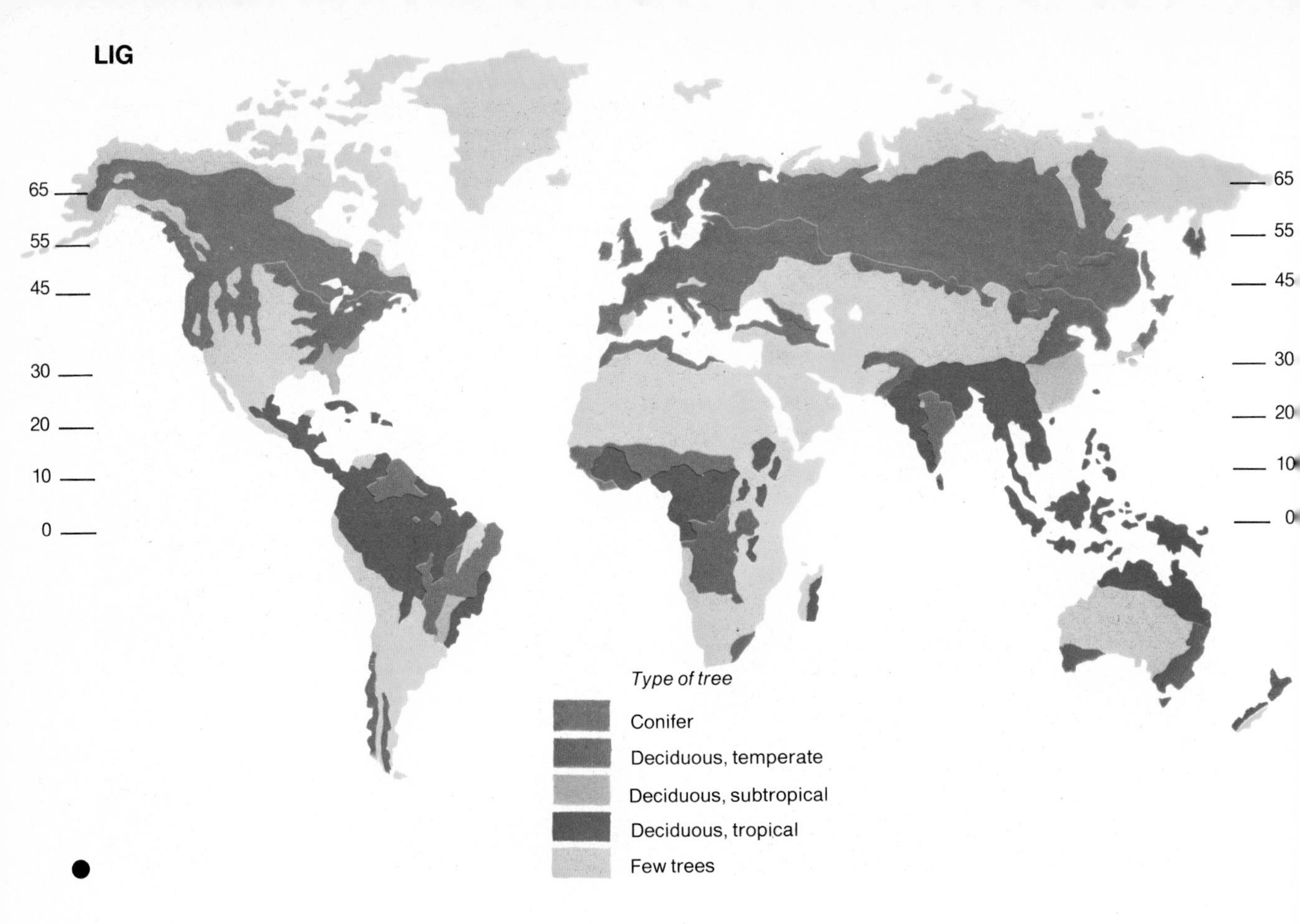

length is also important for some plants, many CULTIVARS of the florist's chrysanthemums will flower only when the days are less than 12 hours long; these are known as short day (or long night) plants. By shortening the day artificially, using blackout materials, chrysanthemums can be had in bloom all the year round. Conversely, such plants as wheat and some sorts of *Iris* require days longer than 12 hours to flower. These are long day (or short night) plants.

LIGNEOUS Woody; strengthened with lignin.

LIGNIN A substance which, when combined with cellulose, forms wood. The chemical nature of lignin is not fully understood.

LIGNUM VITAE *(Guaiacum officinale)*. An evergreen tree to 10m or more in the *Zygophyllaceae* from the W. Indies. It has a smooth white and green, mottled trunk and PINNATE leaves of 4–6, OBOVATE to oval leaflets. The 5-petalled flowers are blue. It provides the hard, heavy, greenish-brown wood and a medicinal resin.

LIGULATE [LIGULAT|US, -A, -UM] A little tongue, e.g. the ray florets of a daisy flower.

LIGULE A collar-like outgrowth at the junction of the sheath and blade of most grasses. It may be like a clerical collar or much taller and in many cases, it is drawn up to a point at the back, e.g. in tufted HAIR GRASS.

LIGULIFLORAE A sub-family of the *Compositae* (daisy family) having all flower heads composed of strap or tongue-shaped florets, e.g. dandelion, chicory.

LILAC *(Syringa)*. A genus of 30 deciduous shrubs or small trees in the *Oleaceae* from S.E. Europe to E. Asia. They show false DICHOTOMOUS branching; the terminal bud of each stem fails to develop and two lateral buds grow on. The leaves are usually OVATE, entire, rarely PINNATE and the small, tubular, 4-lobed flowers are in large terminal THYRSES. Common lilac *(S. vulgaris)* is a large shrub or small tree to 6m with ovate-cordate 50–125mm long leaves and blue-purple, fragrant flowers in 150–200m clusters. Many CULTIVARS are known with blue, purple or rose, single or double flowers.

LILACIN|US, -A, -UM Lilac coloured.

LILY Many plants are called lilies, but strictly the name refers to members of the genus *Lilium*, containing 80, bulbous-rooted species in the *Liliaceae* from the northern hemisphere. They have fragile bulbs composed of small, separate scales and erect, unbranched stems. The flowers may be trumpet, bowl-shaped or like a turk's cap (the 6 TEPALS rolled back on themselves). They grow in shades of red, orange, yellow, white and purple and are often spotted. The common turk's cap *(L. martagon)* grows 0.6–1.2m tall with WHORLS of OBLANCEOLATE leaves and RACEMES of red-purple or white flowers; tiger lily *(L. tigrinum)* grows to 1.2m or so with orange-red flowers 63–100mm across having reflexed (but not rolled) purple-black, spotted tepals; Nankeen lily *(L. X testaceum)* is a hybrid between the Madonna and Chalcedonian lilies with pale, orange-yellow flowers on 1.5m stems; Madonna lily *(L. candidum)* has wide trumpets of pure white, fragrant flowers on 0.9–1.2m stems; Easter lily *(L. longiflorum)* has shapely, fragrant white trumpets to 150mm long on robust,

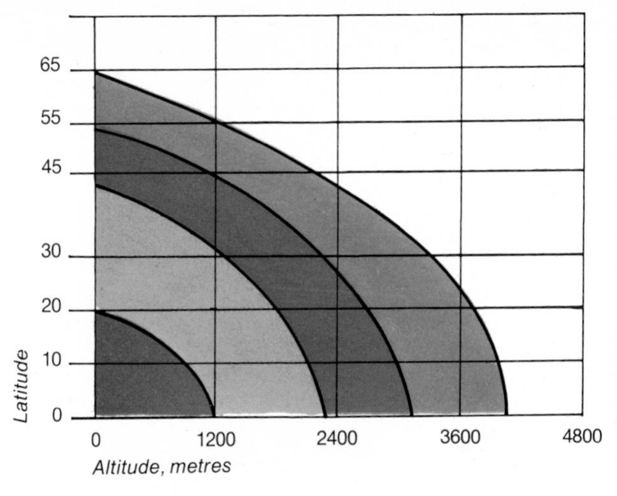

As latitude increases towards the poles the altitude at which the various tree types grow decreases

▲

600–900mm stems; gold-rayed lily *(L. auratum)* has 250mm wide bowl-shaped, fragrant, white flowers, each petal having a central yellow or crimson stripe and deep purple spots.

LILY OF THE VALLEY *(Convallaria majalis)*. A RHIZOMATOUS perennial in the *Liliaceae* from Europe to N.E. Asia. Each shoot has 1–3, 100–200mm long, ovate, lanceolate leaves and an arching tipped raceme of waxy, pure white fragrant bells. The fruit is a red berry.

LIMA BEAN
see BUTTER-BEAN

LIMB The broad, coloured part of a stalked petal.

LIME [LINDEN] *(Tilia)*. A genus of 50 deciduous trees in the *Tiliaceae* from the N. temperate zone, south to Mexico and Indochina. They have broadly OVATE, usually obliquely CORDATE based leaves and pendulous CYMES of 5-petalled flowers. The latter has a wing-like BRACT attached and when the capsules are ripe the whole cluster breaks free in a strong wind. Common lime *(T. X europaea)* is a hybrid between small-leaved lime *(T. cordata)* and large-leaved lime *(T. platyphyllos)*. It grows to 24.4m or more with 50–88mm long leaves and yellow white flowers. Like all limes it is a good bee flower.

● **LIMIT OF TREES** That point of altitude or exposure beyond which trees cannot grow. At the equator it may be many thousand feet up in the mountains, but in Arctic regions is at sea level. Wind and rainfall are important factors in deciding the upper limit.

LIMOS|US, -A, -UM Growing in muddy places.

LINDEN
see LIME

LINEAR [LINEAR|IS, -E] Narrow, with the sides parallel or almost so.

LINEATE [LINEAT|US, -A, -UM] Marked with fine parallel lines.

LING
see HEATHER

LINGULAT|US, -A, -UM [LINGUIFORM|IS, -E] Tongue-shaped.

LINIFOLI|US, -A, -UM Flax *(Linum)* leaved.

▲ **LINNAEUS, CARL [CARL VON LINNÉ] (1707–78)** Swedish botanist and naturalist born at Reashult, Småland. He first studied theology but soon turned to botany and medicine, eventually graduating as M.D. at Leyden in Holland. In 1732 he journeyed in Lapland and later wrote *Flora Lapponica*. From 1741 until his ceath he was professor of botany at Uppsala where he became an inspiring teacher of great influence, several of his students later becoming botanical leaders in their own right. Linnaeus is best remembered as a pioneer in the systematic classification of all objects in the natural world. As early as 1735 he produced *Systema Naturae*, the first of a series of encyclopedic works which described and classified in a systematic way all the plants (also insects and minerals) known to him at that time. Plants were classified according to his sexual system which meant grouping by the number of stamens and pistils in each flower. Although essentially an artificial system, it was successful in distinguishing plant entities we now call genera and species. In the course of time the sexual system became modified and then replaced by more natural classifications.

His major contribution to biology was the binomial system (*see* NOMENCLATURE), which gave each plant or animal a mere two names in place of the existing unwieldy Latin sentences. Linnaeus's fame as a writer and teacher made Uppsala a mecca for the botanical students of Europe; he was created a Knight of the Polar Star in 1753 and ennobled in 1761, taking the name von Linné. His epitaph in Uppsala cathedral reads simply 'Princeps Botanicorum' (Prince of Botanists).
Main books: *Systema Naturae* (1753; facsimile: Vol. 1, London 1956, Vol. 2, Weinheim 1964); *Bibliotheca Botanica, Fundamenta Botanica* (1736); *Critica Botanica, Flora Lapponica* (1737); *Flora Suecica* (1745); *Flora Zeylanica* (1747); *Species Plantarum* (1753; facsimile: London 1957–70); *Systema Vegetabilum* (1774).

LINT The natural 'wool' in the cotton seed pod.

LIP (1) An alternative name for the LABELLUM of orchids. (2) The two divisions at the mouth of a LABIATE flower; e.g. the tubular flowers of the dead-nettle are divided into a hooded upper part which protects the stamens and a lower, flattened lobe as an alighting platform for insects.

LIQUIDAMBAR A genus of 6 deciduous trees in the *Altingiaceae* from N.E. America, Turkey, S.E. China and Taiwan. They yield a useful timber and a fragrant gum resin and are grown as ornamentals. Common liquidambar or sweet gum *(L. styraciflua)* grows to 14.4m or more with 5–7 lobed, PALMATE, 75–100mm wide leaves which turn to shades of yellow, red and purple in autumn. The petal-less flowers are MONOECIOUS, the males in short catkins, the females in globular heads which become aggregate fruits of narrow, woody capsules.

LIQUORICE *(Glycyrrhiza glabra)*. An erect perennial in the *Leguminosae* from S. Europe to W. Asia. It is a RHIZOMATOUS plant with stems to 1.2m bearing PINNATE leaves and AXILLARY RACEMES of small, purple-blue pea-flowers; liquorice is obtained from the roots by boiling and steaming.

LITCHI [LEECHEE, LITCHEE, LYCHEE] *(Litchi chinensis)*. An evergreen tree to 12.2m in the *Sapindaceae* from S. China. It has pinnate leaves of 5–9 LANCEOLATE leaflets and terminal PANICLES of 5-petalled, whitish, starry flowers. The ovoid, 25–38mm long, red-brown fruits have a shell-like, prickly, warty rind, a jelly-like whitish pulp (the aril) with combined flavours of grape and melon, and a glossy-brown seed.

LITMUS *(Rocella tinctoria)*. A lichen from the rocky coasts of the Mediterranean, W. Europe, the Canary Is. and S. America. It is tufted in habit with slender, cylindrical, brownish, sparingly branched fronds. It has been used since Roman times to produce a purple dye. When treated with ammonia, it yields litmus, the indicator which turns red in acid and blue in alkaline solutions.

LITTORAL|IS, -E Growing on the sea-shore.

LIVE-LONG
see STONECROP

LIVERWORT *(Hepaticae)*. Primitive spore-bearing plants which, with the mosses *(Musci)*, make up the order *Bryophyta*. There are two distinct sorts: those with flattened, often lobed stem-like structures (thallose) and those with prostrate thread-like stems bearing minute leaves rather like mosses (foliose). In both cases they represent the GAMETOPHYTE generation and are equivalent to the PROTHALLUS of the fern. The fertilised ovum remains attached to the plant, and produces a SPOROPHYTE composed of a stalk and a spore capsule. The latter release their spores by splitting back into four flaps or valves. A commonly seen thallose liverwort is *Marchantia polymorpha* with waved and branched ribbon-like stems which form mats in damp, shady spots and can be a nuisance smothering the soil of pot plants grown outside. They have star-shaped spore capsules and tiny cup-like organs (gemmae cups) which bear tiny green seed-like objects called gemmae. These are the equivalent of BULBILS in higher plants and quickly grow into new plants.

LIVID|US, -A, -UM Originally a dull, bluish-grey, but now also blues and reds darkened with grey or black.

● **LIVING STONES** A name given to various SUCCULENT plants from the stony deserts of S. Africa which mimic their surroundings. Best known is the genus *Lithops* in the *Aizoaceae* with 50 species, some clump-forming, some solitary. Each shoot or plant is made up of two grossly swollen leaves which are virtually fused together and are coloured to look just like a pebble. During the hottest months, the leaf pair withers and is later replaced from within by another pair which burst through the dried remains. Large daisy flowers are borne from between the leaf pair.

LIZARD ORCHID *(Himantoglossum hircinum)*. A tuberous ground-dwelling plant in the *Orchidaceae* from Europe and N. Africa. It has elliptic-oblong 75–150mm leaves and a 200–375mm stem bearing a dense RACEME of greenish and whitish flowers with purple spots and streaks. The LABELLUM of each flower is bizarrely shaped with two slender side lobes like legs and a very long, thin middle lobe from between them like a tail. The remaining PERIANTH segments arch forwards like a 'head'.

LOBE A section of a leaf, bract, sepal or petal which stands out like a cape or isthmus from the main part of the organ. In some leaves which are deeply and regularly lobed, the organ may be entirely taken up with lobing. It is used also for the petal-like divisions at the mouth of a tubular flower.

LOCULICIDAL Splitting down the middle of each CARPEL in the fruit when ripe.

LOCULUS A section of the ovary; usually representing one CARPEL in a SYNCARPOUS ovary.

LOCUST or BLACK LOCUST TREE
see FALSE ACACIA

LODICULE
see GRASS

LOGANBERRY *(Robus X loganobaccus)*. A scrambling, thorny shrub in the *Rosaceae*, reputedly a hybrid between the American blackberry *(R. vitifolius)* and a garden raspberry. It is said to have arisen in the garden of Judge Logan in California in 1881. It has TRIFOLIATE leaves with OVATE, toothed or lobed leaflets to 100mm long, and small PANICLES of 5-petalled, whitish or pinkish 25mm wide flowers. The bluntly conical dark red fruits are aggregates or DRUPELETS.

LOMENTUM A legume-type seed pod constricted between the seeds and breaking up into one-seeded sections when ripe.

LONGAEV|US, -A, -UM Long-lived.

LONG AND SHORT SHOOTS The characteristic of certain trees and shrubs to form two distinct sorts of growths; rapidly extending shoots with well-spaced leaves, and short slow-growing ones having more densely borne leaves. Fruit trees are an example of this where the short shoots form the fruiting spurs. The pine provides a different example, the short shoots bearing the leaves, the long extending the size of the plant.

LOOFAH *(Luffa cylindrica* syn. *L. aegyptica)*. A climbing annual in the *Cucurbitaceae* from the tropics. It has rounded, often angular or lobed leaves about 100mm wide, and MONOECIOUS, yellow flowers like those of a cucumber. The cylindrical to club-shaped fruits are 150–300mm long and have a tough, fibrous skeleton which is rather like a sponge when cleaned of pulp and is also known as both sponge or dish-cloth gourd.

LOOSESTRIFE Two unrelated, herbaceous perennials. Yellow loosestrife *(Lysimachia vulgaris)* is a member of the *Primulaceae* from Europe and Asia. It is a RHIZOMATOUS plant with erect stems 0.6–1.2m high bearing WHORLS of 3–4, LANCEOLATE, 50–100mm long leaves and terminal PANICLES of 5-petalled, yellow flowers. Purple loosestrife *(Lythrum salicaria)* belongs to the *Lythraceae* and is native to the north temperate zone. Has stems to 1.2m, opposite pairs of SESSILE, lanceolate leaves and a tapering spike of dense whorls of small, purple flowers. Both are natives of wet soils, often by water.

LOQUAT *(Eriobotrya japonica)*. An evergreen tree to 9.0m in the *Rosaceae* from China, but much grown in other sub-tropical countries. It has leathery, glossy, 150–300mm long lanceolate leaves, strongly ribbed and woolly beneath, and panicles, of 5-petalled 19mm wide, cream flowers on woolly stalks. The small, pear-shaped fruits are yellow and may be eaten raw, cooked or made into jam, wine or liqueur.

LORATE Strap-shaped, e.g. leaves with parallel margins.

LORDS AND LADIES [CUCKOO PINT, WILD ARUM] *(Arum maculatum)*. A tuberous-rooted perennial in the *Araceae* from Europe and N. Africa. It has a cluster of 75–200mm, triangular-HASTATE, long-stalked leaves direct from the tuber, and a pale green SPATHE to 250mm long with a club-shaped dull purple (rarely yellow) SPADIX.

● **LOTUS [SACRED LOTUS]** *(Nelumbo nucifera* syn. *Nelumbium nuciferum).* A robust, aquatic plant in the *Nelumbonaceae* from tropical Asia. It has a thick, tuberous RHIZOME which is edible and 0.6–1.8m long leaf stalks bearing circular, PELTATE blades to 600mm across. The fragrant cup-shaped pink or pink and white flowers can be 300mm across. This is a sacred plant in India and China and traditionally the flower in which Buddha sits. It is often confused with the sacred lotus of Egypt which is considered to be *Nymphaea lotus*, a true water lily.

LOUSEWORT *(Pedicularis sylvatica).* A small perennial plant in the *Scrophulariaceae* from W. and C. Europe. Of tufted growth with decumbent stems and LINEAR, PINNATISECT leaves. The tubular, 2-lipped, 25mm long pink flowers are carried in short, erect RACEMES.

LOVAGE *(Ligusticum scoticum).* A bright green perennial in the *Umbelliferae* from the rocky coasts of N.W. Europe. It has TERNATE or biternate leaves with OVATE, 32–50mm leaflets, the tips toothed. The tiny greenish-white flowers are born in UMBELS on 150–750mm tall stems.

LOVE APPLE
see TOMATO

LOVE-IN-A-MIST *(Nigella damascena).* Sometimes called by the curiously opposite name Devil-in-a-bush, this is an annual plant in the *Ranunculaceae* from the Mediterranean region. It has 300–600mm, wiry, erect stems and leaves cut into thread-like segments. The terminal, cup-shaped blue flowers are almost hidden by long INVOLUCRAL BRACTS as finely cut as the leaves.

LOVE-LIES-BLEEDING
see AMARANTH

LUCENS Glistening, shining as if polished.

LUCERNE [ALFALFA] *(Medicago sativa).* A bushy, erect or ascending perennial in the *Leguminosae* probably from the Mediterranean and W. Asia, but much grown elsewhere as stock feed. It grows 300–600mm or more tall, with TRIFOLIATE leaves having narrowly OBOVATE leaflets and dense, 19mm long RACEMES of small purple pea-flowers. The pod is wound into a two to three times spiral.

LUCID|US, -A, -UM Clear, transparent, shiny.

LUNAT|US, -A, -UM [LUNULAT|US, -A, -UM] Half-moon or crescent-shaped.

LUNGWORT *(Pulmonaria).* A genus of 10 species of perennials in the *Boraginaceae* from Europe and W. Asia. They are RHIZOMATOUS with large, LANCEOLATE to OVATE basal leaves and tubular flowers in CYMES which open pink and age to blue. Common lungwort *(P. officinalis)* has broadly ovate pubescent leaves, often white spotted and 9–13mm long flowers.

▲ **LUPIN [LUPINE]** *(Lupinus).* A genus of 200 species of annuals, perennials and shrubs from N. and S. America and the Mediterranean. Some species are naturalised elsewhere. They have DIGITATE leaves with up to 15 narrow leaflets and terminal RACEMES of pea-shaped flowers. The pods split with a violent twist scattering the seeds afar. Tree lupin *(L. arboreus)* is a bush to 1.8m with the leaves having

▲

6–9 leaflets and 100mm racemes of blue, yellow or purple fragrant flowers. The popular lupins of gardens are hybrids between this and the herbaceous *L. polyphyllus*, a 0.9–1.5m species with 10–17 leaflets and racemes 300–600mm long.

LUPULIN|US, -A, -UM Hop-like.

LURID|US, -A, -UM Dirty brown to smoky yellow.

LUTESCENS Yellowish, becoming yellow.

LUTE|US, -A, -UM Deep or buttercup yellow.

LYME GRASS *(Elymus arenarius)*. An erect, 0.9–1.8m tall plant in the *Gramineae* from the north temperate zone. A rhizomatous species from coastal sand dunes, with broad, rigid, GLAUCOUS leaves and dense 150–300mm long flower spikes not unlike those of wheat.

LYRATE Lyre-shaped; leaves with a broad, rounded apex and several small, lateral lobes towards the base.

MACASSAR OIL [YLANG-YLANG] *(Cananga odorata)*. An evergreen tree to 24.5m in the *Annonaceae* from Malaysia but widely grown in the tropics. It has elliptic to lanceolate, 75–200mm long leaves and 100mm wide yellow, fragrant flowers with 6 narrow, pointed petals. The scented ylang-ylang or macassar oil is extracted from the flowers and used in perfumery and soaps. It was once used as a hair dressing, hence the antimacassars of Victorian times.

MACE
see NUTMEG

MACRANTH|US, -A, -UM Having large flowers.

MACROPHYLL|US, -A, -UM Having large leaves.

MACULAT|US, -A, -UM Spotted or blotched.

MADDER *(Rubia tinctoria)*. A scrambling 1.2–1.8m perennial in the *Rubiaceae* from the Mediterranean and W. Asia. It has 4-angled stems and WHORLS of 4–6 rigid LANCEOLATE leaves at each NODE. The small yellow flowers have 5 pointed lobes and are followed by red-brown, berry-like fruits. The dried and powdered roots yield the dye turkey red, still used by artists but superseded elsewhere by a synthetic substitute.

MADWORT *(Alyssum)*. A genus of 150 annuals and perennials in the *Cruciferae* from the Mediterranean region and Asia. Most are low-growing with simple grey or green leaves and RACEMES of small, 4-petalled yellow, white or purplish flowers. Several are cultivated as ornamentals: sweet alyssum *(Lobularia maritima* syn. *Alyssum maritimum)* is a 75–150mm bushy annual with white, lilac or purple flowers; gold dust *(A. saxatile)* is a sub-shrubby, 200–250mm tall perennial with spathulate, grey-hairy leaves and abundant, tiny golden flowers.

MAGNOLIA A genus of 80 species of deciduous and evergreen shrubs and trees in the *Magnoliaceae* from east N. America, and east Asia, south to Borneo and Java. They have LANCEOLATE, OVATE or OBOVATE leaves and erect

●

or nodding, bowl or chalice-shaped flowers. The cylindrical fruits are made up of many CARPELS each with 2 red-ARILLED seeds: southern magnolia or bull bay *(M. grandiflora)* is an evergreen tree to 24.5m in its native S.E. America with rusty-backed, glossy, leathery, 150–250mm long elliptic to obovate leaves. The creamy-white, fragrant, cup-shaped flowers can attain 250mm in width; yulan *(M. denudata)* is a deciduous tree to 12.2m with obovate 75–150mm leaves and erect, chalice-shaped white flowers before the leaves. Crossed with the shrubby, purple-flowered *M. liliflora*, it has produced *M. X soulangiana*, a popular garden plant with white and purple flowers; Chinese magnolia *(M. sinensis)* grows to 6m with oval, 75–175mm leaves and nodding 75–100mm wide white, bowl-shaped flowers bearing crimson stamens.

MAHOGANY *(Swietina mahagoni)*. An evergreen tree to 30m in the *Meliaceae* from Central America, W. Indies and Florida. It has PINNATE leaves and panicles of tiny, yellowish or red-flushed, 5-petalled flowers in AXILLARY PANICLES. The heavy, hard wearing red-brown wood is used for quality furniture. Several other timbers of similar appearance are passed off as mahogany.

MAIDENHAIR FERN *(Adiantum)*. A genus of 200 evergreen and deciduous perennial ferns in the *Adiantaceae* of cosmopolitan distribution, particularly in tropical America. They are tufted in growth with very slim, wiry leaf stalks, often black and glossy, and pinnate to quadripinnate fronds, composed of numerous stalked, asymmetric leaflets. The SORI are carried on the margins of the leaflets (PINNAE): common maidenhair *(A. capillus-veneris)* has bipinnate fronds to 300mm with irregularly fan-shaped pinnae: northern maidenhair *(A. pedatum)* has 300–600mm palmately forking fronds, each 'finger' composed of broad, crescent-shaped pinnae.

MAIDENHAIR TREE *(Ginkgo biloba)*. A primitive deciduous tree in the *Ginkgoaceae* reaching 30m; it is allied to the conifers and comes from China. It has stalked, 50–75mm wide fan-shaped leaves and DIOECIOUS flowers, the males in short catkins, the females consisting of two naked OVULES only. The fruit is yellowish with a single nut-like seed containing an edible fruit. This tree relies upon man for its survival and is no longer convincingly wild even in its native China.

● **MAIZE [INDIAN CORN, SWEET CORN]** *(Zea mays)*. An erect, unbranched member of the *Gramineae* from Central America, but widely grown in warm countries. It grows to 2.4m with alternate pairs of broad leaves 50–100mm wide. The flowers are MONOECIOUS, the male spikelets forming a large terminal PANICLE, the female aggregated into cigar-shaped spikes (cobs), covered with sheathing leaves and set low down on the plant. From the tip of each cob hangs a tassel of stigmas which catch the wind-blown pollen. There are several distinct sorts of maize: pop corn, with small hard grains having a soft centre which explode on heating; flint maize, with larger and only slightly less hard grains, used for poultry feeding; dent and flour maizes are similar having a soft starchy centre and are used for milling; sweet corn has a sweet endosperm and is cooked and eaten before it matures, as corn on the cob. There are also CULTIVARS with coloured grains or variegated leaves grown as ornamentals.

MAJAL|IS, -E Flowering in May.

▲ **MALLOW** A name given to several members of the mallow family *(Malvaceae)*. Musk mallow *(Malva moschata)* is a 450–750mm perennial from Europe and N. Africa. It has 50–75mm wide PALMATELY lobed leaves, the upper deeply so, and 5-petalled, rose-pink flowers 32–50mm wide: marsh mallow *(Althaea officinalis)* grows by the sea, usually in marshes, with stems 0.6–1.5m tall, rounded, shallowly 3–5 lobed leaves and funnel-shaped, 5-petalled, rose-purple flowers, 25–38mm across and borne in leafy RACEMES, the roots yielding a mucilage used in confectionery and as a medicine; tree mallow *(Lavatera arborea)* is a biennial with a woody stem persisting the first winter. In bloom it can reach 2.4–3.0m, with rounded, CORDATE, shallowly-lobed leaves to 200mm wide, and widely funnel-shaped 5-petalled, 38mm wide rose-purple flowers having darker veins. It comes from the Mediterranean and Atlantic coasts of Europe and N. Africa.

MAMMEE APPLE *(Mammea americana)*. An evergreen tree to 18.2m in the *Guttiferae* from tropical America and the W. Indies. It has 125–200mm long, OBOVATE leaves and fragrant white 38mm wide flowers with 4–6 petals and many stamens. The globose, 75–150mm diameter fruits have a rough, russet skin and a bright yellow sweet and scented flesh; also known as St Domingo apricots, they are used to prepare the liqueur Eau de Creole.

MAMMOTH TREE
see BIG TREE

MANDIOC
see CASSAVA

MANDRAKE *(Mandragora officinarum)*. A rosette-forming perennial in the *Solanaceae* from S. Europe and S.W. Asia. It has broadly lanceolate, somewhat wrinkled leaves to 300mm or more long and purplish, 5-lobed, bell-shaped flowers. The latter open with the young leaves in the spring and are followed by yellowish berries.

MANGEL-WURZLE
see SUGAR BEET

MANGO *(Mangifera indica)*. An evergreen tree to 18.2m in the *Anacardiaceae* from S.E. Asia. It has narrowly oblong-lanceolate leaves to 200mm and terminal panicles of tiny, 4 or 5-petalled whitish flowers. The 75–100mm long ovoid or slightly kidney-shaped fruits are yellow-green, sometimes red-tinged with orange-yellow, sweet juicy flesh, but are often somewhat fibrous and have a large seed. There are several CULTIVARS.

MANGOLD
see SUGAR BEET

MANGOSTEEN *(Garcinia mangostana)*. An evergreen tree to 13.7m in the *Guttiferae* from Malaysia. It has elliptic-oblong to ovate 150–200mm long leaves and small clusters of red flowers with 4-cupped petals. The globular brown-purple fruits are the size of small oranges with a similar thick rind. The whitish pulp is divided into about 5 segments and has a somewhat treacly flavour.

MANGROVES Forests that form over the tidal mud of river estuaries in the tropics and subtropics. Several species are involved, but all are characterised by having either aerial roots from trunks or branches, or breathing roots standing out of the mud. In many cases, the seeds germinate in the fruit and fall in the water with a well-developed root ready to grow quickly once lodged in a suitable place. The most important genus is *Rhizophora*, with 7, evergreen species in the *Rhizophoraceae*. They have ovate leaves, 4-petalled flowers and aerial roots from trunk and branches. *Avicennia* has 14 species in the *Avicenniaceae*, some of them forming mangroves on the edge of the subtropics, e.g. *A. resinifera* in New Zealand. They have small, yellowish 5-petalled flowers and erect, breathing roots.

MANIOC
see CASSAVA

MAN ORCHID *(Aceras anthropophorum)*. A tuberous-rooted perennial in the *Orchidaceae* from Europe and N. Africa. It has a cluster of OBLONG-LANCEOLATE leaves 75–125mm long, and a 200–375mm RACEME of small, greenish-yellow flowers often tinged red-brown. The hooded petals of each blossom form a head and the hanging, barely 13mm long LABELLUM looks just like a cut-out doll, with narrow, pointed arms and legs.

MANZANITA *(Arctostaphylos manzanita)*. An evergreen shrub to 3.6m or more in the *Ericaceae* from California. It is of rather crooked growth with leathery, ovate to oblong 25–63mm long leaves. The white or pink, urn-shaped flowers are carried in pendent, terminal PANICLES and the 8mm berries are reddish brown. Several other *Arctostaphylos* species are known as manzanita.

MAPLE *(Acer)*. A genus of 200 deciduous trees (a few bushy or semi-evergreen) in the *Aceraceae* from the northern hemisphere. Leaves may be OVATE, entire or PALMATELY 3–13 lobed; a few species have PINNATE leaves. The small yellow, purple, red or green flowers are borne in RACEMES, CORYMBS or UMBELS. In some species they have petals. The fruits are formed of two winged CARPELS known as keys which split apart when ripe and spin in a strong wind like propellers. Field maple *(Acer campestre)* grows to 7.5m or more with 38–75mm wide bluntly 5-lobed leaves and corymbose clusters of yellow-green flowers; Japanese maple *(A. palmatum)* grows 6m or more with 50–100mm wide leaves having 5, slender pointed lobes and purple flowers in small corymbs; Norway maple *(A. platanoides)* grows to 27.3m with 100–175mm wide leaves having 5 broad, widely-toothed lobes and showy, bright green-yellow flowers in CORYMBS; sugar maple *(A. saccharum)* has similar leaves on a tree to 30m or more and petal-less flowers in small clusters on thread-like stalks before the leaves; the sap is tapped in spring and yields maple syrup after boiling; red maple *(A. rubrum)* grows to the same height with 3–5 lobed leaves each 75–100mm long and small red flowers before the leaves; sycamore maple *(A. pseudoplatanus)* grows to above 27m with 5-lobed, 100–200mm wide leaves and narrow, pendulous PANICLES of yellow-green flowers; box elder *(A. negundo)* grows to 18m or more with PINNATE leaves, of 5, OVATE, toothed leaflets and yellow-green flowers in pendulous RACEMES.

MARE'S TAIL *(Hippuris vulgaris)*. An aquatic perennial in the *Hippuridaceae* from Europe, Asia and N. Africa. It has a stout, creeping RHIZOME and erect, unbranched stems to 750mm or occasionally more. The LINEAR leaves are carried in WHORLS of 6–12 at each node, those on submerged stems are longer and translucent. Tiny, APETALOUS flowers are borne on aerial stems.

MARGINAL|IS, -E [MARGINAT|US, -A, -UM] Having a distinctive margin.

MARIGOLD A name given mainly to several members of the daisy family *(Compositae)* but primarily to the common or pot marigold *(Calendula officinalis)*, a bushy annual from S. Europe. It grows to 900mm with oblong-ovate leaves and orange daisy-flowers heads 50–75mm wide. Double flowered CULTIVARS are often seen in gardens. *See also* AFRICAN, FRENCH, BUR, CORN and MARSH MARIGOLDS.

MARIN|US, -A, -UM Growing close to or in the sea.

MARIPOSA LILY *(Calochortus)*. Also known as star lily, this is a genus of 60, mainly woodland, bulbous plants in the *Liliaceae* from west N. America and Mexico. They are slender, tulip-like plants with LINEAR to narrowly LANCEOLATE leaves and erect or nodding, cup-shaped or globular flowers. Each flower is composed of 6 perianth segments; in some species the outer 3 are narrow and sepal-like, in others they are coloured as petals. The larger inner segments are often beautifully marked and fringed.

MARITIME PINES
see PINE

MARITIM|US, -A, -UM Growing near the sea.

MARJORAM Several species of *Origanum*, perennials in the *Labiatae* from Europe, the Mediterranean and N. Africa. They have opposite pairs of OVATE leaves and spikelets of small, tubular, 2-lipped flowers in PANICLES and CORYMBS: garden, knotted or sweet marjoram *(O. marjorana)* is sub-shrubby and grows to 600mm with purple or white flowers in oblong spikelets; pot marjoram *(O. onites)* is sub-shrub to 300mm with erect, seldom branched stems and whitish flowers in ovoid spikelets which are arranged in corymbs. This and garden marjoram are used as culinary herbs for flavouring stuffings, etc.; common marjoram *(O. vulgare)* is used in herbal medicines. It is a herbaceous perennial to 750mm with oblong spikelets of rose-purple flowers in panicles.

MARMORAT|US, -A, -UM Marbled; irregularly striped or veined with a contrasting colour.

● **MARRAM GRASS** *(Ammophila arenaria)*. A RHIZOMATOUS, perennial plant of sand dunes in the *Gramineae* from the coasts of western Europe. It has 0.6–1.2m stems and rolled, quill-like leaves. The spikelets form stout, spike-like panicles 63–150mm long. A common sand-binding grass.

MARROW
see CUCURBITS

MARSH A wet area such as may be found on the margins of a lake or a large pond. The underlying soil is of a mineral nature, not peat as is the similar FEN. Where the land is covered with water, reeds flourish and form reedswamp, but where the water drops, at least in summer, characteristic marsh plants are great tussock SEDGE, yellow IRIS, MEADOWSWEET, purple LOOSESTRIFE, MEADOW-RUE, MARSH CINQUEFOIL, MARSH FERN, etc. If not grazed, marshland becomes colonised with such trees as grey and white WILLOWS, BIRCH and ALDER, and becomes wet woodland.

MARSH CINQUEFOIL *(Potentilla palustris* syn. *Comarum palustre)*. A rhizomatous perennial in the *Rosaceae* from the north temperate zone. Frequents marshes, fens, bogs and wet heaths and grows 150–450mm in height. It has PALMATE or PINNATE leaves with 5–7 oblong to elliptic, coarsely toothed, 32–56mm long leaflets. The flowers are borne in terminal CYMES. All parts of the flowers are purple or deep purple including the 5 petals, sepals and EPICALYX segments.

MARSH FERN *(Thelypteris palustris)*. A RHIZOMATOUS fern in the *Thelypteridaceae* from wet land in the north temperate zone, S. Africa and New Zealand. The 30–120cm pinnate, erect fronds are light green and emerge singly or rarely in loose tufts. The LANCEOLATE PINNAE are PINNATIFID, each lobe of the fertile ones bearing a continuous row of sori each side of the mid-vein.

MARSH MARIGOLD [KINGCUP] *(Caltha palustris)*. A perennial plant of wet ground in the *Ranunculaceae* from the north temperate zone. Has long-stalked, rounded to almost triangular-ovate leaves and golden yellow, 19–20mm wide, cup-shaped flowers of 5–8 TEPALS on stems of 300mm or so.

MARSH ORCHIDS Several species of *Dactylorhiza* (syn. *Dactylorchis*), natives of marshes, fens and damp meadows, members of the *Orchidaceae* from Europe, Asia, N. Africa and the Canary Is. They have fingered, tuberous roots and lanceolate, sometimes dark-spotted leaves which become narrower up the erect stems. The flowers are carried in dense terminal spikes, each one having 3 spreading sepals, 2 arching petals and a 3-lobed, spurred LABELLUM: common spotted orchid *(D. fuchsii)* varies from 150–500mm in height with blotched leaves and white or pink flowers, the labellum marked with red veins.

MARSH PENNYWORT *(Hydrocotyle vulgaris)*. A slender, creeping perennial of wet ground in the *Hydrocotylaceae* from Europe and N. Africa and naturalised in New Zealand. It has circular, PELTATE, round-toothed leaves and stalked, 2–5 flowered UMBELS of tiny, pinkish-green flowers from the leaf AXILS. *See also* NAVELWORT.

MARVEL OF PERU
see JALAP

MAST A vernacular name for the nuts of beech, oak and sweet chestnut.

MASTIC TREE *(Pistacia lentiscus)*. An evergreen shrub or small tree to 6m in the *Anacardiaceae* from S. Europe. It has PINNATE leaves with 4–12, narrowly OBLONG to OBVATE leaflets to 38mm long and small, dense catkin-like PANICLES of tiny red flowers. The resin mastic is exuded when the bark is cut or wounded. It is used in medicine and varnish-making. American or Peruvian mastic, also known as the Californian pepper tree *(Schinus molle)*, is an evergreen tree to 15m in the same family, from S. America and much planted for its elegant weeping habit. The 125–225mm long, pinnate leaves have linear-lanceolate leaflets and the tiny yellow-green flowers are borne in 25–63mm panicles, to be followed by pea-like red fruits.

MATÉ [YERBA MATÉ, BRAZILIAN TEA] *(Ilex paraguariensis)*. An evergreen tree to 6m in the *Aquifoliaceae* from Brazil. It has 50–125mm coarsely toothed, obovate leaves, small AXILLARY clusters of tiny greenish flowers and globular red berries. The leaves are smoked, dried, roasted and ground, then brewed like ordinary tea. It is an important beverage in Brazil.

MAT GRASS *(Nardus stricta)*. A tufted, wiry perennial in the *Gramineae* from Europe, N. Asia and Greenland. It has rough, bristle-like leaves, and slender stems to 300mm with a one-sided spike of very narrow spikelets. It is rejected by sheep because of its harsh foliage.

MATRIMONY VINE Another vernacular name for the DUKE OF ARGYLL'S TEA TREE.

MATRONAL|IS, -E Pertaining to the Roman matrons whose festival was on 1 March.

MAW SEED
see POPPY (opium).

MAY
see HAWTHORN

MAY APPLE *(Podophyllum peltatum)*. A perennial plant in the *Podophyllaceae* from N. America. It has erect, 300–450mm stems bearing at the tip 2, PALMATE, stalked leaves between which sits a 6-petalled, white, bowl-shaped, 38–50mm wide flower. The ovoid, yellow or red, 25–50mm long fruits are edible. Non-flowering stems have PELTATE, lobed leaves to 300mm across.

MAYWEED A name used for several members of the *Compositae* with white, daisy-like flowers. Best known is scentless mayweed *(Tripleurospermum maritimum)*, an annual or perennial from Europe and W. Asia. It grows to 450mm or more with oblong leaves, bipinnately cut into many LINEAR segments. The stems are smooth and angular and the flower heads are 25–38mm across: *T.m.* ssp. *maritimum* is a prostrate plant which grows by the sea; stinking mayweed or chamomile *(Anthemis cotula)* is similar but with a foetid smell and sparsely hairy stems; similar again is scented mayweed or wild chamomile *(Matricaria recutita)* but it has a pleasant, aromatic smell and the DISKS of the flower heads are markedly conical; rayless mayweed or pineapple weed *(M. matricarioides)* is easily distinguished by its strong pineapple smell and lack of RAY florets.

MEADOW
see PASTURE

MEADOW-GRASS *(Poa pratensis)*. A variable perennial species in the *Gramineae* from Europe, Asia, N. Africa and N. America. It is loosely to densely tufted with flat leaves and stems to 600mm or more bearing broadly to narrowly pyramidal PANICLES of tiny, purplish or greyish-green spikelets. An important pasture grass. A GLAUCOUS-leaved form is cultivated in the USA as Kentucky blue grass.

MEADOW-RUE *(Thalictrum)*. A genus of 150 perennials in the *Ranunculaceae* from the north temperate zone and mountains of S. America, Africa and S.E. Asia. Most species have COMPOUND leaves, pinnately or ternately divided, and large panicles of small flowers. The latter are APETALOUS, but in some species coloured sepals are present. Several species are cultivated as ornamentals. Common meadow rue *(T. flavum)* is 45–105cm tall with bi- or tripinnate leaves and flowers with bright yellow stamens; lesser meadow-rue *(T. minus)* varies from 150mm to 1m in height and has equally variable tri- or quadri-pinnate leaves that resemble maidenhair fern. The tiny flowers are yellowish or purplish-green.

MEADOW SAFFRON
see COLCHICUM

MEADOWSWEET *(Filipendula ulmaria)*. A perennial plant of wet soil in the *Rosaceae* from Europe and Asia and escaped in N. America. It grows 0.6–1.2m tall with pinnate leaves composed of 4–10 main OVATE, toothed leaflets, a larger 3-lobed, terminal one and several very small ones on the stalk. The tiny, creamy-white 5-petalled flowers are borne in large CYMOSE PANICLES. A similar but smaller plant of well-drained grassland is dropwort *(F. vulgaris* syn. *F. hexapetala)* with more numerous PINNATIFID leaflets and flowers having 6 petals.

MECONOPSIS A genus of 43 perennial or MONOCARPIC plants in the *Papaveraceae* from Asia (one in W. Europe). They form tufts or rosettes of usually hairy, often lobed leaves and erect stems bearing one to many cup-shaped poppy-like flowers composed of four or more broad petals in shades of blue, purple, red or yellow. Most species are alpine or sub-alpine and several are grown as choice ornamentals. The blue poppy *(M. betonicifolia* syn. *M. baileyi)* grows 0.9–1.5m tall with long-stalked, oblong, CRENATE leaves, the blade 100–150mm long, and 50–75mm wide sky-blue flowers (mauve-blue on limy soils); *M. grandis* is similar with narrower leaves and larger, darker flowers; lampshade poppy *(M. integrifolia)* has OBLANCEOLATE, densely hairy leaves and globular, 50mm wide, clear yellow flowers; Welsh poppy *(M. cambrica)* is the only European species and has deeply and irregularly lobed leaves and yellow or orange 38mm wide flowers.

MEDIAL [MEDIAN] The middle, e.g. a central leaf vein or stripe.

MEDICK *(Medicago)*. A genus of 100 annual and perennial plants in the *Leguminosae* from Europe, W. Asia and N. Africa and naturalised elsewhere. They have TRIFOLIATE leaves, tiny pea-shaped flowers in ovoid RACEMES and spirally coiled pods often with prickles. Black medick *(M. lupulina)* is a procumbent or ascending annual or short-lived perennial, having downy OBOVATE 6–18mm long leaflets and 9mm racemes of bright yellow flowers. *See also* LUCERNE.

● **MEDLAR** *(Mespilus germanica)*. A large, sometimes spiny, deciduous shrub or small tree to 6m in the *Rosaceae* from S.E. Europe and S.W. Asia. It has 50–113mm long OBLANCEOLATE leaves and solitary 25–38mm wide white flowers with 5 petals and long, leafy sepals. The brown, flattened-globose 19–32mm wide fruits are only edible when bletted (allowed to rot).

MEDULLA Loose textured cell tissue known as pith, found mainly in the centre of young stems.

MEDULLARY RAYS Vertical plates of PARENCHYMATOUS cells that extend radially from the MEDULLA to the CAMBIUM and which serve to transport water and foodstuffs.

MEGASPORE [MACROSPORE]
see HETEROSPOROUS

MEIOSIS
see CELL DIVISION

MELIA [AZEDERACH, BEAD TREE] *(Melia azadirachta)*. A deciduous tree to 15m in the *Meliaceae* from S.E. Asia. It has PINNATE leaves to 375mm long comprising 9–15, LANCEOLATE leaflets and 150–225mm AXILLARY PANICLES of small, white 5–6 petalled, honey-scented flowers. The oval, purple fruits contain bony seeds used as beads.

MELICK *(Melica)*. A genus of 70 perennial species in the *Gramineae* from all temperate countries, except Australia. They are mostly of tufted, slender habit with panicles of smooth, oval to oblong spikelets. Wood melick *(M. uniflora)* grows 300–600mm tall with sparse panicles of 3mm long, purplish spikelets each with a fertile floret.

▲ **MELILOT** *(Melilotus)*. A genus of 25 annuals, biennials and short-lived perennials in the *Leguminosae* from Europe, Asia and N. Africa. They are mostly erect, slender plants with TRIFOLIATE leaves and AXILLARY RACEMES of small, yellow or white pea-flowers. The small, ovoid pods are INDEHISCENT. Several species have the sweet smell of coumarin (new mown hay). Common melilot *(M. officinalis)* is a 0.6–1.2m biennial with oblong-elliptic 13–19mm long leaflets and 19–50mm long racemes of yellow flowers.

MELON
see CUCURBITS

MEMBRANE A thin film of tissue with microscopic pores of molecular size. *See* OSMOSIS.

■ **MENDEL, GREGOR JOHANN (1822–84)** Austrian biologist born in Silesia. He joined an Augustinian monastery at Brünn in 1843 as a monk, later becoming Abbot. For about fifteen years he taught natural sciences at the school in Brünn and carried out experiments in the monastery's garden from which he developed his now famous laws of HEREDITY. Mendel published the results of his experiments in 1868 in a somewhat obscure journal and it failed to gain recognition in his lifetime. His further, more ambitious experiments were unfortunately with plants now known to be APOMICTIC and consequently they did not bear out his earlier results. It is said that because

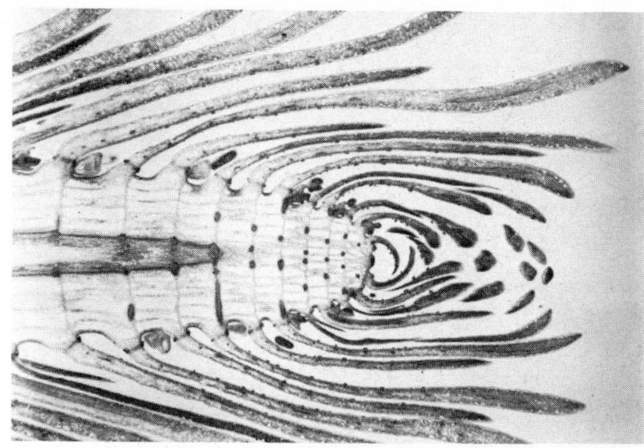

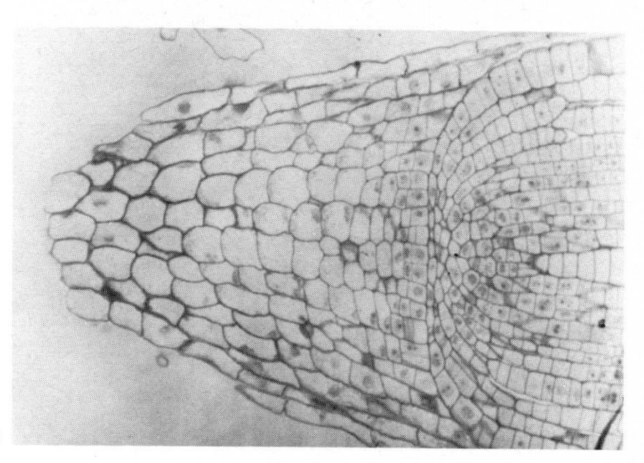

of this he died a disappointed man at a time when DARWIN'S evolutionary theories of the origin of species had become generally accepted, replacing the old idea that each species was a separately created unchangeable entity. Mendel's work, which formed the basis of much later (and continuing) research, pointed the way to the discovery of the means by which evolutionary change takes place. His article is translated as 'Experiments in Plant Hybridisation' and is found in J. A. Peters: *Classic Papers in Genetics* (1959).

MENTUM A projection or chin at the base of the column in certain orchid flowers.

MERCURY *(Mercurialis)*. A genus of 8 species of annuals and perennials in the *Euphorbiaceae* from Europe and Asia. They have elliptic to ovate leaves and DIOECIOUS or MONOECIOUS flowers with 3, PERIANTH segments: annual mercury *(M. annua)* grows from 100–500mm with crenate, serrate leaves 13–50mm long and small, green dioecious flowers, the males in slender, axillary spikes, the female in almost stalkless clusters; dog's mercury *(M. perennis)* has far-creeping RHIZOMES, 150–375mm erect, unbranched stems and flowers similar to those of annual mercury. GOOD KING HENRY is also known as mercury.

MERICARP A one-seeded section of a usually dry fruit, derived from a SYNCARPOUS ovary, e.g. the 'seeds' of mallow or cranesbill.

MERIDIONAL|IS, -E This is used of two distinct occurrences; (1) of midday, referring to plants whose flowers open around midday and (2) south or southern.

○ **MERISTEM** The extreme tip of a stem or root where the cells are actively dividing and growing.

MESCAL BUTTON [DUMPLING CACTUS] *(Lophophora williamsii)*. A small, spineless, flattened, globe-shaped plant in the *Cactaceae* from Texas and Mexico. It is about 50mm across, blue-green with a small tuft of hairs at each AREOLE. The 19mm flowers are pink or whitish and somewhat daisy-like. The flesh contains the hallucinogenic drug, mescalin.

MESOCARP
see ENDOCARP

MESOPHYLL The layer of photosynthetic cells between the upper and lower EPIDERMAL surfaces of a leaf. The upper part of this layer is the pallisade tissue composed of cylindrical cells containing numerous CHLOROPLASTS. The lower part is the spongy mesophyll made up of irregularly rounded cells with fewer chloroplasts, often only in contact by their protruberances and with large air spaces. The STOMATA open into this layer which is thus in contact with the atmosphere and gases can pass in and out.

MESOPHYTES Average or ordinary plants, those which flourish under average soil and climatic conditions.

MESQUITE GRASS
see BOUTELOUA

METABOLISM The chemical processes that take place as the plant grows, involving such things as PHOTOSYNTHESIS, RESPIRATION, TRANSPIRATION.

METAPHASE
see CELL DIVISION

METAXYLEM
see XYLEM

MEXICAN POPPY *(Argemone mexicana)*. A 300–600mm annual plant in the *Papaveraceae* from Central America. It has prickly stems and PINNATIFID, spiny-toothed, GLAUCOUS leaves and pale yellow or orange 50–63mm wide, cup-shaped flowers composed of 4–6 petals.

MEZEREON
see DAPHNE

● **MICHAELMAS DAISY** Several species of *Aster*, herbaceous perennials in the *Compositae* from N. America. They are clump-forming plants with rigid stems, LINEAR to LANCEOLATE leaves and small, daisy-flowers in large CORYMBOSE PANICLES. The commonest Michaelmas daisy is *A. novi-belgii* a 1.2m plant with 25–38mm wide, purple-rayed flower heads. It has given rise to many hybrid CULTIVARS in shades of purple, blue, red, pink and white and in a height range from 150mm to 1.5m.

MICRANTH|US, -A, -UM Having small flowers.

MICROPHYLL|US, -A, -UM Having small leaves.

▲ **MICROPYLE** The canal into the EMBRYO SAC in the OVULE through which the pollen tube grows. Also present as a tiny pore in the testa of a mature seed.

MICROSPECIES A unit of classification below the species level, used to separate very variable species into recognisable entities, but mainly of concern to professional botanists. Thus, the common blackberry *(Rubus fruticosus)* has been split up into almost 400 microspecies on such small characters as presence or absence of stalked glands, size, shape and number of leaflets, prickles, etc.

MICROSPORE
see HETEROSPOROUS

MIDRIB The central main vein, present in many leaves.

■ **MIGNONETTE** Several species of *Reseda*, annuals, biennials and perennials in the *Resedaceae* from Europe, the Mediterranean region and E. Africa. They have mainly narrow leaves, either entire, lobed or PINNATISECT, and small flowers in terminal RACEMES; common or garden mignonette *(R. odorata)* can be annual or a sub-shrubby perennial to 600mm, with LANCEOLATE or TRIFID leaves and very fragrant, yellow-white flowers with prominent orange stamens. Wild mignonette *(R. lutea)* is a 300–600mm perennial with PINNATIFID leaves and greenish-yellow flowers in compact conical racemes.

MILDEW Minute parasitic fungi, often serious pests of crop and ornamental plants. They are recognisable as a fine, white downy, or powdery coating on leaves and shoots caused by thread-like HYPAE which send out HAUSTORIA into the cells of the host plant, absorbing the food material. As a result, the leaves of the host plant may turn yellow and die or become discoloured and distorted.

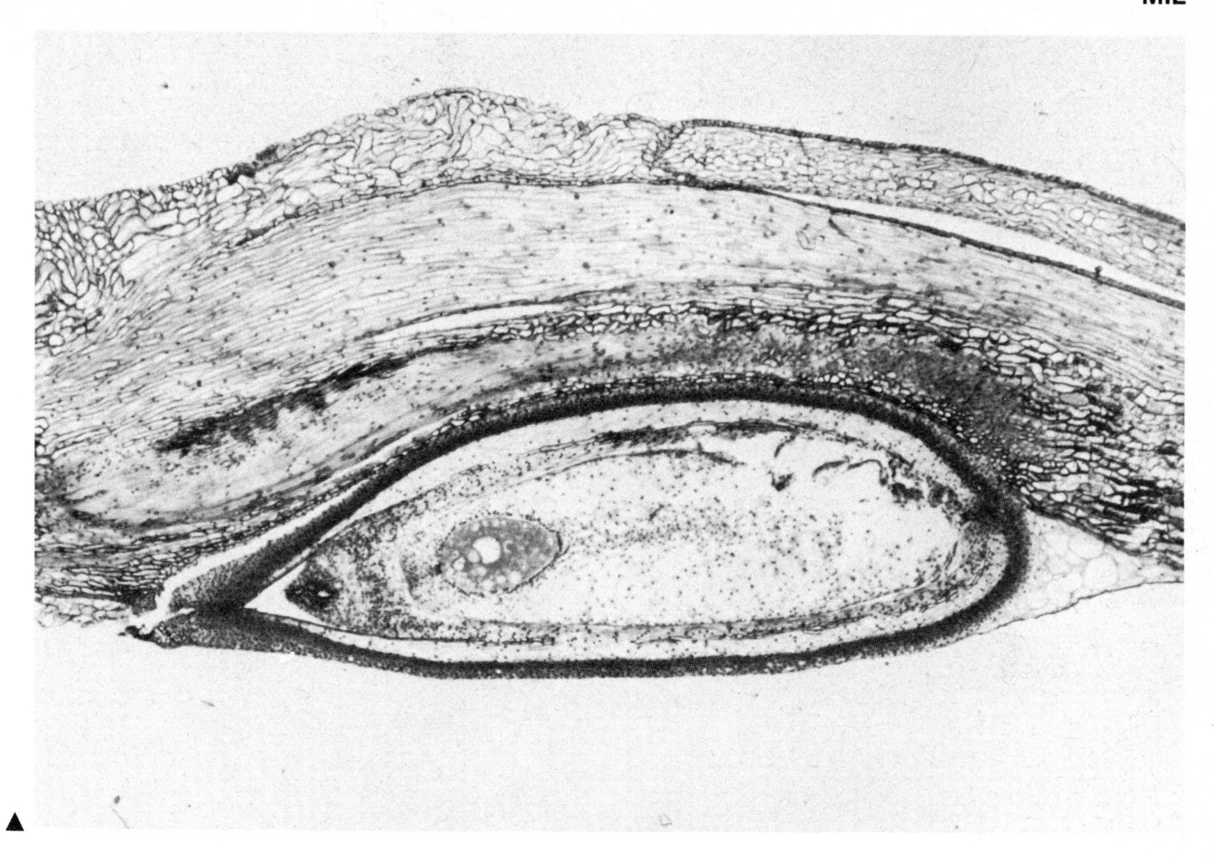

▲

MILFOIL
see YARROW; WATER MILFOIL

MILITARY [SOLDIER] ORCHID *(Orchis militaris)*. A perennial ground-dwelling plant in the *Orchidaceae* from Europe, Turkey and Siberia. It has a tuft of elliptic, oblong leaves to 100mm and a 200–450mm spike of reddish-violet flowers. The sepals arch forwards to form a hood or head and the LABELLUM fancifully resembles the body of an old-fashioned, red-coated soldier.

MILK PARSLEY *(Peucedanum palustre)*. An erect biennial in the *Umbelliferae* from Europe to the Urals. It grows 45–150cm tall, with bi- to quadripinnate, long-stalked leaves and 32–75mm wide UMBELS of tiny, 5-petalled, white flowers.

MILK THISTLE *(Silybum marianum)*. An annual or biennial plant in the *Compositae* from Europe and the Mediterranean region to the Caucasus, and introduced elsewhere. It grows 45–120cm tall, with large, SINUATELY lobed and wavy-margined leaves, having a white vein pattern. The thistle-like flower heads have long, spiny bracts and red-purple florets.

MILK-TREE
see COW-TREE

MILK VETCH Several species of *Astragalus*, a large genus of annuals, perennials and shrubs in the *Leguminosae* of cosmopolitan distribution except Australia. They have PINNATE leaves and AXILLARY RACEMES of small, pea-flowers. Common milk vetch *(A. glycyphyllos)* is a prostrate or ascending perennial 900mm long with 100–200mm long leaves of oblong, elliptic leaflets and creamy-white, 13mm long flowers.

MILK WEED *(Asclepias)*. A genus of 120 perennial and shrubby plants in the *Asclepiadaceae* from N. and S. America. They have OVATE to LANCEOLATE leaves in opposite pairs or WHORLS of three, and small flowers in lateral or terminal UMBELS. Each flower has 5 petals, 5 sepals and 5 stamens united into a tube at the top of which are 5 petal-like hoods. The fruit is a pod-like follicle and the seeds have tufts of hair adapted for wind dispersal.

MILKWORT Several species of *Polygala*, a large genus of annuals, perennials and shrubs in the *Polygalaceae* of world-wide distribution except New Zealand and the Arctic. They have LINEAR to OVATE leaves and flowers usually in spikes or RACEMES. The flowers superficially resemble those of the *Leguminosae*. The two innermost of the 5 sepals are large and petal-like forming 'wing' petals, the 3 petals are united with the 8 stamens to form a tube, analagous to the KEEL of a pea-flower. Common milkwort *(P. vulgaris)* is a tufted plant 100–300mm high with narrow, OBOVATE leaves and blue, pink or white flowers, 6mm long.

MILLET A general name for several tropical, small-grained cereal grasses in the *Gramineae*. The most important is great millet or sorghum, also called guinea and Kaffir corn *(Sorghum vulgare)*. It is a variable plant, 0.9–4.5m tall with broad leaves like maize and dense, terminal PANICLES of tiny florets. There are many CULTIVARS, with red to white and sweet to bitter seeds. It is

an important foodcrop in parts of Africa, India and China, but in the USA and elsewhere it is grown for livestock feeding: finger millet *(Eleusine coracana)* is a smaller, tufted plant with narrow leaves and grain carried in clusters of about 5 finger-like spikes. It is a staple food crop in parts of S. India, Sri Lanka and Africa; bulrush millet *(Pennisetum typhoideum)* is like sorghum in growth, but the grain is carried in dense, cylindrical panicles. There are brown and white seeded CULTIVARS, the latter known as pearl millet. This is the most drought-resistant of all the millets and is important in the arid areas of East Africa, India and Pakistan; common millet *(Panicum miliaceum)* has long been a food crop; it was the 'milium' of the Romans and is mentioned in the Old Testament (Ezekiel 4:9). It grows 0.9–1.2m tall with a compact, terminal PANICLE shaped rather like an old-fashioned broom. It is widely grown in India, Asia, S. Europe and parts of N. America: foxtail or Italian millet *(Setaria italica)* probably originated in Africa. It grows 0.9–1.5m high and has dense, cylindrical panicles to 300mm (well known to bird fanciers in Britain, where whole ears can be obtained in pet shops). There are white, yellow, red, brown and black seeded cultivars grown in all warm temperate and subtropical countries.

● **MIMOSA** A genus of up to 500 perennials and shrubs in the *Leguminosae* from tropical and subtropical America, Africa and Asia. They have bipinnate leaves which in several species are sensitive to the touch, folding rapidly downwards when knocked. The tiny tubular flowers are aggregated into globose heads or cylindrical spikes. Best known is the humble or sensitive plant *(M. pudica)*, a prickly, short-lived perennial to 900mm with very sensitive leaves and rose-purple flowers. For the yellow mimosa of florists, *see* ACACIA (dealbata).

MIND-YOUR-OWN-BUSINESS *(Soleirolia soleirolii,* syn. *Helixine soleirolii)*. A mat-forming perennial in the *Urticaceae* from Corsica, Sardinia and the Balearic Is. It has thread-like, much branched stems and almost round leaves up to 6mm wide. The minute green flowers are petalless.

MINERS' LETTUCE
see CLAYTONIA

MINT *(Mentha)*. A genus of 25 perennial species and many hybrids in the *Labiatae* from the north temperate zone, Australia and S. Africa. Many species are aromatic and used for flavouring. They have opposite pairs of OVATE to LANCEOLATE leaves and tiny tubular flowers borne in dense WHORLS in the upper leaf AXILS: apple mint *(M. rotundifolia)* grows to 600–900mm with broadly-ovate, white-hairy leaves to 38mm long and lilac-pink flowers (the best for mint sauce); spear mint *(M. spicata)* is the most commonly grown as a herb, with oblong-lanceolate leaves 38–75mm long and lilac flowers; peppermint *(M. X piperita)* is used to flavour many sorts of confectionery and is a hybrid between water mint and spearmint; water mint *(M. aquatica)* grows in wet places with ovate leaves 57mm long and the WHORLS of lilac flowers condensed to form a terminal head; pennyroyal *(M. pulegium)* is a prostrate plant with oblong leaves 9–19mm long and semi-erect flowering stems with lilac blossoms.

▲ **MIRROR ORCHID [MIRROR OF VENUS]** *(Ophrys speculum)*. A ground-dwelling perennial in the *Orchidaceae* from around the Mediterranean. It has narrowly oblong basal leaves and stems 100–300mm tall, bearing 2–6 bee-like flowers. These have green and maroon lateral sepals and an upper hooded one, very small maroon petals and a lobed pear-shaped, 13–19mm long LABELLUM. This is an iridescent blue, edged yellow and heavily fringed with blackish-red.

MISTLETOE *(Viscum album)*. An evergreen, semi-parasitic shrub in the *Loranthaceae* from Europe, Asia and N. Africa. It grows to about 900mm long with repeatedly forking, olive-green, semi-woody stems and opposite pairs of yellow-green, OBOVATE leaves. The tiny unisexual flowers have SEPALOID, greenish petals and are followed by white, mucilaginous berries. The sticky seeds are distributed by birds. Several other members of the *Loranthaceae* are known as mistletoe in different parts of the world.

MITIS Mild or innocuous, e.g. a plant without spines where these might normally be expected.

MITOSIS
see CELL DIVISION

MOCCASIN FLOWER
see LADY'S SLIPPER

MOCK ORANGE *(Philadelphus)*. A genus of 75 shrubs in the *Philadelphaceae* from the north temperate zone. They have opposite pairs of strongly veined, OVATE to LANCEOLATE leaves and cup-shaped, 4-petalled white flowers, sometimes purple-blotched and fragrant. Several species and hybrids are grown as ornamentals and called syringa. The original mock orange is *P. coronarius*, a 2.4–3.6m species with strongly fragrant 25mm wide flowers.

MOLL|IS, -E Soft, usually meaning softly hairy.

MONANDROUS Flowers having one stamen only.

MONEYWORT [CREEPING JENNY] *(Lysimachia nummularia)*. A prostrate perennial in the *Primulaceae* from Europe. It has rounded leaves 13–25mm across and cup-shaped, 13–19mm wide, yellow, 5-petalled blooms.

MONKEY BREAD The fruit of BAOBAB.

MONKEY FLOWER *(Mimulus)*. A genus of 100 annuals, perennials and subshrubs in the *Scrophulariaceae* of cosmopolitan distribution. They have opposite pairs of narrowly lanceolate to ovate leaves and solitary, AXILLARY or RACEMOSE red or yellow flowers. These are tubular, expanding to 5 broad, spreading lobes, 3 pointing down, 2 up. This stigma has 2 sensitive flaps which close when touched. Common musk *(M. moschatus)* is a spreading, sticky, woolly perennial with bright yellow flowers. Until about 1914 it had a sweet musky odour, but since then it has disappeared, a botanical mystery that has never been satisfactorily explained.

■ **MONKEY ORCHID** *(Orchis simia)*. Very similar to the MILITARY ORCHID, but with a white, rose or crimson LABELLUM, the lower part of which is forked into two leg-like lobes with a small tooth or tail between them.

● **MONKEY PUZZLE [CHILE PINE]** *(Araucaria araucana)*. An evergreen, coniferous tree to 30m in the *Araucariaceae* from Chile and S. Argentina. It has WHORLS of horizontal branches covered with overlapping, OVATE, hard, leathery spine-tipped leaves. The cones are almost globular to 175mm and contain large, edible seeds.

MONKSHOOD
see ACONITE

MONOCARPIC Of plants which flower once then die. Annuals and biennials are monocarpic, but the term is usually applied to plants which live for several years before flowering, and then die, e.g. some agaves, bell-flowers and meconopses.

MONOCHLAMYDEOUS
see HETEROCHLAMYDEOUS

MONOCOTYLEDONS One of the two classes that make up the *Angiospermae*. It is characterised by the seedlings having only one cotyledon and the flowers having their parts (petals and sepals) in WHORLS of three.

MONOECIOUS Having unisexual flowers on the same plant, e.g. hazel.

MONOGYNOUS Having one STYLE only to each flower.

MONOPODIAL Referring to a stem which continues to extend from the same growing point, e.g. pine and fir trees with one main stem only.

MONOTYPIC A genus containing one species only.

MONSTERA A genus of 50 evergreen, root climbers in the *Araceae* from tropical America and the W. Indies. They have large, leathery, OVATE leaves, often PINNATIFID and curiously perforated, and arum-like boat-shaped flowering spikes. *M. deliciosa* is often grown as a house plant and then has pinnatifid or slotted leaves to 300mm long. In the wild the leaves reach 600mm with numerous perforations and the stems climb to 12m or more.

MONTBRETIA *(Crocosmia)*. A genus of 6 cormous-rooted perennials in the *Iridaceae* from S. Africa. They have fans of sword-shaped leaves and spikes of tubular flowers with 6 spreading lobes. Several hybrid CULTIVARS are grown as ornamentals: *C. crocosmiflora* grows 300–900mm with deep orange, red-suffused flowers 25–50mm across.

MONTICOL|US, -A, -UM Growing on the mountains.

MOON DAISY
see OX-EYE DAISY

MOON FLOWER *(Calonyction aculeatum)*. A twining perennial to 3m in the *Convolvulaceae* from the tropics. It has OVATE-CORDATE or HASTATE 75–200mm long leaves and fragrant, trumpet-shaped, 75–150mm wide white flowers which open only at night.

▲ **MOONWORT** *(Botrychium)*. A genus of 40 ferns in the *Ophioglossaceae* of cosmopolitan distribution. The solitary fronds are PINNATE to tripinnate and the SPORANGIA

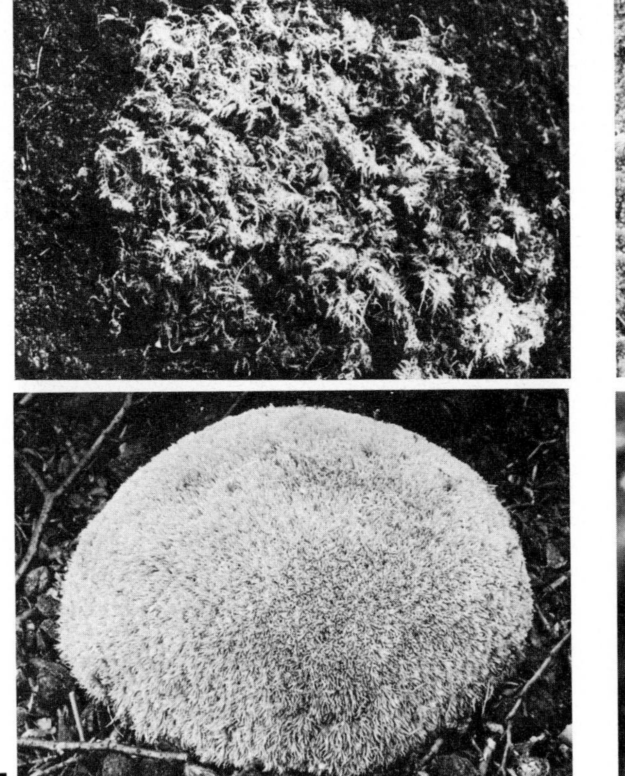

are borne in stalked PANICLES appearing as if united with the leaf base. Common moonwort *(B. lunaria)* grows 50–150mm or more in height with pinnate fronds of fan-shaped leaflets.

MOORLAND
see HEATHLAND

MOREL *(Morchella esculenta)*. A curious *Ascomycete* fungus from temperate, deciduous woodland, hedgerows and grassland, usually on rich soils. It has a yellowish, somewhat grooved stalk (stipe) and an irregularly ovoid and honeycombed yellow and brown cap 57–113mm long.

MORNING GLORY Several climbing species of *Ipomoea* and *Pharbitis* members of the *Convolvulaceae* from the tropics and subtropics. The best known is *Pharbitis tricolor* (syn. *Ipomoea tricolor*), an annual twiner to 3m with ovate-cordate leaves and 75–125mm wide trumpet-shaped flowers of red, purple and sky-blue.

MORPHOLOGY The study of the form or make-up of things. The form which a plant takes often reflects the conditions under which it lives, e.g. cushion plants, water plants, cacti.

MOSCHATEL [TOWNHALL CLOCK] *(Adoxa moschatellina)*. A small, RHIZOMATOUS perennial in the *Adoxaceae* from the north temperate zone. Has small tufts of long-stalked, BITERNATE leaves, and the 13–50mm long leaflets usually trilobed. Slender 50–100mm tall stems bear 5 green, 4-petalled 6mm wide flowers arranged like the four faces of the town hall clock with one on top.

MOSCHAT|US, -A, -UM Having a smell of musk.

■ **MOSS** *(Musci)*. Primitive SPORE-bearing plants which, with the LIVERWORTS *(Hepaticae)* form the order *Bryophyta*. They form mats or cushions of erect or prostrate stems covered with frequently overlapping, scale-like or awl-shaped leaves of one cell thickness. The moss plant represents the GAMETOPHYTE generation and is equivalent to the PROTHALLUS of the fern. The fertilised ovum remains attached to the plant and produces a SPOROPHYTE consisting of a stalk and a capsule. The latter mainly release their spores through a hole in the top of the capsule, though some open by four splits and the sphagnums have an explosive mechanism. Mosses are found in a wide variety of habitats: on exposed rocks and walls, on trees, on soil, and particularly in moist and shady places, and in water.

MOTHER-IN-LAW'S-TONGUE *(Sansevieria trifasciata)*. A familiar house plant in the *Agavaceae* from tropical W. Africa. It has thick RHIZOMES and forms clumps of 0.3–1.2m tall rigid, erect, LINEAR-LANCEOLATE leaves. They are somewhat fleshy with a pattern of pale cross-bands on a dark green background; 6-petalled, greenish-white flowers are followed by 8mm wide, orange berries.

MOTHER-OF-THOUSANDS
see SAXIFRAGE

MOTILE Capable of independent movement, e.g. the sperms of algae and ferns.

MOULD A rather general term for several small, thread-like fungi which live on dead and decaying plant and animal remains, causing greenish, bluish or blackish areas of HYPHAE and spores. The commonest is *Mucor mucedo*, the blue-green bread mould. Some moulds are of value to man, particularly species of *Penicillium* which yield the drug penicillin and are also used to ripen the various kinds of blue cheese.

MOUNTAIN ASH An alternative name for ROWAN.

MOUNTAIN EVERLASTING *(Antennaria)*. A genus of 100 species of evergreen mat-forming plants in the *Compositae* from the temperate areas of both hemispheres. They have entire, grey-green, narrowly *spathulate* to LANCEOLATE leaves and clusters of small flower heads, each surrounded by chaffy BRACTS. The commonest species is *A. dioica*, with 150mm tall stems and clusters of white-tipped, pink bracts.

MOUNTAIN PLANTS
see CUSHION PLANTS

MOUSE-EAR [MOUSE-EAR CHICKWEED] Several annual and perennial members of the genus *Cerastium*, small plants in the *Caryophyllaceae* mostly from the north temperate zone, but some cosmopolitan. They have opposite pairs of entire, usually narrow leaves and dichasial CYMES of small white flowers with notched or bilobed petals. Common mouse-ear *(C. holosteioides* syn. *C. fontanum)* has lax stems to 300mm with elliptic to oblanceolate leaves and petals hidden by 6mm sepals.

MOUSETAIL *(Myosurus minimus)*. A tiny annual plant in the *Ranunculaceae* from Europe, S.W. Asia and N. Africa, naturalised in N. America and Australia. It has a rosette of LINEAR, grassy leaves and solitary greenish flowers of 50–125mm stems. Each flower has a tall, cylindrical PISTIL of many carpels, 5 narrow oblong sepals and 5 tubular petals.

● **MUCILAGE [MUCILAGINOUS]** A sticky or slimy fluid, often secreted by glandular hairs. This gives a clammy feel to leaves and helps to cut down water loss from soft, young growth. CARNIVOROUS PLANTS such as sundews and butterworts secrete mucilage to trap insect prey. The water-holding capacity of mucilage is utilised by succulent plants and the seed-coats of some seeds which swell up when wetted.

MUCRONATE [MUCRONAT|US, -A, -UM] Leaves or bracts abruptly terminating in a firm, often sharp point.

MUGWORT
see ARTEMISIA

MULBERRY Several species of *Morus*, deciduous trees in the *Moraceae* from the north temperate zone. They have OVATE, often lobed leaves and tiny green, unisexual flowers in small catkins. The CALYX of each flower becomes fleshy as the seeds ripen and each flower spike becomes one, multiple fruit. Common or black mulberry *(M. nigra)* grows to 9.1m with a broad head of branches. The 75–150mm long leaves are 2–5 lobed and coarsely toothed. The 25mm fruit clusters are dark red.

▲ **MULLEIN** *(Verbascum)*. A genus of 360 biennials, perennials and shrubs in the *Scrophulariaceae* from Europe and Asia, and naturalised elsewhere. The biennial and perennial species are best known, having large rosettes of ovate to obovate leaves and terminal spikes or PANICLES of mainly yellow, but sometimes white or purple flowers. The latter have 5, petal-like lobes and the stamen filaments have long, woolly hairs. Great mullein or Aaron's rod *(V. thapsus)* is a thickly white woolly plant to 1.8m with a dense spike-like RACEME of 19–32mm yellow flowers.

MULTIFID|US, -A, -UM Cleft into many parts, e.g. leaves.

MULTILOCULAR Bearing many parts or CARPELS to the OVARY.

MUNG BEAN [GREEN GRAM] *(Phaseolus aureus)*. An annual pulse crop in the *Leguminosae* from India, but grown in subtropical Asia, Africa and America. It is a rather floppy, bushy plant with TRIFOLIATE hairy leaves and yellow pea-flowers. The slender pods contain up to 15 small beans. These are moistened and germinated in the dark to produce the bean sprouts of Chinese cookery.

MURAL|IS, -E Growing on a wall.

MURICAT|US, -A, -UM Covered with small, pointed, wart-like outgrowths.

MUSCI
see MOSS

MUSCOS|US, -A, -UM Moss-like

■ **MUSHROOM** Several species of *Agaricus, Basidiomycetes* fungi from temperate grassland and woodland. They have convex caps and short, sturdy STIPES with a narrow, membranous ring which usually soon falls: common or field mushroom *(A. campestris)* has a silky-white, 50–100mm cap with pink gills that darken on maturity; the similar cultivated mushroom *(A. bisporus)* is brown flushed in its wild state, but its cultivated form is *A.b. albida* which is white. Field mushroom has 4 spores to each BASIDIUM; cultivated mushroom has two only.

MUSK MALLOW
see MALLOW

MUSK PLANT
see MONKEY FLOWER

MUSTARD In a general sense referring to several species and genera in the *Cruciferae*, but mainly to the black mustard *(Brassica nigra)*. This 300–900mm tall annual comes from Europe and is much grown elsewhere for its seeds which yield the condiment mustard. It has LYRATE-PINNATIFID, bristly leaves to 150mm long and erect RACEMES of bright yellow, 4-petalled flowers. Mustard is also obtained from white mustard *(Sinapis alba)*, a similar plant but with shorter pods having long, flattened, curved beaks. The seedlings of both mustards are used in the partnership of mustard and cress. *See also* GARLIC and HEDGE MUSTARDS.

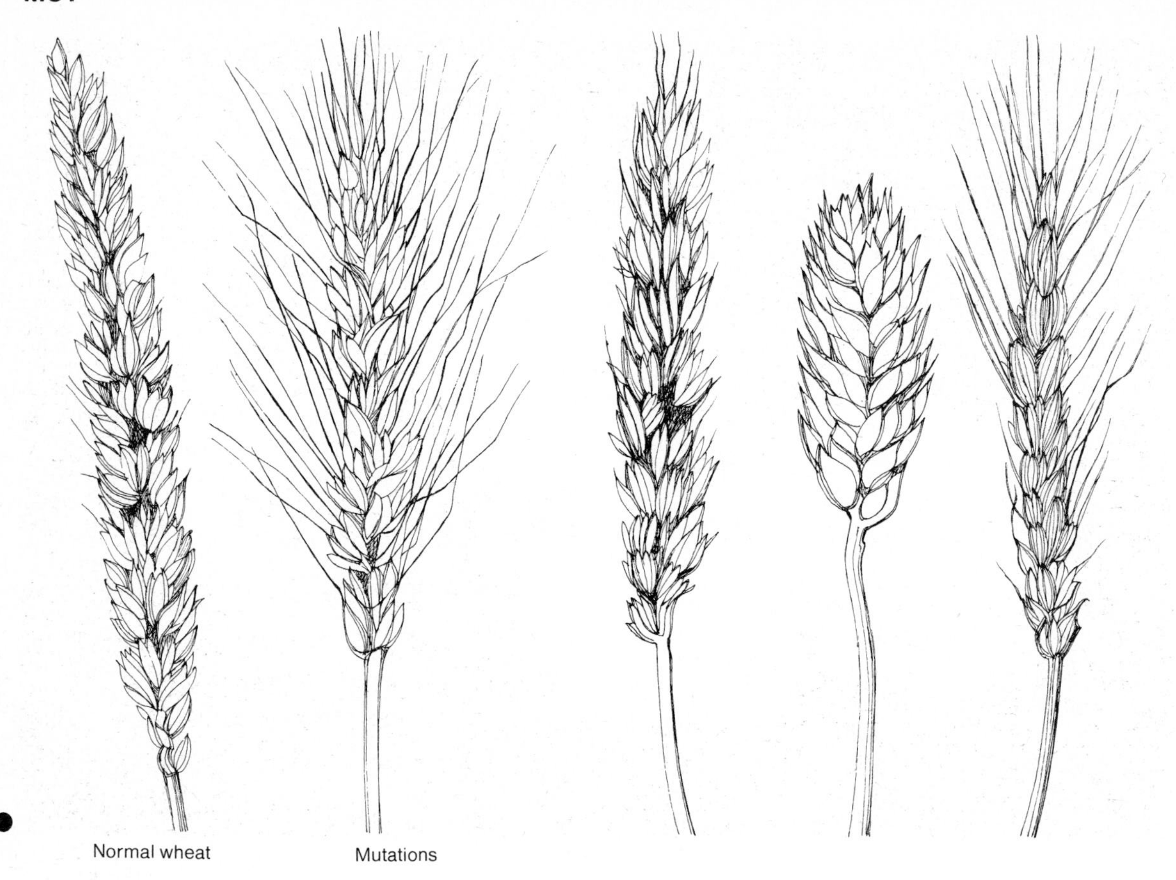

MUTABIL|IS, -E Changeable, e.g. the colour of flowers.

● **MUTANT [MUTATION]** Spontaneous changes that take place in plant organs or in whole plants. Often these are small, but sometimes they are large and startling, as when a flower changes colour or a normally tall plant stays very dwarf. A series of mutational changes can result in the formation of new species and is the basis of evolution.

MYCELIUM A collective term for a mass of HYPHAE. *See* FUNGI.

MYCORRHIZA The association of fungal HYPHAE with the roots of higher plants. It is a common phenomenon in the plant world, e.g. many trees, orchids, heaths and heathers could not grow properly or at all without a fungus associate. Mycorrhizal partnerships occur in soils deficient in nitrogen and the fungus is able to obtain this from decaying organic matter, passing some on to the plant in return for sugars. There are two kinds of mycorrhiza, ectotrophic and endotrophic. The latter penetrate the cells of the cortex of the roots and in heathers, penetrate the whole plant, even the seeds. Ectotrophic mycorrhiza forms a mainly surface covering to roots and is the commonest type in forest trees where it also partially or wholly replaces the tree's root hairs. The fungi involved in these associations are mainly *Basidiomycetes*, e.g. species of cep *(Boletus)* and agaric *(Amanita)*.

MYRMECOPHILY The association of plants with ants of various species. The ants nest in plant organs such as the large hollow thorns of *Acacia* or the aerial tubers of *Myrmecodia*.

MYROBALAN PLUM
see PLUM

MYRRH *(Commiphora myrrha)*. A spiny shrub in the *Burseraceae*, mainly from Somalia. It has small, OBOVATE leaves, tiny unisexual flowers and yields the resin myrrh used for medicine, incense and embalming.

MYRTLE *(Myrtus communis)*. An evergreen aromatic shrub in the *Myrtaceae* from S. Europe to W. Asia. It has opposite pairs of OVATE to LANCEOLATE 25–50mm long leaves and fragrant, white, 19mm wide 5-petalled flowers with numerous slender stamens. The 13mm long fruits are purple-black. *See also* BOG MYRTLE.

▲ **MYXOMYCETES** Popularly known as slime moulds, these are curious fungi which seem to combine the characters of primitive animals and plants. They start off as slime-like masses of protoplasm with numerous nuclei known as plasmodia. These move slowly along, ingesting food particles, usually decayed plant and animal remains, and gaining in size. Finally they crust over and produce spores, often in intriguingly shaped and coloured SPORANGIA.

NAN|US, -A, -UM Dwarf of squat.

NAPELL|US, -A, -UM Like a little turnip, e.g. root or tuber.

NAPIFORM Turnip-shaped.

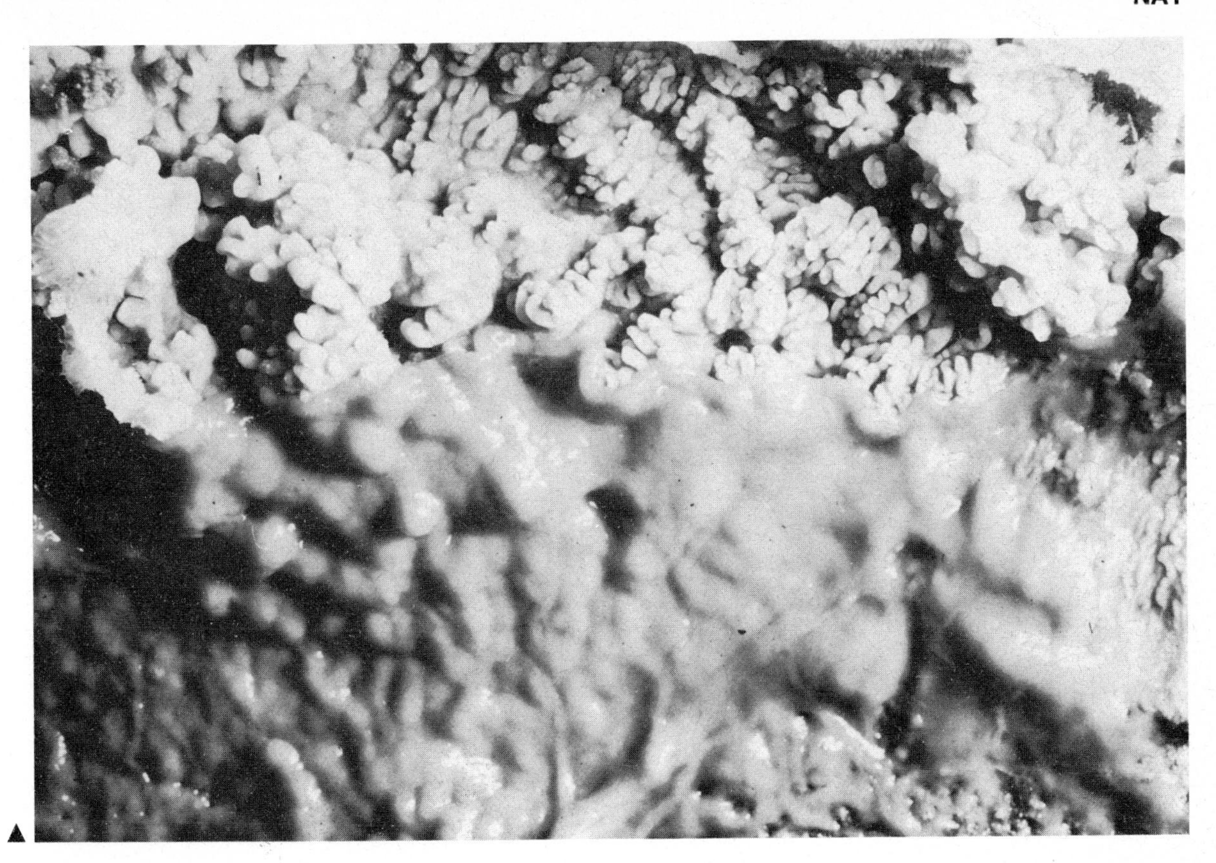

▲

NARCISSUS A genus of 60 species of bulbous plants in the *Amaryllidaceae* from Europe, the Mediterranean region and W. Asia. They have LINEAR leaves and small UMBELS or solitary flowers. Each bloom is tubular with 6, coloured, petal-like lobes (the PERIANTH) and a cup or trumpet-shaped organ called a corona. This corona is formed from a flap of tissue that grows out from the base of the perianth lobes and it may be of a contrasting colour. Daffodil is a name applied to narcissi with trumpet-shaped coronas as long as or longer than the perianth lobes. If the corona is more than one-third as long as the lobes, but less than equal to them, it is known as a large-cupped narcissus, if less than one-third, a small-cupped narcissus: common daffodil or lent lily *(N. psuedonarcissus)* grows 250–350mm tall, each flower having a 25mm long yellow corona and paler petals; *N. tazetta* has UMBELS of 4–8 white flowers, 25–50mm wide with small cup-shaped coronas; pheasant's eye *(N. poeticus)* has solitary white, fragrant flowers and very shallow, yellow, red-margined coronas. Several species and many hybrid cultivars are grown in gardens.

NARD GRASS
see MAT GRASS

NASTIC Movement in plants in response to such stimuli as light intensity, changes in temperature and contact. Many flowers close (crocus and dandelion are examples) or leaves fold and droop at night (as in clover), being said to sleep. The most marked nastic movement is displayed by the various sensitive plants which, when knocked, fold their leaves and appear to wilt. These movements are facilitated by the presence of somewhat swollen, joint-like structures at the base of the leaflets and leaf stalks known as pulvini. These pulvini are held firm by water pressure and any change in this pressure will alter the position of the leaves.

NASTURTIUM *(Tropaeolum)*. A genus of 90 annuals and tuberous-rooted perennials in the *Tropaeolaceae* from Mexico and S. America. Many are climbers, twisting their leaf stalks around a support. Some are trailers. They have entire to deeply lobed, PELTATE leaves and basically cup or trumpet-shaped flowers with 5 petals and a long spur: common nasturtium or Indian cress *(T. majus)* is a climber to 1.8m or more with entire, peltate leaves and 38–50mm wide orange flowers. Non-climbing cultivars are grown and semi-double forms in colours from yellow to red. The leaves and flowers can be used in salads and the green seeds pickled as a substitute for capers; flame creeper *(T. speciosum)* climbs to 3m with deeply 5–6 lobed leaves, scarlet flowers with stalked petals and blue seeds; *T. polyphyllum* forms large grey mats on the Andean screes, the tip of each stem wreathed with yellow or orange flushed blooms; *T. tricolorum* is a slender climber to 600mm with deeply lobed leaves and quaint inflated flowers of orange, yellow and purple.

NATANS Floating on or under the water.

NATURALISED Plant introductions from abroad whether by accident or design, which now live and reproduce as though native.

NATURAL SELECTION
see EVOLUTION

NATURAL SYSTEM The prevailing approach to classification, which seeks to arrange plants in a sequence showing their natural relationships.

● **NAVELWORT [PENNYWORT]** *(Umbilicus rupestris)*. An erect, succulent perennial in the *Crassulaceae* from Europe, the Mediterranean and the Azores. It has fleshy, orbicular, PELTATE leaves 13–63mm wide and terminal RACEMES of 9mm whitish, tubular flowers.

NAVICULAR [NAVICULAR|IS, -E] Boat-shaped, e.g. bracts or leaves.

NEBULOS|US, -A, -UM Cloud-like.

NECROSIS Brown or blackish areas of dead cell tissue usually resulting from the attack of a parasitic fungus.

NECTAR A sugary liquid secreted by certain cells in the nectary.

NECTARINE A smooth-skinned MUTANT of the peach.

NECTARY A small gland which secretes nectar, usually found at the base of flowers and serving to attract pollinating insects. Extra floral nectaries occur in some plants, e.g. on the leaf stalks of cherry and passion flower; their purpose cannot be satisfactorily explained.

NEEDLES The slender, rigid leaves of many coniferous trees and ericaceous plants.

NEMORAL|IS, -E [NEMOROS|US, -A, -UM] Growing beneath the shade of trees.

NERVOS|US, -A, -UM With conspicuous nerves or veins.

NETTLE *(Urtica)*. A genus of 50 annual and perennial plants in the *Urticaceae*, mainly from the temperate regions. They are typified by having stinging hairs on leaves and stems and opposite pairs of LANCEOLATE to OVATE leaves. The tiny green flowers are carried in CYMES or PANICLES, sometimes in dense, rounded heads. Common stinging nettle *(U. dioica)* grows 45–150cm with ovate-acuminate leaves and panicle-like flower clusters. *See also* DEAD NETTLE.

NETTLE TREE *(Celtis australis)*. A somewhat elm-like tree to 21m in the *Ulmaceae* from S. Europe, N. Africa and Turkey. It has a smooth grey trunk, 50–125mm ovate-lanceolate leaves and small clusters of petal-less green flowers. The 8–13mm wide, globose fruits are sweet and edible.

NEW ZEALAND FLAX *(Phormium tenax)*. A tufted, evergreen perennial in the *Agavaceae* from New Zealand. It has tough, LINEAR leaves to 1.8m or more long, and erect stems to 3m bearing PANICLES of tubular, dull red 38–63mm long flowers. The leaves yield a tough fibre used for rope and twine making.

colour	
ruber	red
viridis	green
caeruleus, azureus	blue
purpureus	purple
luteus, flavus	yellow
roseus	pink

hairiness	
lanatus	woolly
hirsutus	hairy
pubescens	downy
glabrus	smooth

shape	
lanceolatus	lance-shaped
ensatus	sword-shaped
orbiculatus	rounded/circular
acicularis	needle-like
ovatus	egg-shaped

geography	
europaeus	European
asiaticus	Asian
africanus	African
americanus	American
australasicus	Australasian
antarticus	Antarctic
arcticus	Arctic
britannicus	British
chinensis	Chinese
guatemalensis	Guatemalan
kewensis	from Kew (England)

discoverer	
delavayi	J.M. Delavay
forrestii	G. Forrest
florindae	Florinda Ward (F. Kingdon Ward's first wife)
hookeri	W.J. & J.D.
hookerianus	Hooker
wilsoniae	Mrs E.H. Wilson
wilsonii	E.H. Wilson

■

NEW ZEALAND SPINACH *(Tetragonia expansa)*. A prostrate, somewhat succulent plant in the *Aizoaceae* from the coasts of Tasmania, New Zealand, the Pacific Islands and S. America. It has OVATE to triangular-ovate 50–100mm long leaves with a crystalline texture, and small, yellow petal-less flowers. The leaves are eaten as spinach.

NID|US, -A, -UM Nest or nest-like.

NIDUS-AVIS Resembling a bird's nest, e.g. bird's nest orchid.

NIG|ER, -RA, -RUM Black or almost so.

▲ **NIGHTSHADE** A name applied mainly to several members of the *Solanaceae*, annuals, perennials and sub-shrubs from temperate climates: black nightshade *(Solanum nigrum)* is an annual of cosmopolitan distribution, 15–120cm tall with angular-ovate to almost diamond-shaped, sinuately toothed leaves. The 9mm wide white flowers have five, REFLEXED, pointed lobes and a cone of yellow stamens. They are followed by 6mm globular, black berries. *See also* DEADLY NIGHTSHADE, BITTERSWEET [Woody nightshade], ENCHANTER'S NIGHTSHADE.

NIGRICANS, NIGRESCENS Blackish.

NIPPLEWORT *(Lapsana communis)*. An erect annual in the *Compositae* from Europe, Asia and N. Africa and naturalised in N. America. It has LYRATE-PINNATIFID leaves, the large terminal leaflet, coarsely SINUATE toothed, and PANICLES of 13–19mm wide yellow flower heads entirely composed of LIGULATE florets.

NITENS [NITID|US, -A, -UM] Shining or polished.

NITRIFYING BACTERIA Nitrogen-fixing, soil-living bacteria which produce nitrogen compounds. *Nitrosomonas* oxidises the ammonia salts from decaying plant and animal remains to nitrite, while *Nitrobacter* oxidises these nitrites to nitrates, a form of nitrogen usable by plant roots.

NIVAL|IS, -E Associated with snow, or snow-white.

NIVE|US, -A, -UM The colour of snow.

NODE That part of a stem, often slightly swollen, where the leaf is joined.

■ **NOMENCLATURE** It is obviously difficult to discuss or classify any object which has no name, and plants are no exception. Those with food and medicinal values were probably the first to acquire vernacular names, and plants with colourful and appealing flowers would obviously have been identified early; but the majority remained nameless. As soon as botanists started to catalogue and describe the plants around them, the need for a uniform, meaningful system of naming became apparent. The early attempts at Latin descriptive names were cumbersome (all scientific work was then carried on in Latin), some plants being named by a whole descriptive sentence. Finally LINNAEUS devised an abbreviated method of naming which became known as the binomial system, each plant being given two names. They can be likened to a human surname (the first, generic name) and Christian name (the second or specific name).

For a plant to be authentically named, it must be described in Latin and published in an approved journal or book. In the original description and in scientific papers on plants, also in serious botanical books such as floras, the name of the author follows the specific epithet of the plant. For example, the common daisy is '*Bellis perennis* Linnaeus', Carl Linnaeus being the author. To avoid making names too cumbersome, however, there is an internationally accepted abbreviation for each author's name. Linnaeus, being the originator of the binomial system and known to all botanists, has his name abbreviated to L. In general, other abbreviations are less drastic. Where a small f. follows the author's name it indicates son (fils), e.g. 'Hook.' stands for Sir W. J. Hooker, 'Hook. f.' for his son, Sir J. D. Hooker. Quite often, further study and research makes it necessary to re-classify a genus, placing some of the species in other or newly created genera. When this is done, the name of the original author is placed in parenthesis followed by that of the re-classifying author. For example, Linnaeus described the tree we know as Norway spruce as '*Pinus abies* L.'. Later, G. K. W. Karston placed it in the spruce genus *Picea* and thus it became '*Picea abies* (L.) Karst.' The scientific names used for plants are often derived from other languages than Latin, but are always given the grammatical form of that language. In general they are descriptive, commemorative or geographic. Descriptive names usually highlight a major character of the plant, its flower colour, hairiness (or lack of hair), leaf shape, etc. Commemorative names honour the discoverer, a relative or distinguished colleague. The endings of these specific names are a key to their meanings. Those derived from a personal name have a terminal *-i* or *-ii*; these tell that the discoverer was a man, *-ae* or *-iae*, a woman. If *-iana* or *-ana* ends the name, it honours the memory of a person usually deceased. Geographical names most often end in *-icus* or *-us*.

NORFOLK ISLAND PINE *(Araucaria heterophylla* syn. *A. excelsa).* An elegant, coniferous tree to 60m in the *Araucariaceae* from Norfolk Island in the Pacific. It has an erect, central trunk and regular, well-spaced WHORLS of horizontal branches clad with firm, incurving, awl-shaped, 6mm long leaves. The somewhat flattened, globular cones are 75–113mm wide. Young specimens are often grown as pot plants in the home and have bright green, spreading leaves to 13mm long.

NUCELLUS The tissue surrounding the EMBRYO SAC in the OVULE.

NUCLEOLUS Small, granular, densely-staining bodies in the nucleus and containing proteins and nucleic acid.

NUCLEUS
see CELL; CELL DIVISION

NUDIFLOR|US, -A, -UM Naked flowers; i.e. borne before the leaves.

NUD|US, -A, -UM Naked, devoid of an expected covering.

NUTANS Nodding or pendent.

NUTATION The slow, lateral, often circular movement that takes place as the tip of a stem elongates.

NUTLETS Small, nut-like ACHENES, typical of the dead nettle family *(Labiatae).*

NUTMEG *(Myristica fragans).* An evergreen tree to 9.1m in the *Myristicaceae* from the Moluccas. It has 50–150mm long aromatic, oblong-lanceolate leaves and DIOECIOUS petal-less, yellow flowers with bell-shaped, 3-lobed waxy CALYCES. The broadly pear-shaped, fleshy fruits are 50–88mm long with a central seed, the nutmeg of commerce. Surrounding the seed is a red, LACINIATE ARIL known as mace and used in savoury dishes and pickles. The nutmeg itself is used to flavour milk dishes, cakes and drinks. Both contain a narcotic substance which is poisonous in quantity.

NUTS Hard INDEHISCENT fruits usually containing one seed and sometimes surrounded by protective bracts, e.g. HAZEL, BEECH, SWEET CHESTNUT.

● **OAK** *(Quercus).* A genus of 450 evergreen and deciduous trees and shrubs in the *Fagaceae* from the northern hemisphere, S. America and S.E. Asia. They vary in the shape, size and lobing of the simple leaves, but are easily distinguished by the fruits known as acorns. These are ovoid nuts, borne in cup-shaped INVOLUCRES which may be very shallow or deeply goblet-shaped. The tiny petal-less flowers are MONOECIOUS, the males in slender catkins, the females solitary or in small clusters. Several species have valuable timber and bark used for tannin. Cork is the bark of *Q. suber,* an evergreen tree to 18m or more with 25–63mm long OVATE leaves, grey-felted beneath, from S. Europe and N. Africa. The 50–100mm thick spongy cork-bark can be stripped off every 8–10 years; common or pedunculate oak *(Q. robur)* grows to 30m or more with 50–125mm OBOVATE, lobed leaves and stalked acorns; sessile oak *(Q. petraea)* is very similar but has almost sessile acorns; evergreen or holm oak *(Q. ilex)* reaches 27.4m or more, with 25–63mm long, LANCEOLATE to OVATE, almost entire to prickle-toothed leaves; Turkey oak (*Q. cerris)* can attain 36.4m with 50–100mm deciduous leaves and the acorns in mossy cups; scarlet and red oak *(Q. coccinea* and *Q. borealis)* are deciduous, N. American trees to 45m, having obovate 75–150mm long leaves with prominent, pointed lobes and bright autumn colour; American evergreen or live oak *(Q. virginiana)* grows to 18.2m with entire, leathery, 50–100mm elliptic, oblong leaves. *See also* SILKY OAK.

OAK APPLE Spongy, red-flushed, wrinkled, crab-apple-like galls found on oak twigs in spring. Not to be confused with the commoner, smooth, hard pale brown marble galls.

OAK FERN *(Gymnocarpium dryopteris* syn. *Thelypteris dryopteris).* A RHIZOMATOUS perennial in the *Aspidiaceae* from the north temperate zone. It forms colonies of individual long-stalked fronds each 100–300mm high, bi- or tri-pinnate and triangular in outline. The tiny SORI are borne near the margins of the PINNAE.

OAR WEED Several species of brown seaweed, but mainly cuvie *(Laminaria hyperborea)* in the *Laminariaceae* from cold and temperate waters. It has a 30–120cm, rigid, cylindrical stem or stipe and a fan-shaped leaf 300–600mm wide divided into strap-shaped segments which are shed

annually. Very similar is tangle *(L. digitata)* with a flexible stipe. Sea belt *(L. saccharina)* has a 300mm stipe and a 0.6–1.5m blade about 100–150mm wide with frilled margins. When dry it has a sugary deposit and is sometimes known as sugar wrack. It is also called poor man's weather glass, the dried fronds becoming limp with the approach of rain.

OAT *(Avena sativa)*. A cereal-producing annual in the *Gramineae* probably from E. Europe. It is a tufted grass to about 1.2m with broad, flat leaves and loose wide PANICLES of nodding, narrowly ovoid spikelets to 25mm long. It is thought to be derived from the wild oat *(A. fatua)*, a similar plant but with awned spikelets and hairy grains tightly enclosed in the lemma and palea.

OAT GRASS Several species of *Helictotrichon*, tufted perennials in the *Gramineae* from Europe and Asia: hairy or downy oat grass *(H. pubescens)* is a loosely tufted plant, 300–900mm tall with hairy leaf sheaths and narrow, erect PANICLES of glistening green or purplish awned spikelets; meadow oat grass *(H. pratense)* is densely tufted with smooth leaf sheaths.

OBCORDATE Reversed heart-shaped.

OBLANCEOLATE The reverse of lanceolate: the stalk at the narrow end of the leaf, which widens to the tip.

OBLONG A leaf, bract or petal with the sides parallel for some distance, the ends tapering.

OBOVATE The reverse of OVATE: the stalk at the narrow end, the leaf broadening to the tip.

OBSOLETE Referring to plant organs which are rudimentary or missing, e.g. petals, stamens.

OBTUS|US, -A, -UM With a blunt or rounded end.

OCCIDENTAL|IS, -E From the west, usually applying to America.

OCEANIC|US, -A, -UM In or near the sea.

OCHROLEUC|US, -A, -UM Whitish-yellow or pale bluff.

OCOTILLA [COACH WHIP] *(Fouquieria splendens)*. A curious desert shrub in the *Fouquieriaceae* from S. W. America and Mexico. It has several 1.8–6.0m tall, spiny, erect, cane-like stems branching from near ground level. In the AXILS of each spine is a cluster of 13–25mm long fleshy, OBOVATE leaves which only appear after the spring and autumn rains and soon fall. The scarlet, tubular, 19–25mm long flowers are borne in narrow PANICLES.

OCTANDROUS Flowers with 8 stamens.

OCTOGYNOUS Flowers with 8 styles.

OCTOPETALOUS [OCTOPETAL|US, -A, -UM] Flowers with 8 petals.

ODORAT|US, -A, -UM [ODOR|US, -A, -UM] Fragrant.

OFFICINAL|IS, -E Used of plants with medicinal value.

OFFSET A modified shoot borne at the end of a short stalk from the base of a plant, mainly occurring in houseleeks. They readily become detached and grow into new plants.

OIL Many plants produce oils as food reserves, mainly in seeds and storage organs such as tubers. They are formed as minute droplets in the cytoplasm or cell sap. Oil is also secreted by special glands on leaves and stems, particularly in plants from warm, dry climates. These often aromatic oils are volatile and seem to slow down water loss on still, hot days. Some of the commercially important oil producing plants are; oil palm, cotton, olive, sesame, ground nut, soy bean, sunflower, rape, maize, flax (linseed), and safflower. In all these examples the seeds are used.

OLD MAN
see ARTEMISIA

OLD MAN'S BEARD
see TRAVELLER'S JOY

OLEANDER *(Nerium oleander)*. An evergreen shrub to 5.4m in the *Apocynaceae* from the Mediterranean region and much grown in other warm countries. It has leathery, LINEAR-LANCEOLATE 100–150mm long leaves and terminal CYMES of tubular flowers which open out to 5 flat lobes. The wild plant has 38mm wide pink of white blooms but CULTIVARS range up to 75mm and occur also in shades of crimson and purple.

OLEASTER *(Elaeagnus angustifolia)*. A deciduous shrub or tree to 6m in the *Elaeagnaceae* from S. Europe and W. Asia. It has spiny stems, glistening silvery-scaly when young and 38–88mm long, narrowly oblong to LANCEOLATE leaves. The silvery, fragrant, bell-shaped flowers are 4-lobed and followed by 13mm oval, yellowish berries with silvery scales.

OLERACE|US, -A, -UM Kitchen garden plants, usually vegetables, but sometimes the weeds that grow with them.

● **OLIVE** *(Olea europaea)*. A long-lived, round-headed, evergreen tree to 12.2m in the *Oleaceae* from the Mediterranean region and grown elsewhere. It has pairs of 38–75mm long narrowly OBOVATE leaves, grey-green above, white beneath and tiny, white 4-petalled flowers in AXILLARY RACEMES. The oval, oily fruits are plum-like, 19mm long, purplish when ripe. They are pickled both green and ripe or pressed for oil.

ONION *(Allium cepa)*. A bulbous-rooted plant in the *Alliaceae* probably from W. Asia. It has two ranks of tubular 300–600mm leaves and 600–900mm stems bearing large, spherical UMBELS of greenish-white, 6-tepalled flowers. There are many CULTIVARS with white, yellowish or reddish-skinned bulbs that range from flagon-shaped to flattened globes. They are grown annually from seeds or tiny bulbs called sets. The Spanish onion is a white skinned sort; Egyptian or tree onion *(A. cepa aggregatum)* forms clumps of narrow bulbs and bears heads of BULBILS instead of flowers. *See also* WELSH ONION.

▲

▲

ONTOGENY The life history of any particular plant from seed to flowering and fruiting at maturity.

OOGONIUM A specialised cell that produces one or more eggs, mainly in certain filamentous algae and fungi, e.g. mildew *(Peronospora)*.

OPERCULUM The lid of a moss spore capsule.

OPPOSITE [OPPOSITIFOLI|US, -A, -UM] Of two leaves that arise on opposite sides of the same point on a stem.

ORACHE *(Atriplex)*. A genus of 200 annuals or shrubs in the *Chenopodiaceae* from temperate and subtropical regions. The leaves may be toothed or lobed sometimes entire, and often more or less triangular or RHOMBOID, though occasionally LINEAR. The tiny, unisexual flowers are petal-less, the female enclosed by two persistent BRACTEOLES which enlarge in fruit. The orache of gardens is *A. hortensis*, a 0.9–1.8m annual with 100–125mm long oblong leaves and is used as a spinach substitute; *A.h. atrosanguinea* has red leaves; the wild or common orache, *A. patula*, is an erect or semi-prostrate annual spreading to 900mm with 38–75mm linear oblong leaves.

ORANGE
See CITRUS

ORANGE PEEL FUNGUS *(Aleuria aurantia)*. A common cup fungus in the *Ascomycetes*. It forms clusters of irregularly shaped or split-cup-like fruiting bodies, orange within and downy without.

ORBICULAR|IS, -E Rounded or disk-shaped.

▲**ORCHID** All members of the *Orchidaceae*, highly specialised *Monocotyledons* of cosmopolitan distribution. Many are EPIPHYTIC with swollen stems known as PSEUDOBULBS, and AERIAL roots. Others are ground-dwelling with or without root tubers. The leaves are simple, ranging from oval to linear, those of the epiphytes sometimes fleshy. The flowers are often large and showy and very complex, adapted to be pollinated by a wide range of insects, some of which they mimic. Basically, each flower has 3 sepals, frequently petal-like in appearance and colouring, 2 often similar petals and a modified third petal known as the LABELLUM or lip. The stamens are combined with the PISTIL in various ways to form a central organ called the column. There is only one, sometimes two stamens, each lobe bearing a club-shaped sticky mass of pollen grains called a pollinium. These pollinia are readily carried by insects from flower to flower, often stuck onto them in such a way that they can only be rubbed off on to the stigma. The labellum may be a fairly simple alighting platform for insects or a trap to ensure that visitors carry out pollination, e.g. slipper orchids. In many cases, the labellum is shaped like the body of the insect pollinator, and in its attempt at copulation with this model, pollen is transferred to the stigma. Orchids are closely associated with fungal MYCORRHIZA and the seeds will not germinate properly, or sometimes not at all, without their presence. Horticulturalists have overcome this problem by culturing the seeds in sterile flasks of agar jelly, sugar and minerals. *See* BEE, BIRD'S NEST, EARLY PURPLE, FLY, LADY, LADY'S SLIPPER, LIZARD, MAN, MARSH, MILITARY, MIRROR, MONKEY, SPIDER ORCHIDS.

ORDER The unit of classification above the family, each one containing few to many closely related families. Names of orders end in *-ales,* e.g. *Ranales* contains the buttercup family *(Ranunculaceae)*, magnolia family *(Magnoliaceae)*, and water lily family *(Nymphaeaceae)*.

OREGON CEDAR
see CYPRESS (Lawson).

OREGON GRAPE *(Mahonia aquifolium)*. A suckering, evergreen bush to 1.8m in the *Berberidaceae* from west N. America. It has PINNATE leaves with 38–75mm long OVATE, glossy, prickly marginal leaflets and terminal clusters of short RACEMES bearing yellow, bowl-shaped, 6-petalled flowers. These are followed by blue-black edible berries like small grapes.

ORIENTAL|IS, -E Eastern. From the Orient (Asia).

ORNITHOPHILY Flowers pollinated by birds.

ORPINE
see STONECROP

ORRIS ROOT
see IRIS

ORTHOTROPOUS
see OVULE

OSAGE ORANGE [BOW WOOD] *(Maclura pomifera)*. A thorny, deciduous tree to 12.2m in the *Moraceae* from N. America. It has ovate-acuminate, 38–100mm long leaves and tiny green, unisexual flowers, the males in catkin-like clusters the females in globose heads. The 63–100mm wide orange-like fruit is an aggregate one, being composed of numerous small DRUPELETS, one from each flower.

OSMOSIS Process occurring when a solution of water and sugar, or dissolved minerals, is divided from pure water by a membrane which allows water only to pass through (a semi-permeable membrane), and flow into the sugar solution. This water flow is osmosis and it sets up what is termed osmotic pressure. Combined with the pull of TRANSPIRATION, from the leaves above, it enables water to reach the topmost twig of the highest trees. The root hair membranes are not totally semi-permeable, however, and allow certain dissolved soil minerals to move in with the water.

OSTIOLE
see CONCEPTACLE

OSTRICH FERN *(Matteuccia struthiopteris)*. A stoloniferous fern in the *Aspidiaceae* from the north temperate zone. It forms elegant vase-shaped rosettes of LANCEOLATE, bipinnate, sterile fronds to 1.5m in height. The fertile fronds are smaller, with the linear pinnae largely occupied by SORI.

OSWEGO TEA
see BERGAMOT

OVAL|IS, -E Oval or elliptic; broadest in the middle, narrowing to rounded ends.

OVARY A CARPEL or fused group of carpels at the base of the flower and in which the ovules are borne.

OVATE [OVAT|US, -A, -UM] Egg-shaped in outline, with the stalk at the broadest end.

OVOID Egg-like as some fruits, the stem at the broad end.

OVULE A minute egg-shaped body, consisting mainly of PARENCHYMATOUS cells called the NUCELLUS and attached to the ovary by a short stalk or funicle. The nucellus is sheathed by one or two protective coats or layers known as integuments, except for the extreme tip where a narrow opening is left. This is the MICROPYLE, the way through for the pollen tube. Lying within the nucellus is the embryo sac, the vital female organ containing the egg cell or ovum, which, after fertilisation, becomes the embryo plant within the seed. Ovules are carried in the ovary in three basic positions. In the majority of cases they are anatropous, the FUNICLE bending sharply over so that ovule and micropyle face downwards. Orthotropous ovules are erect with the micropyle furthest from the funicle. In the campylotropus ovule, the nucellus is curved, also bringing the micropyle to a down-facing position.

OVULIFEROUS SCALE The scales which form the female 'flower' or strobilus of a *Gymnosperm* and upon which are borne the ovules.

OVUM
see OVULE

● **OXALIS** A genus of 800 annuals, perennials and shrubs in the *Oxalidaceae*, of cosmopolitan distribution, but chiefly from S. America. They are typified by having TRIFOLIATE leaves, 5-petalled flowers and seeds with elastic ARILS which turn violently inside out, shooting the seed several feet away. Some of the perennial species have tubers or bulbils and are troublesome weeds in gardens and orchards. In this category is Bermuda buttercup *(O.pes-caprae)* with stems to 300mm, bearing nodding clusters of 25mm long, bright canary yellow, funnel-shaped blooms. A native of S. Africa, it is now a pest of many warm countries. *See also* WOOD SORREL.

OX-EYE DAISY [MARGUERITE] *(Chrysanthemum leucanthemum)*. An erect, 300–900mm tall clump-forming perennial in the *Compositae*. It has OBOVATE to SPATHULATE, boldly CRENATE, toothed or lobed leaves and 38–50mm wide daisy-flower heads with yellow disk and white ray florets.

OXLIP
see PRIMULA

OXYPETAL|US, -A, -UM With pointed petals.

OXYPHYLL|US, -A, -UM With pointed leaves.

OYSTER MUSHROOM *(Pleurotus ostreatus)*. A parasitic *Basidiomycetes* fungus found on beech and other trees in temperate climates. The 38–150mm wide, lop-sided, bracket-like cap is dark blue-grey when young, brownish when older, and the gills are white or yellow-tinged. It is edible but rather tasteless.

OYSTER PLANT [NORTHERN SHOREWORT] *(Mertensia maritima)*. A slightly succulent, maritime perennial in the *Boraginaceae* from Atlantic coasts of Europe north to Iceland. It has prostrate stems to 600mm, GLAUCOUS, OVATE to OBOVATE, 19–50mm long leaves in two rows, and terminal CYMES of 9–13mm long, tubular flowers which open pink and change to blue and pink.

PACHYCARP|US, -A, -UM Thick fruited.

PACHYPHYLL|US, -A, -UM With thick, often fleshy leaves.

PACHYRHIZ|US, -A, -UM With thick roots or RHIZOMES.

PADDY Flooded fields for growing rice.

PAGODA TREE *(Sophora japonica)*. An elegant, deciduous tree to 24.4m in the *Leguminosae* from China, Korea and much grown in Japan. It has 150–250mm PINNATE leaves of 9–17 ovate leaflets and PANICLES of 13mm long white pea-flowers.

PAIGLE
see PRIMULA (cowslip)

PALAEOBOTANY The study of fossil plants.

PALATE The inflated or projecting part at the mouth of a 2-lipped, tubular flower, partially or wholly closing it, e.g. snapdragon.

PALEA [PALE]
see GRASS

PALLENS [PALLID|US, -A, -UM] Pale, usually of colour.

PALLISADE TISSUE
see MESOPHYLL

PALM All members of the *Palmae*, a well-defined, MONOCOTYLEDON family of 2500 trees, shrubs and climbers in 217 genera, mainly from the tropics and subtropics. Typically they have erect, unbranched stems or trunks and terminal heads of PALMATE (fan palms) or PINNATE (feather palms) leaves. There are exceptions, the doum palm and some others having branches and the climbing rattan palms have well-spaced leaves. The large, branched, RACEMOSE inflorescences are often pendulous and known as SPADICES (singular: spadix). They are protected, at least when young, by large bracts called spathes. The often yellow flowers have a PERIANTH of 6 tepals, 6 stamens and a one- or three-celled ovary. The fruits are drupes or berries, often of large size. Palms have many uses and are very important in the tropics providing oil, sugar, fruit, timber, etc: coconut *(Cocos nucifera)* is a 24–30m tree with feather leaves to 6m and seeds (nuts) enclosed in a thick fibrous husk which is removed before importation to Britain. The sun-dried 'meat' of the nut is copra and provides oil for margarine, etc. The fibre is used for matting, ropes, etc; date palm *(Phoenix dactylifera)* is a stiff tree to 30m with grey-green, feather leaves and DIOECIOUS flowers. The cylindrical-ovoid fruits are plum-like before processing; oil palm *(Elaeis guineensis)* grows

● to 15m tall with feather leaves to 4.5m and up to 200, ovoid, orange fruits in each rounded cluster; sugar palm *(Arenga saccharifera)* grows to 12.2m with feather leaves having 0.9–1.5m long leaflets, silver beneath. When the flowers appear, the spadix is cut, the freely flowing sap collected and boiled down to obtain molasses and sugar; it is also fermented to produce palm toddy. Several other palms yield a sweet sap, notably the palmyra palm *(Borassus flabellifer)*, a fan-leaved tree to about 21m with large, 3-lobed nuts which yield a refreshing drink and can be eaten when young; toddy is also brewed from the fish-tail or wine palm *(Caryota urens)*, a distinctive 30m species with leaves to 6m long composed of oblique, fish-tail-like leaflets; sago or swamp palm *(Metroxylon sagu)* has large feather leaves and lives for about 15 years before it flowers and dies (MONOCARPIC). Just before flowering, the trunk is full of starch and is then felled, the pith scraped from it and the starch washed out. Fishtail and sugar palms also produce sago (*see also* CYCAD); thatch and palmetto palms *(Sabal)* grow 12–24m with large fan-leaves used for thatching in the W. Indies and elsewhere; ivory nut palm *(Phytelephas macrocarpa)* has a creeping stem with feather leaves to 6m high and seeds almost as hard as ivory and used to carve figurines, etc.; rattan palms (several species of *Calamus*) have slender, climbing stems sometimes 30m or more long and feather leaves. The stems are used for baskets, chair battens, furniture and walking sticks; royal palm *(Roystonia regia)* grows to 18m or more with a smooth grey trunk often curiously bulged about the middle and feather leaves to 3m. It is often grown as an ornamental; cabbage palm *(R. oleracea)* can reach 36.6m with 3–4.5m leaves; the central growing point is tender and edible. Many other palms have edible hearts, e.g. coconut; betel-nut palm *(Areca catechu)* is a slender tree, rarely to 30m with feather leaves to 1.8m and ovoid, red or orange 38–50mm long seeds which are chewed with betel pepper leaves in S.E. Asia and E. Africa; doum palm *(Hyphaene)* is curious in having trunks that fork into two, once to several times, forming an open, branched tree to 12.2m with a tuft of fan-shaped leaves at each branch tip. Fan palms are numerous, the best known being species of *Chamaerops* and *Trachycarpus*. *C. humilis* is the only native European palm, a bushy species to 3m, but often less with grey-green, fan leaves 450mm wide; raffia palm *(Raphia ruffia)* has a trunk to 7.5m and feather leaves to 18.2m in length. The fibres from these form the tying material used by gardeners and also are used for mats, ropes, etc.

PALMATE Used for leaves shaped like outspread hands, usually with 5 or more leaflets, the mid-veins of each one uniting at the stalk.

▲ **PALMATIFID** Leaves lobed in a palmate fashion to half their length.

■ **PALMATISECT** Lobed palmately almost to the base.

PALMETTO
see PALM

PALUDOS|US, -A, -UM [PALUSTRIS, -E] Growing in marshes, swamps or bogs.

PALYNOLOGY
see POLLEN ANALYSIS

▲ ■

PAMPAS GRASS *(Cortaderia)*. A genus of 15, densely tufted, large, perennial grasses in the *Gramineae* from temperate S. America and New Zealand. They have long, tapered, arching leaves and tall stems bearing large PANICLES of small, DIOECIOUS spikelets. The pampas grass of gardens is *C. selloana* (syn. *C. argentea, Gynerium argenteum*) having GLAUCOUS leaves 0.9–1.8m long and flowering stems to 3m with silvery-white pyramidal panicles, sometimes tinged red or purple.

PANICLE An inflorescence made up of branched RACEMES or CYMES.

PANNOS|US, -A, -UM Felted, densely covered with hairs.

PANSY [HEARTSEASE] *(Viola tricolor)*. An annual or perennial semi-erect plant in the *Violaceae* from Europe and Asia. It has oval to OVATE-CRENATE leaves with prominent, lobed stipules. *Viola tricolor* is the wild pansy with 5-petalled, violet-blue, yellow or bicoloured flowers. The lower petals bear an upturned spur. The garden pansies are derived from it, mainly as hybrids with the mountain pansy *(V. lutea)*.

PAPER MULBERRY *(Broussonetia papyrifera)*. A deciduous small tree to 9.1m in the *Moraceae* from E. Asia and Polynesia. The broadly, OVATE-CORDATE, toothed leaves are sometimes bi- or tri-lobed. The flowers are unisexual, the males in cylindrical catkins to 900mm and the females in 13mm wide, globose heads. Later the small red fruits fuse together to form a multiple fruit. Tapa cloth is made by the Polynesians from the beaten bark and a good paper can also be made from it.

PAPILLAE Small, soft nipple-like protuberances on a leaf, stem, fruit, etc.

PAPPUS In a general sense, the tuft of hairs on a fruit or seed to assist distribution by wind, e.g. willow-herb. In the strict botanical sense, it is used for the bristles, hairs or scales which represent the CALYX on the top of fruits belonging to members of the *Compositae*. These are often a very efficient parachute, e.g. in dandelion.

PAPYRIFER|US, -A, -UM Used for paper-making, or bearing papery bark.

PAPYRUS *(Cyperus papyrus)*. A 2.4–3.0m tall, aquatic grass-like perennial in the *Cyperaceae* from N.E. Africa. The main part of the plant consists of clumps of dark green stems, triangular in cross-section and tiny yellowish flowering spikelets carried in large, terminal, spherical heads. The ancient Egyptians made paper from it by slicing the pith from the middle of the stems, laying the strips edge to edge then pressing and drying them.

PARADISIAC|US, -A, -UM From paradise, or sometimes of gardens.

PARADOX|US, -A, -UM Contrary to what is expected.

PARAMO Alpine grassland areas in the northern Andes of S. America. Characteristic plants are the frailjones *(Espeletia)*, members of the *Compositae* with a remarkable resemblance to the giant groundsels of the tropical African mountains.

PARAPHYSIS Sterile, multicellular hairs found in the ANTHERIDIAL and ARCHEGONIAL cups of mosses and the reproductive CONCEPTACLES of certain seaweeds, e.g. wrack. The sterile BASIDIA produced in certain mushrooms and toadstools are also known as paraphyses.

PARASITE A plant which gains all or some of its food supply by robbing others. Many fungi and bacteria are common examples and some flowering plants also, including BROOMRAPES and TOOTHWORT. These lack CHLOROPHYLL and are therefore unable to make their own food, being total parasites, relying entirely upon the host for a food supply. Mistletoe has green leaves and can elaborate its own sugars and starches and is a semi- or hemi-parasite. Some hemiparasites only supplement their food supply from a host plant, e.g. EYEBRIGHTS, YELLOW RATTLE, BARTSIA (red), etc. which attach HAUSTORIA to grass roots but also have their own root systems. All these live on or beside their host plants with haustorial attachments (exoparasites). Parasitic fungi live mostly within the host, sending HYPHAE between or into the cells; these are endoparasites.

PARASOL MUSHROOM *(Lepiota procera)*. A *Basidiomycete* fungus from grassy places amongst scrub or at the edges of woods in temperate climates. It grows to about 175mm tall with a 100–200mm wide convex cap raised in the centre like an old-fashioned parasol. It is grey-brown with shaggy, darker brown scales and a membranous double ring on the stipe (stalk). It is edible with a rich mushroom flavour.

PARDALIN|US, -A, -UM Spotted like a panther.

PARENCHYMA Simple unmodified cells which form the background tissue of all plants.

PARIETAL PLACENTATION
see PLACENTATION

PARIPINNATE
see PINNATE

PARROT'S BILL [KAKA-BEAK] *(Clianthus puniceus)* An evergreen, scandent shrub to 3.6m in the *Leguminosae* from New Zealand. It has 75–150mm long PINNATE leaves and AXILLARY clusters of pendulous, 88–125mm long bright red pea-flowers with a long, cued, pointed keel. A white form is known.

PARSLEY *(Petroselinum crispum)*. An erect biennial in the *Umbelliferae*, probably from S. Europe but long cultivated. It has long-stalked, tripinnate leaves somewhat triangular in outline with CUNEATE, lobed leaflets. The much branched, 450–600mm stems bear 25–50mm UMBELS of tiny, yellowish flowers. The garden CULTIVARS have the leaflets tightly curled.

PARSLEY FERN *(Cryptogramma crispa)*. A tufted mountain fern in the *Cryptogrammataceae* from Europe and Siberia. It has 50–150mm tall, tripinnate sterile fronds, triangular-ovate in outline and longer stalked fertile ones. The leaflets of the latter have rolled margins and continuous bands of SORI.

PARSNIP *(Pastinaca sativa)*. A strong-smelling, erect biennial in the *Umbelliferae* from Europe and introduced into N. and S. America, Australia and New Zealand. It has a rosette of PINNATE leaves having OVATE, lobed and SERRATE leaflets. The 0.6–1.5m branched stems bear 38–100mm wide umbels of tiny yellow flowers.

PARTHENOCARPY The formation of fruits without seeds. This often happens naturally as in bananas and 'Conference' pear, but can be induced artificially.

PARTHENOGENESIS The formation of viable seeds without fertilisation, a characteristic of certain plants, e.g. dandelions and blackberries.

PARTIAL PARASITE
see PARASITE

PARTRIDGE BERRY (1). *(Gaultheria procumbens)*. A creeping evergreen shrub in the *Ericaceae* from east N. America with elliptic to obovate leaves, nodding urn-shaped white or pink-flushed flowers and bright red, globose berries.
(2) *(Mitchella repens)*. A prostrate evergreen shrub in the *Rubiaceae* from N. America, with pairs of ovate, pale-veined leaves, twinned white tubular flowers and double, red berries.

PARV|US, -A, -UM [PARVI-] Small or puny; parvifolius — of leaves; parviflorus — of flowers.

PASCUAL Growing in pastures.

● **PASQUE FLOWER** *(Pulsatilla vulgaris)*. A tufted, herbaceous perennial in the *Ranunculaceae* from Europe and W. Asia. It has bipinnate leaves with the leaves cut into LINEAR segments and 100–250mm stems bearing a WHORL of segmented bracts. The solitary, nodding to erect, violet-purple flowers have 6 petal-like PERIANTH segments and are followed by plumed ACHENES. Reddish and white forms are grown in gardens.

PASSAGE CELLS Those cells in the ENDODERMIS of a root opposite the PROTOXYLEM groups. They do not become thickened with cellulose like the rest.

PASSION FLOWER *(Passiflora)*. A genus of 500 species, mainly tendril climbers in the *Passifloraceae*, the majority from S. America, also from N. America, Australasia and Malagasy. The leaves vary from OVATE to PALMATE, often deeply lobed, and some have curiously bilobed leaves shaped like a boomerang or swallow-tail. The shallowly bowl-shaped or flat flowers have 5 sepals (sometimes coloured) 5 petals and the pistils and stamens on a central ANDROPHORE. Around the base of the androphore is a ring of filaments or thread-like outgrowths of the RECEPTACLE, called the corona. Missionaries to S. America saw the flowers as symbolical of Christ on the cross, hence the name. The pistil is the cross, the stigmas the nails and the corona the crown of thorns. Passion fruits are the 50–75mm long, ovoid purple or yellow berries of *P. edulis*, a vigorous plant with 100–150mm trilobed leaves and 75–100mm white fragrant flowers having purple-banded coronas. Granadilla is the name given to the fruits of

several *Passiflora* species, including *P. edulis*, but best belonging to *P. quadrangularis*, a robust climber with ovate-cordate, 100–200mm long leaves and 113mm wide bowl-shaped white or pink-flushed flowers. These have very long, wavy coronas cross-banded blue, purple and white. The yellow-green fruits are ovoid. Water lemon, belle apple or Jamaica honeysuckle *(P. laurifolia)* is similar, but the orange-yellow fruits are only 50–75mm long; common passion flower *(P. caerulea)* is almost hardy outside in Britain and is often grown as a pot plant. It has deeply lobed, palmate leaves and 50–75mm white or pinkish flowers with blue and purple coronas. The 25–32mm orange-yellow fruit are broadly OVOID.

PASTURE A meadow regularly grazed by livestock.

PATCHOULI *(Pogostemon cablin)*. A 0.9–1.2mm sub-shrub in the *Labiatae* from the Philippines, but much cultivated in Sumatra, Seychelles, Malagasy and Brazil. It has pairs of OVATE leaves, and whitish tubular flowers in small WHORLS. It yields a scented, volatile oil used in perfumery, soap, hair tonics and tobacco.

PATELLIFORM|IS, -E Shaped like a dish or plate.

PATENS Spreading or opening out widely.

PAUCIFLOR|US, -A, -UM Few-flowered.

PAUCIFOLI|US, -A, -UM Few-leaved.

▲ **PAW PAW** *(Carica papaya)*. An unbranched tree-like perennial in the *Caricaceae* from Central America, but much planted in the tropics. Has a stem to 6m with a tuft of long-stalked, 300–750mm wide, palmate leaves having 7–11 lobes, each lobe irregularly PINNATIFID. The creamy or yellow flowers are DIOECIOUS, the tubular 25mm long males with 5 lobes, the females with 5 separate lobes. There is an hermaphrodite cultivar 'Solo'. The yellow or orange fruits are ovoid to pear-shaped, about 75–300mm long. The leaves and unripe fruits contain papain, a protein ferment used in digestive salts. Mountain paw-paw *(P. cundinamarcensis)* from Ecuador is a smaller, hardier, branched plant with angular yellow fruits.

PEA Used as a suffix, this refers to several members of the *Leguminosae*. More specifically the garden pea *(Pisum sativum)*, a climbing annual probably native of W. Asia but long cultivated in S. Europe and now in all temperate climates. It has PINNATE leaves with 4–6 OVATE leaflets and 2 large SESSILE, CORDATE STIPULES. The flowers are white, reddish or purplish and are followed by oblong, cylindrical pods of spherical seeds (peas). It is a very variable plant ranging in height from 0.3–1.5m. The common garden pea can be classified into two main groups: those with sweet seeds that are wrinkled when dry, the marrow fats, and the more floury tasting, rounded seeded sorts. There are also sugar peas or mangetout CULTIVARS, the young pods of which are eaten whole. 'Petit Pois' has very small seeds of fine flavour. *See also* CHICK, COW, EVERLASTING, GLORY, PIGEON and SWEET PEAS.

PEACH *(Prunus persica)*. A deciduous tree to 7.5m or so in the *Rosaceae* from China, but much cultivated elsewhere. It has 75–150mm long slender-pointed, finely toothed, LANCEOLATE leaves and 5-petalled cup-shaped,

pink flowers before the leaves. The finely pubescent yellow or red flushed fruits are 50–75mm wide DRUPES with a large fissured stone. Nectarine is a smooth, hairless peach. Red, white, pink and double-flowered CULTIVARS are grown as ornamentals, as is David's or China peach *(P. davidiana)*, a similar tree with white or rose blooms.

PEA-FLOWER The hallmark of the *Leguminosae* or pea family. It is a ZYGOMORPHIC flower with 5 petals; the 2 lower ones fit together around the PISTIL and stamens and are known as the keel, the lateral ones (wings) usually point forwards on either side of the keel and the much larger upper or standard petal stands erect behind providing the main floral impact. Flowers of this form are found only in the *Papilionoidae* division of the *Leguminosae*. The two other divisions *(Mimosoidae* and *Caesalpinioidae)* having regular or slightly zygomorphic flowers without a keel.

PEAR *(Pyrus communis)*. A deciduous tree to 15m in the *Rosaceae* from Europe and Asia. It has broadly OVATE, 38–88mm leaves with rounded or CORDATE bases and CORYMBS of 5-petalled, cup-shaped white flowers as the leaves unfurl. The fruits may be rounded to oval or tapered in the typical pear shape, yellow to russet, sometimes flushed reddish. The pear has been a popular fruit for many centuries and at least 39 different sorts were known in Roman times. Today there are many hundreds of CULTIVARS. *See* AVOCADO PEAR (Alligator pear) and CACTUS (Prickly pear).

PEARL MILLET Another name for bulrush millet.
see MILLET.

PEARLWORT *(Sagina)*. A genus of about 30 species of small annuals and perennials in the *Caryophyllaceae* from the north temperate zone. They are prostrate or ascending, rarely erect plants with slender stems and LINEAR to LINEAR-LANCEOLATE leaves in pairs. The tiny flowers are usually solitary in the leaf AXILS with 4 or 5 sepals. Procumbent pearlwort *(S. procumbens)* is a prostrate perennial with linear 6–13mm long leaves and flowers with 4 hooded sepals and tiny white (or no) petals.

● **PEAT** The soil of bogs and fens composed largely of dead plant remains in a state of arrested decomposition due to lack of oxygen. Moss peat is largely composed of ▲ sphagnum moss; sedge peat, of the roots and leaves of sedges *(Carex)*.

PECTINATE, [PECTINAT|US, -A, -UM] Having closely and regularly placed leaves, leaflets or lobes like the teeth of a comb.

PEDATE, [PEDAT|US, -A, -UM] A deeply lobed, PALMATE leaf, the lobes again divided.

PEDICEL The stalk of an individual flower.

PEDICELLAT|US, -A, -UM Having stalked flowers.

PEDUNCLE The main stem of an inflorescence, usually bearing branches and several flowers, but in the case of solitary blooms, e.g. tulip, one bloom only.

PEEPUL
see BO TREE

● **PELARGONIUM** A genus of 250 shrubs and sub-shrubs in the *Geraniaceae,* mainly from Africa. They have OVATE, ORBICULAR or PALMATE leaves and 5-petalled flowers in UMBEL-like, stalked clusters. The fruit is like a stork's bill and splits into 5, one-seeded segments, each one having a sliver of the STYLE, bearing a plume of hairs. Some species have tuberous roots. Others are succulent, several are aromatic (an attar of roses substitute is obtained from one). Several special and many hybrids are grown as ornamentals, mainly under the name of *Geranium*, a genus now confined to the cranesbills. *P. zonale* and its hybrids *P. X hortorum* are the popular bedding and pot plant geraniums. They grow to 1.8m with orbicular, long-stalked, shallowly lobed or CRENATE leaves, often carrying a bronze-red horseshoe pattern. The flowers cover many shades of red, pink, purple and white, some cultivars having double blooms: regal geraniums or pelargoniums *(P. X domesticum)* are also of hybrid origin. They grow to 900mm or so with palmate, plain green, lobed and toothed leaves and larger flowers in smaller clusters in shades of red, white, purple, maroon, often as striking bicolours.

PELLITORY-OF-THE-WALL *(Parietaria judaica* syn. *P. officinalis, P. diffusa).* A perennial of rocks, walls and hedgebanks in the *Urticaceae* from Europe. It has ovate to lanceolate leaves to 63mm long and AXILLARY, CYMOSE clusters of tiny greenish unisexual flowers.

PELLUCID [PELLUCID|US, -A, -UM] Transparent or almost so.

▲ **PELOR|IA, -IC** Curiously symmetrical mutant flowers that arise on such plants as foxglove and toadflax with normally ZYGOMORPHIC flowers.

PELTATE [PELTAT|US, -A, -UM] An ORBICULAR leaf with the stalk attached at a point beneath the leaf, not at the edge as is usual.

PENCIL CEDAR
see JUNIPER

PENDUL|US, -A, -UM, [PENDULIN|US, -A, -UM] Drooping, hanging down.

PENICILLIN
see MOULD

■ **PENNYCRESS** Several species of *Thlaspi*, annual, biennial or perennial plants in the *Cruciferae* from the north temperate zone. The common field pennycress is an overwintering annual with a rosette of OBLANCEOLATE leaves and an erect 150–600mm leafy stem, often branched at the top with RACEMES of small, white, 4-petalled flowers. The flattened seed pods (siliculae) are almost circular, and 10–19mm wide.

PENNY ROYAL
see MINT

PENNYWORT
see MARSH PENNYWORT, NAVELWORT

PENSTEMON A genus of 252 annual, perennial and sub-shrubby plants in the *Scrophulariaceae*, all but 2 from N. America. Many are alpine or sub-alpine, others from semi-arid regions. They have simple leaves from LINEAR to OVATE, and RACEMES of ZYGOMORPHIC, tubular flowers with 5 lobes rather like a foxglove, in shades of red, purple and blue. The flowers are sometimes hairy within, giving them the vernacular name beard tongue. Several species are grown as ornamentals.

PENTAGYNOUS With 5 styles.

PENTAMEROUS With parts in fives, used of flowers.

PENTANDROUS With 5 stamens.

PEONY *(Paeonia)*. A genus of 33 herbaceous perennials and deciduous shrubs in the *Paeoniaceae* from the north temperate zone. They have pinnately or ternately lobed leaves and large saucer-shaped flowers with 5–10 broad petals and a boss of numerous stamens. Most of the species and many hybrid CULTIVARS are grown as ornamentals: common red peony *(P. officinalis)* is often seen in gardens in its double form; most of the herbaceous hybrids are derived from *P. lactiflora*, a 600mm species with biternate leaves and 75–100mm white, fragrant flowers.

PEPPER (1) Black or white pepper *(Piper nigrum)*. A twining, climbing plant to 100–150mm or more in the *Piperaceae* from S. E. Africa. It has slender-pointed 100–150mm broadly ovate leaves and spikes of tiny, petal-less flowers followed by red then black berries. The whole dried fruits are ground to give black pepper, the seeds only to give white pepper.
(2) Red and green peppers *(Capsicum annuum)*. A variable annual to short-lived perennial in the *Solanaceae* from the tropics. It grows to 300–900mm or more in height with OBLONG to OVATE leaves and nodding white flowers with 5 pointed lobes and a cone of yellow stamens. *C. a. acuminatum* includes chilli, cayenne and paprika with red, slenderly conical and often twisted fruits, 100–250mm long. *C. a. grossum* includes the large salad, red and green fruited peppers with ribbed, almost oblong fruits 75–125mm long and wide.

PEPPERWORT Several species of *Lepidium*, annual, biennial and perennial plants in the *Cruciferae* from Europe, some in Asia. Common pepperwort *(L. campestre)* is an annual or biennial with grey-green, LYRATE basal leaves and AMPLEXICAUL stem ones. The erect 200–600mm stems produce a cluster of short RACEMES of tiny, 4-petalled white flowers. *See* GARDEN CRESS.

PERENNIAL Any plant that lives for several to many years, including trees and shrubs. In a more restricted sense, an herbaceous perennial (*see* HERB).

PERFECT Applied to flowers with functional pistils and stamens (HERMAPHRODITE).

PERFOLIAT|US, -A, -UM Leaves or bracts which appear to have the stems growing through the middle because the leaf base extends around the stem and fuses with that of its opposite pair.

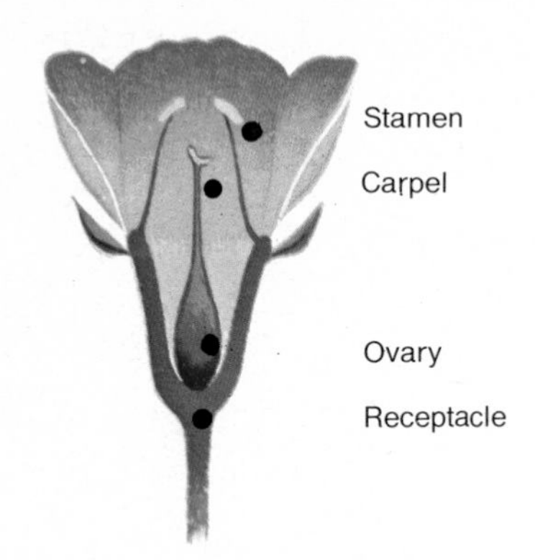

PERFORAT|US, -A, -UM Leaves or bracts that have small holes or translucent dots.

PERIANTH The two outer protective WHORLS of a flower, the sepals or CALYX and petals or corolla. Perianth is often used as a descriptive term when the sepals and petals look alike, e.g. tulip and narcissus.

PERIBLEM The layer or zone of cells in a root tip which lies between the inner core or plerome and the outer skin or dermatogen. *See also* MERISTEM.

PERICARP The wall of an ovary when it has matured into a fruit, e.g. an ACHENE or nut or the fleshy coat of a berry.

PERICYCLE (1) The PARENCHYMATOUS tissue that surrounds the XYLEM and PHLOEM (the vascular system) in the central core or stele of a root.
(2) Similar tissue on the outside of the vascular bundles in a stem, often modified to strengthening fibres.

PERIDERM Another name for the cork CAMBIUM (phellogen) and its derivatives which form the bark covering of a tree.

PERIGONE Another name for the PERIANTH.

● **PERIGYNY [PERIGYNOUS]** Flowers with the OVARY borne in a cup formed out of a flattened RECEPTACLE, e.g. cherry and plum. Although the ovary is below the stamens and petals it is still on top of its receptacle and is said to be superior (*See* HYPOGYNOUS).

PERISTOME The ring of fine teeth around the top of a moss spore capsule. They fold inwards in wet weather, preventing the spores from escaping. In dry weather they open jerkily, flicking away any spore entangled with them.

PERITHECIUM The complex fruiting bodies of certain *Ascomycetes* fungi, formed of a woven mass of HYPHAE around a fertile ascogonium which later gives rise to ASCI and ascospores.

▲ **PERIWINKLE** Several species of *Vinca*, evergreen shrubs and perennials in the *Apocynaceae*. They have OVATE to LANCEOLATE leaves and tubular flowers with 5 broad, asymmetric lobes: lesser periwinkle *(V. minor)* is a prostrate soft-stemmed shrub with lanceolate 25–38mm leaves and 25mm wide, blue-purple flowers; white, purple, red-purple and double forms are cultivated. The fruit is a slender forked FOLLICLE about 25mm long; greater periwinkle *(V. major)* has arching stems to 600mm and prostrate ones, with 25–63mm long ovate leaves. The purple-blue flowers are 25–28mm wide.

PERSIAN LILAC
see MELIA

PERSIMMON *(Diospyros kaki)*. A deciduous tree to 12.2m in the *Ebenaceae* from China and Japan and grown elsewhere. It has 75–200mm long oval leaves, 25–38mm wide yellow-white flowers and globose orange fruits about 75mm wide known as Chinese or Japanese persimmons or date plums; American persimmon or date plum *(D. virginiana)* grows to 30m, but with leaves, flowers and fruits only half the size.

PETAL
see COROLLA

PETALOIDS [PETALODY]
see DOUBLE FLOWERS

PETIOLE The stalk of a leaf; the petiolule is the stalk of a leaflet in a compound leaf.

PETRAE|US, -A, -UM Growing among rocks.

■ **PETTY WHIN [NEEDLE FURZE]** *(Genista anglica)*. A low-growing twiggy shrub rarely above 450mm in the *Leguminosae* from W. Europe. It has slender, straight spines, ovate, GLAUCOUS leaves to 8mm and yellow pea-flowers of the same length in the upper leaf AXILS.

△ **PETUNIA** A genus of 40 annual and perennial plants in the *Solanaceae* from N. and S. America. They have entire, often sticky-hairy leaves and AXILLARY funnel-shaped flowers: *P. integrifolia* is a sub-shrubby perennial of spreading growth with LANCEOLATE, 50mm long leaves and rose to purple flowers; *P. nyctaginiflora* is an annual with semi-prostrate stems to 600mm, ovate-oblong leaves and large, white flowers. Both are S. American and have clammy foliage. Crossed together they have produced the popular garden petunias with 50–125mm wide, single and double flowers in shades of pink, red, mauve, purple-blue, white, cream and pale yellow.

PHANEROGAM
see SPERMAPHYTA

PHANEROPHYTE Woody plants (shrubs and trees) with buds more than 250mm above soil level; the first of a 7-category system of classification of life forms.

PHELLOGEN
see CORK

PHENOTYPE
see GENOTYPE

PHILODENDRON A genus of 275 evergreen climbers and small trees in the *Araceae* from tropical America. They have small to very large leaves in a wide variety of forms from LANCEOLATE and HASTATE to deeply PINNATE or bipinnatifid. The tiny, petal-less flowers are borne in spathes. Many species and hybrids are grown as house plants. Most commonly seen is *P. oxycardium* (syn. *P. scandens* or *P. cordatum*), a climber with broadly OVATE-CORDATE slender pointed leaves, 95–150mm long on potted specimens but 300mm in the wild when mature; *P. selloum* is tree-like to 3m or so with long-stalked, deeply pinnate lobed leaves 600–900mm wide when mature.

PHLOEM
see VASCULAR BUNDLE

○ **PHLOX** A genus of 67 perennial and annual plants in the *Polemoniaceae* from N. America and Mexico, one in N. E. Asia. Some are prostrate alpines, others erect herbaceous plants. They have opposite pairs of linear to ovate leaves and terminal CORYMBS of tubular flowers with 5 petal-like lobes, often notched at the tip. Several species are cultivated as ornamentals: border phlox *(P. paniculata)* grows 0.6–1.2m with 50–125mm long oblong to elliptic leaves and large terminal PANICLE-like clusters of 10–32mm wide, fragrant violet-purple flowers. Many cultivars are known with pink, blue, white and bicoloured flowers; moss phlox *(P. subulata)* is a mat-forming evergreen with linear, 13–19mm long leaves and small corymbs of pink, lilac, blue-purple or white flowers.

PHOTOPERIODISM
see LIGHT

PHOTOSYNTHESIS The formation or synthesis of organic substances from water and carbon dioxide via CHLOROPHYLL and the energy of sunlight. The end-products of this very elaborate process (which occurs in all the green parts of the plant) are sugars (later changed to starch), fats and proteins.

PHOTOTAXIS The attraction to light of motile, single-celled plants such as *Chalmydomonas*.

PHOTOTROPISM The response of plants to light. Most stems grow towards a light source and are said to be positively phototropic. Most roots are insensitive to light but some, e.g. the aerial roots of ivy, grow away from it and are said to be negatively phototropic. Most leaves are said to be diaphototropic as they assume a more or less horizontal position, the broad, upper surface at right angles to the light source.

● **PHYLLOCLADE** A short flattened stem having the form and function of a leaf, e.g. butchers' broom.

▲ **PHYLLODE** A flattened leaf stalk (petiole) having the form and function of a leaf.

PHYLLODY The transformation of plant organs such as petals and sepals into leaf-like organs.

PHYLLOTAX|Y, -IS The arrangement of leaves on the stem in such a way as to ensure maximum illumination. In some cases they are borne in opposite pairs, each pair at right angles to the one below (decussate). Other plants have WHORLS of 3 or more leaves at each NODE, e.g. cleavers. The commonest phyllotaxis is the spiral, but varying greatly in its tightness. In the simplest or half-spiral form, the single leaves appear to arise one above the other on alternate sides of the stem: (formerly known as alternate). Tighter spirals have leaves arising at set fractions of the stem circumference, e.g. $\frac{1}{3}$, $\frac{2}{5}$, $\frac{3}{8}$, $\frac{5}{13}$, etc.

PHYLOGENY The evolutionary history of a plant or plant group.

PHYTOLOGY An alternative name for botany; the study of plants.

PICK-A-BACK-PLANT [YOUTH-ON-AGE] *(Tolmiea menziesii)*. An evergreen perennial in the *Saxifragaceae* from N. W. North America. It has long stalked, 38–125mm long PALMATE, shallowly lobed and toothed leaves and stems to 600mm, bearing slightly arching RACEMES of small greenish or purplish flowers with 5 filiform petals. Mature leaves produce plantlets at the junction of the leaf blade and the stalk.

PICOTEE Of white or yellow petals with narrow margins of contrasting colour, e.g. some carnations, pinks and sweet peas.

PICT|US, -A, -UM Coloured, as if painted with another colour.

PIGEON BERRY
see POKEWEED

PIGEON PEA *(Cajanus cajana* syn. *C. indicus)*. A short-lived, hairy evergreen shrub 0.9–3.0m in the *Leguminosae*, probably from Africa but much grown in India and elsewhere. It has trifoliate leaves of elliptic, pointed leaflets and AXILLARY RACEMES of small, yellow pea-flowers. The pointed, cylindrical pods are constricted between the 3–4 reddish seeds which are known as red gram and provide the dish known as dhal in India.

PIG LILY
see ARUM LILY

PIGNUT Two perennial members of the *Umbelliferae* from W. Europe.
(1) *Conopodium majus*, also known as earth nut, this plant has a knobbly nut-like, edible tuber; broadly DELTOID, bipinnate leaves having the leaflets cut into LINEAR segments and slender, 300–600mm hollow stems with 38–75mm wide UMBELS of tiny, white flowers.
(2) Great pignut *(Bunium bulbocastanum)* is a similar plant, but with globose tubers and solid flowering stems. In USA, one of the hickories *(Carya glabra)* is known as pignut.

PILEUS The cap-like structure of mushrooms and toadstools which bears the spores.

PILIFEROUS LAYER The outermost layer of cells near the tip of a root which bears the root hairs.

PILLAR ROOTS The aerial roots of such trees as BANYAN which grow straight down from the branches to root in the soil, then thicken and act as props or pillars.

PILOS|US, -A, -UM Covered with long, ascending hairs.

PIMENTO
see ALLSPICE

PIMPERNEL Several species of *Anagallis*, creeping annuals and perennials in the *Primulaceae* from temperate regions. They have OVATE leaves and funnel-shaped or rotate 5-lobed flowers: scarlet pimpernel or shepherd's weather glass *(A. arvensis)* has 13–25mm long leaves and rotate 13mm wide scarlet, purple or blue flowers; bog pimpernel *(A. tenella)* is a bog plant with very slender stems, orbicular to obovate leaves about 5mm long and pink flowers. It is native only to parts of Europe and N. Africa.

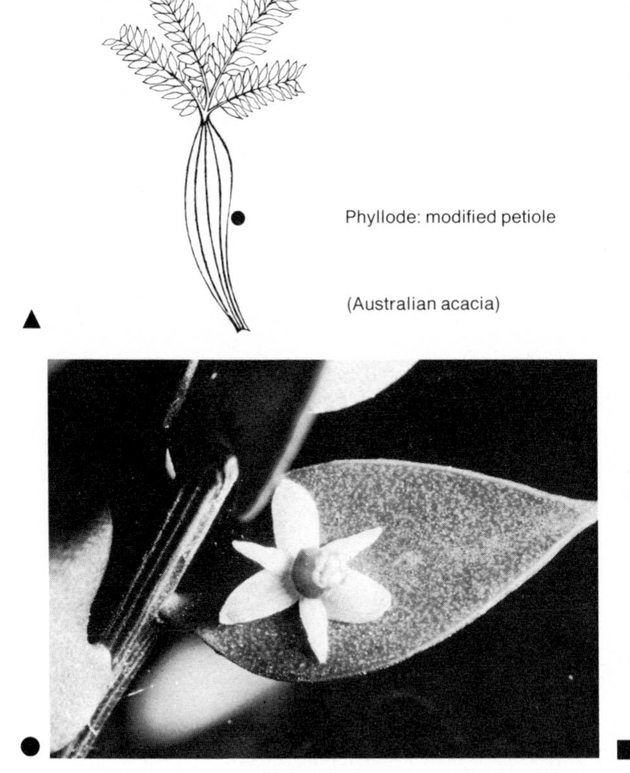

PINE Strictly this name applies to *Pinus*, a genus of almost one hundred evergreen trees in the *Pinaceae* from the northern hemisphere, where they extend from the LIMIT OF TREES in the north to tropical mountains. When young they are typified by having a single, central trunk and WHORLS of horizontal branches at regular intervals. Their true hallmark is the presence of LONG AND SHORT SHOOTS, the LINEAR leaves or needles being borne in groups of 2, 3 or 5 on tiny, stub-like shoots. Pines vary much in size, from the gnarled bushes a few feet tall of dwarf mountain pine *(P. mugo pumilio)*, to the 75m columns of sugar pine. Their flowers or STROBILI are MONOECIOUS, the males in short, cylindrical catkins, the females barrel-shaped like cones in miniature. The mature cones are woody, ranging from under 50mm to 450mm long, pendulous or at right angles to the stems. The nutlet-like seeds are often winged. Pines provided an important source of timber, turpentine and resin; several species have edible seeds: bristle-cone pine *(P. aristata)* from S.W. America and Mexico grows to 12.2m with 38mm leaves in fives and cones to 88mm. This is supposed to be the longest-lived tree known, a specimen having been found of almost five thousand years. However, there is some doubt about the identification of this ancient specimen and it may belong to the closely related *P. longaeva*; Aleppo pine *(P. halepensis)* is the common round-headed pine in the Mediterranean region. It grows to 18.2m with 38–75mm leaves in pairs and 50–113mm cones; maritime or cluster pine *(P. pinaster)* is also Mediterranean in origin, growing to 36m with deeply fissured, red-brown bark, 175–250mm long leaves in pairs and 125–250mm cones. It is the most important of the pines for the production of resin, from which turpentine and rosin are extracted; Austrian and Corsican pines are two of several forms of the black pine *(P. nigra)*, a European species reaching 45m and having 100–150mm long, dark green leaves in pairs and cones, 81mm long; Corsican pine *(P. n. maritima)* is much planted by the Forestry Commission in S.E. England; digger pine *(P. sabiniana)* from the arid foothills of California grows to 24.4m with rather scantily borne grey-green, 225–300mm long leaves in threes and 150–250mm cones; loblolly, slash, swamp or yellow pine *(P. taeda)* from S.E. America grows to 33m with 150–225mm long leaves in threes and cones to 125mm. It is one of the few pines that thrive in really wet soil; long-leaf, pitch or turpentine pine *(P. palustris)* is similar but has longer cones and leaves to 450mm. It has the strongest timber of all pines; Monterey pine *(P. radiata)* from a small area of coastal California grows to 33m or more, with 100–150mm long leaves in threes and cones to 150mm. It is the most widely planted of the pines, millions of acres being grown in S. Africa, Australia and New Zealand; western yellow or pitch pine *(P. ponderosa)* from west N. America grows to 60m or more, with 125–250mm leaves in threes and 75–200mm cones. A common and very important timber tree in America, also yielding resin and turpentine oil. Growing with pitch pine in Oregon and California is sugar pine, *(P. lambertiana)* the largest of all the pines, specimens exceeding 75m. It has 75–100mm leaves in threes and 300–450mm long cones. Wounded trees exude a sugary resin which has been used as a substitute for sugar; Scots pine *(P. sylvestris)* is widespread in Europe, W. and N. Asia and is very variable. *P.s. scotica* is Britain's only native pine. It grows to 30m or more with grey-green 25–100mm long leaves in pairs and 25–75mm cones. An important timber tree known to merchants as red deal, yellow pine and yellow deal; stone pine *(P. pinea)* is a

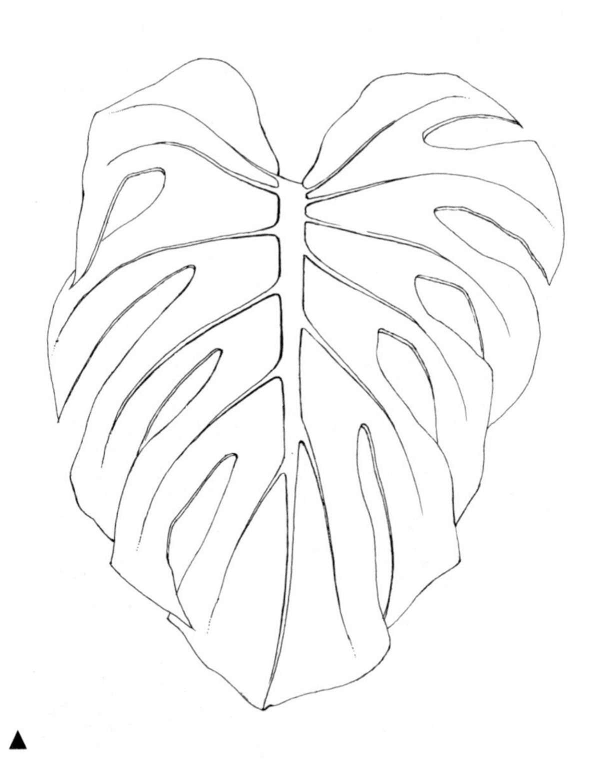

Mediterranean species to 24.4m with broad, umbrella-like crowns of branches, 100–150mm leaves in pairs and glossy 100–150mm cones. The latter have large seeds, the kernels of which are marketed as pignons, pine kernels and pinocchi; Weymouth pine *(P. strobus)* comes from E. Canada and America, but gains its vernacular name from Lord Weymouth who popularised it in the 18th century, planting it on his estate at Longleat, Wiltshire. It grows to 45m or more with blue-green, 75–125mm long leaves in fives and cones to 150mm. Many more pines are grown as ornamentals.

● **PINEAPPLE** *(Ananas comosus)*. A woody-based evergreen perennial in the *Bromeliaceae* from east S. America, but now cultivated throughout the tropics. It has rigid, arching, spiny-toothed, ENSIFORM, grey-green leaves to 900mm or more and cone-shaped spikes of small, 3-petalled blue flowers having a tuft of leaves at the top. Each flower produces a berry-like fruitlet which fuses with its neighbours to form the familiar yellow, aggregate fruit. There are many CULTIVARS, some with fruits to 300mm long or more.

PIN-EYED
see PRIMULA

PINK Applied in a general way to several sorts of *Dianthus*, evergreen perennials in the *Caryophyllaceae* from Europe. They form mats or hummocks of LINEAR, often GLAUCOUS leaves and fragrant flowers with broad, long-clawed petals from a tubular CALYX. The pink of gardens is largely derived from *D. plumarius* a hummock-forming, grey-leaved plant with pink or white, fringed flowers about 25mm wide, sometimes having a crimson centre. It has been crossed with CARNATIONS and other dianthus to produce the popular garden flower; Cheddar pink *(D. gratianopolitanus* syn. *D. caesius)* is similar, with pink, notched petals: maiden pink *(D. deltoides)* is mat-forming, with mid-green leaves and 150–225mm branched stems, bearing 13–19mm wide red-purple to crimson flowers with darker eye markings.

PINK PURSLANE
see CLAYTONIA

PINNA, PINNAE The leaflet(s) of a PINNATE leaf.

PINNATE, [PINNAT|US, -A, -UM] A leaf divided naturally into two ranks of smaller leaves or leaflets on either side of the midrib. Pinnate leaves with an equal number of leaflets in pairs are said to be paripinnate, those with an odd, terminal leaflet are imparipinnate.

▲ **PINNATIFID, [PINNATIFID|US, -A, -UM]** A leaf pinnately divided almost to the midrib (pinnatipartite).

■ **PINNATISECT** A leaf pinnately divided to the midrib, but not rounded off into separate leaflets.

PIP The seeds of orange, lemon, apple, pear and similar fruits. Gardeners sometimes use it for a single floret in an UMBEL of polyanthus, auricula, etc.

PISIFER|US, -A, -UM Pea-bearing, e.g. small fruits likened to peas.

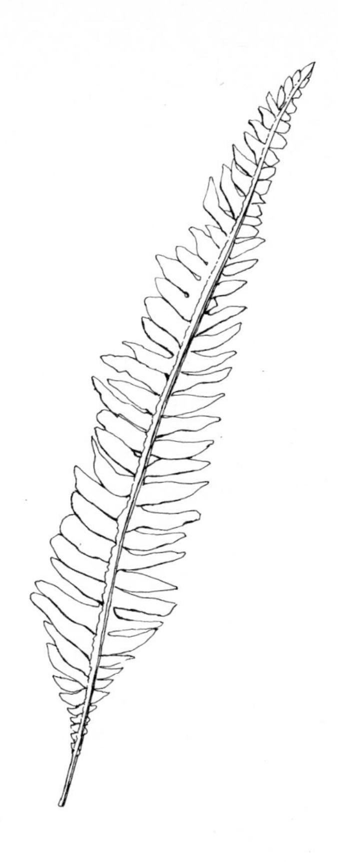

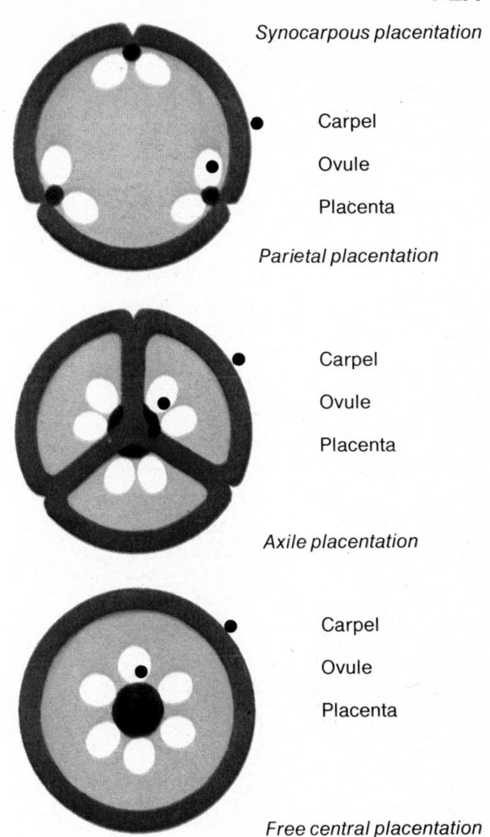

PISTACHIO *(Pistacia vera)*. A deciduous tree to 9m in the *Anacardiaceae* from W. Asia. It has PINNATE leaves of 3–7 ovate to obovate leaflets and small, DIOECIOUS, petal-less flowers in AXILLARY PANICLES. The ovoid, 19–25mm long DRUPES contain ovoid nuts with greenish kernels which are eaten salted and used in confectionery.

PISTIL The femal organ of a flower consisting of an ovary containing the ovules and one or more pollen-receptive organs known as STIGMAS. The latter are usually held up on stalks or styles to catch the pollen-bearing insects or wind-borne pollen.

PITCHER PLANTS A general term for the several CARNIVOROUS plants with the leaf blades formed into a water-holding, pitcher-shaped trap. Insects and other creatures fall into these and drown, their remains providing a source of nitrogen for the plant. Common pitcher-plant *(Sarracenia purpurea)* grows in the bogs of N. America. It forms rosettes of reclining, inflated 100–150mm long leaves with flap-like lids usually veined purple. The nodding, solitary, purple flowers have 5 sepals, 5 petals and a curious umbrella-like out-growth over the STIGMA. Several other species occur, some with slender, erect pitchers called 'trumpet leaf'. Allied to *Sarracenia* is *Darlingtonia* or cobra plant from the bogs of California and Oregon. It has erect, red-netted pitchers to 600mm tall which arch over at the top and bear a forked, crimson and green bract hanging down from the opening. The effect is of a cobra ready to strike. *Nepenthes* is a genus of 67, tropical climbers with basically LANCEOLATE leaves that terminate in stalked, hanging pitchers. These are often brightly coloured and vary from 50–300mm long; the larger ones are capable of drowning small rodents. The tiny, 4-lobed purplish or greenish flowers are borne in RACEMES. Several other kinds of pitcher plants are known.

PITH
see MEDULLA

PITS Simple pits are pore-like openings in the LIGNIN thickening of stone cells and fibres to allow for the passage of water and dissolved food substances. Bordered pits are found in the walls of vessels and have a circular flange of thickening around the opening of each one.

PIXIE MOSS Several species of *Cladonia*, lichens of wide distribution in cool climates, they form flat patches of small, grey-green leafy lobes (squamules) and erect, cup-shaped SOREDIA producing bodies like old-fashioned drinking horns.

PLACENTATION The arrangement of the OVULES in a SYNCARPOUS OVARY. In ovaries formed of carpels joined edge to edge, with the ovules along the placentas at the joins, the arrangement is called parietal, e.g. gooseberry, violet. Where the carpets are folded and joined together at a central point and the ovules arranged in the centre, this is axile, e.g. tulip, lily. In primrose and campion the cross walls of the carpels break down and the ovules sit free of the ovary walls on a spike-like placenta. This is free central placentation.

PLAGIOTROPISM
see GEOTROPISM

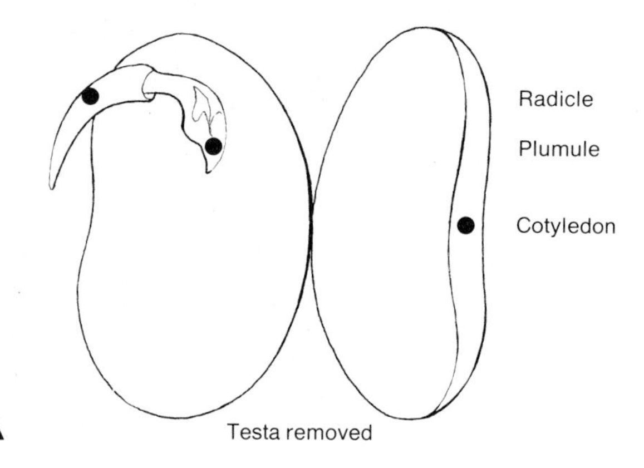

● **PLANE** *(Platanus)*. A genus of 10 deciduous tree in the *Platanaceae* from N. America, Mexico, Europe and Asia to Iran and Vietnam. Except for the Vietnamese plane *(P. kerrii)*, with its elliptic, oblong leaves, they all have large, PALMATELY lobed, maple-like leaves. The tiny, green, unisexual flowers are aggregated into dense spherical heads and are wind-pollinated. The female heads become pale brown seed balls which do not break up until the following spring or summer. The ACHENES have long, bristly hairs. The hybrid London plane *(P. X hispanica*, syn. *P. X acerifolia)* is planted as a street tree in many parts of the temperate world. It grows to 36m with smooth, mottled, flaking bark and 100–250mm wide leaves with 5 broad lobes; oriental plane grows to 24m or more with less obviously flaking bark and deeply 5–7 lobed 75–200mm wide leaves; button wood *(P. occidentalis)* resembles London plane, but with shallowly 3–5 lobed leaves.

PLANTAIN *(Plantago)*. A genus of 265 annuals and perennials in the *Plantaginaceae*, of cosmopolitan distribution. They are mainly rosette-forming plants, though a few are branched and bushy. The LINEAR to OVATE leaves are prominently veined and the tiny, greenish or brownish flowers are carried in dense, erect spikes. Several species are ■ familiar weeds of cultivated land: ribwort *(P. lanceolata)* is found everywhere, from the best kept English garden to the remoteness of Easter Island. It has LANCEOLATE leaves 75–300mm long and black-brown flower spikes with prominent white anthers; greater plantain *(P. major)* has broadly OVATE, 75–250mm long leaves and slender green spikes. In the tropics, certain banana cultivars with a high starch content, and which are cooked like a vegetable, are also known as plantains.

PLANTAIN LILY
see HOSTA

PLANT ASSOCIATION A natural or semi-natural unit of vegetation easily recognisable because the same species are usually found together, e.g. an oakwood on moist soils can be expected to contain hazel, field maple, sallow, dogwood plus smaller perennials such as wood anemone, primrose, dog's mercury. Ecologists split associations into two categories. Where one species is dominant, as in the oak wood, that unit is spoken of as a consociation. Where several tree species are found in equal numbers, as is more usual, this is a true association.

PLANT GEOGRAPHY The distribution of plants on earth in relation to climate and soil, and involving their methods of seed dispersal and migration.

PLASMOLYSIS The reversal of OSMOSIS. If a piece of plant tissue is placed in a solution of sugar or salt and water stronger than the cell sap, water will flow out of the cells and the contents will shrink. Finally the contents part from the cell wall and become reduced to a concentrated globule floating in the sugar and salt solution. If a whole plant is immersed in a strong solution, it soon becomes soft and wilts. Placing the tissue or whole plant in clean water reverses the process back to normal.

PLASTIDS
see CELLS

PLATYCARP|US, -A, -UM Broad-fruited.

■

○

PLATYPHYLL|US, -A, -UM Broad-leaved.

PLEN|US, -A, -UM, [PLENIFORMIS, -E] Full, double; of flowers with more than the usual number of petals.

PLEROME
see PERIBLEM

PLEUROCOCCUS Unicellular algae, commonly seen as green 'powder' on tree trunks and old fences.

PLICATE [PLICAT|US, -A, -UM] Pleated, folded lengthwise.

PLOUGHMAN'S SPIKENARD *(Inula conyza)*. A biennial or perennial in the *Compositae* from Europe and N. Africa. It has ovate-oblong leaves like those of foxglove and erect 30–120cm stems bearing terminal CORYMBS of yellow, shortly cylindrical 9mm wide flower heads of disk florets only.

○ **PLUM** *(Prunus domestica)*. A deciduous tree of 9m or more in the *Rosaceae*, probably originating in S. W. Asia but not known anywhere genuinely wild, and possibly of hybrid origin. It has 38–100mm long OBOVATE or ELLIPTIC leaves and bowl-shaped, white flowers appearing before the leaves. The ovoid to globose fruits vary from yellow and red to purple and blue-black, 19–38mm long in wild plums but up to 75mm in some of the many CULTIVARS: greengage has rounded, 38–50mm long, yellowish-green fruits; bullace *(P. d. institia)* is usually a somewhat thorny shrub with pubescent twigs and globular purple or purple-black 19–25mm fruits; cherry or myrobalan plum *(P. cerasifera)* is not unlike a smooth green-stemmed bullace, but is rarely thorny and often of tree form to 7.5m.

PLUMARI|US, -A, -UM Of feathery appearance.

PLUMBAGINOIDES Like plumbago, the latin name of LEADWORT.

PLUMOS|US, -A, -UM Feathery in appearance as in plumarius.

▲ **PLUMULE** The minute, undeveloped shoot in a seed.

POCKET HANDKERCHIEF TREE
see DOVE TREE

POINSETTIA
see SPURGE

POISON IVY [POISON OAK]
see SUMAC

POKEWEED [PIGEON BERRY, RED INK PLANT] *(Phytolacca americana)*. A clump-forming, herbaceous perennial in the *Phytolaccaceae* from N. America. It has erect, robust stems 0.9–1.8m tall, OVATE leaves to 150mm and dense spikes of small 4-lobed white flowers. The black-purple berries contain poisonous seeds.

POLIT|US, -A, -UM Polished, elegant or neat.

POLLARD A tree cut back to about 2.5m above the ground, either to the trunk or to main branches, to induce

■ the growth of vigorous young stems. Willows are frequently cut this way for basket-making. It is done at this height to prevent the young shoots from being eaten by browsing cattle.

● **POLLEN** The male sex cells in the flowering plants, borne in ANTHER lobes which form the STAMENS. They are single cells of dust-like size, but under the microscope reveal an enormous diversity of shape and form: spherical, elliptical, triangular, etc, often beautifully sculptured or patterned, and in a variety of colours, orange, red, blue, purple though more usually yellow. In many cases the grains are so distinctive that the species from which they have come can be recognised by their pollen alone. Pollen is transported to the stigma mainly by insects or wind, sometimes by gravity or by animals or birds. Pollen can be light and buoyant or heavy and sticky. Each grain has two coats, a thin inner one called the intine and a very tough outer one, the exine, which is remarkably decay resistant. *See* POLLEN ANALYSIS.

POLLEN ANALYSIS The resistant nature of the coat of a pollen grain and its distinctive shape and patterning are used to date and reconstruct past floras. Pollen, particularly from wind-pollinated trees, often coats the ground at flowering time and under the right conditions (in bogs in particular) can become gradually buried by organic debris. Centuries or millennia later it can be recovered by special processes, identified, and the layer of mud and peat where it was found dated by comparison with other sequences and by carbon^{14} dating. Thus a picture can be built up of past plant communities and their changes.

▲ **POLLINATION** The landing of pollen grains on the STIGMA. *See also* FERTILISATION.

POLLINIA
see ORCHID

POLYANDROUS Flowers with many stamens, e.g. poppy.

POLYARCH
see DIARCH

POLYCARPIC Flowering and fruiting more than once, i.e. most perennial plants, trees and shrubs which do so annually. *See also* MONOCARPIC.

POLYEMBRYONY The formation of more than one EMBRYO in each OVULE, sometimes as a result of the division of the fertilised egg cell (e.g. pine), or by a process of budding from the pro-embryo.

POLYGAMY A plant species with separate male, female and hermaphrodite flowers in various combinations on one or several plants.

POLYMORPH|IC, -ISM Plants which have more than one kind of leaf, stem or flower on the same plant, e.g. ivy has climbing stems with PALMATE leaves and non-climbing flowering stems with LANCEOLATE to OVATE leaves.

POLYPETALAE A division of the *Archichlamydae* which has flowers with separate petals.

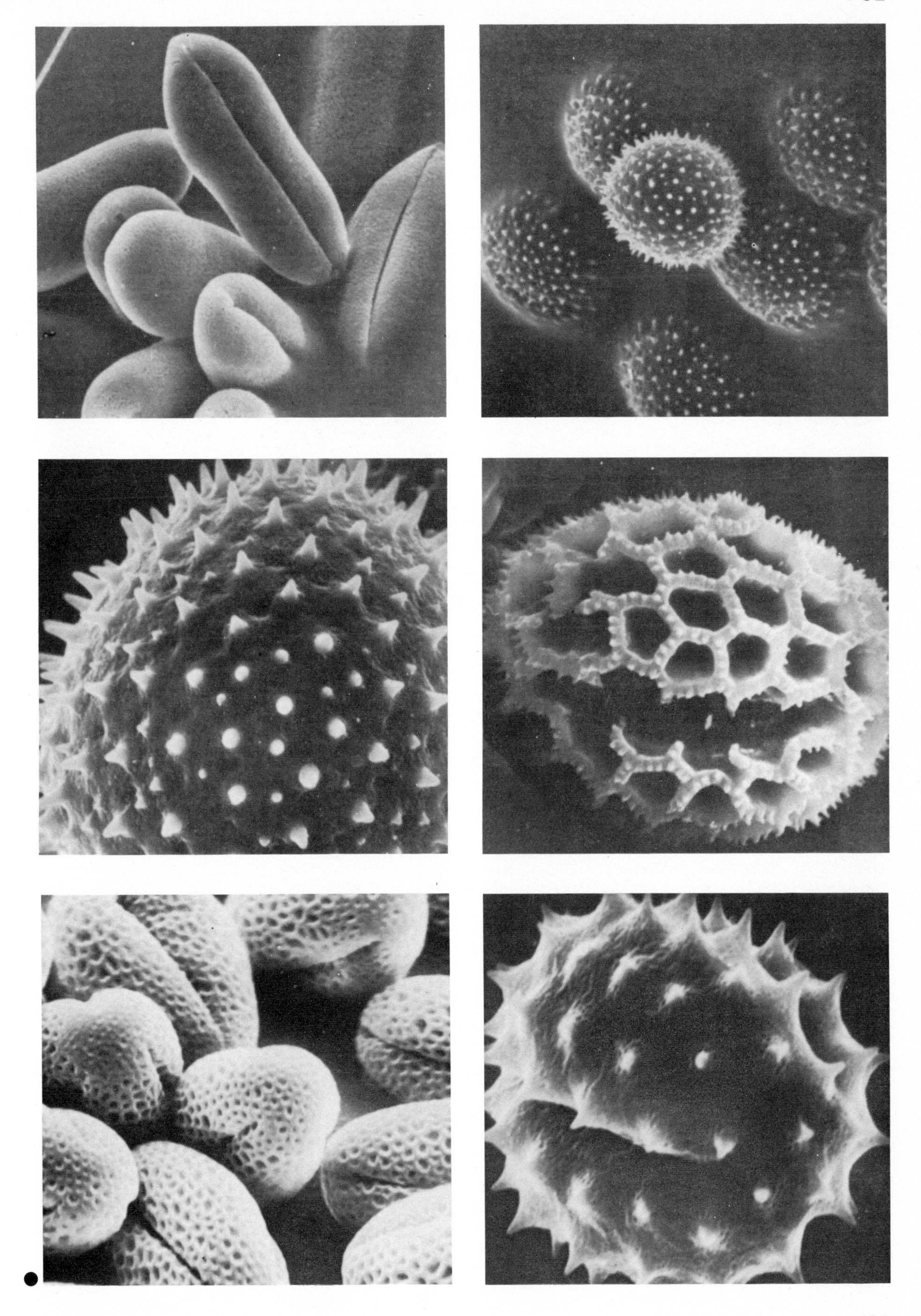

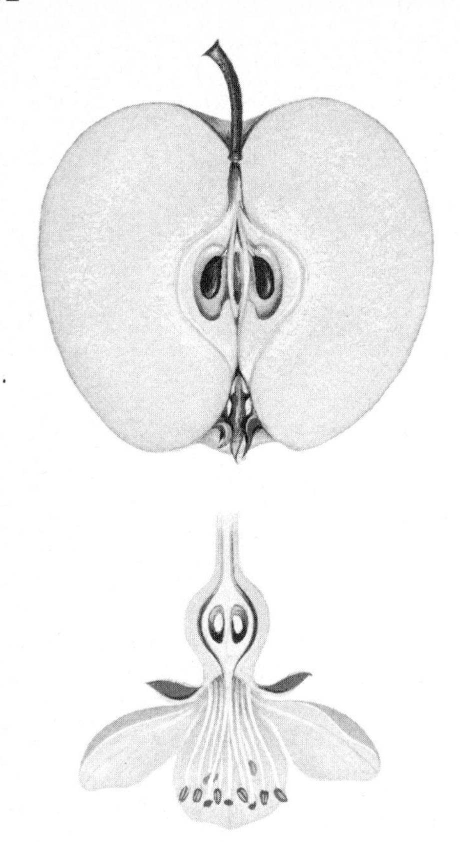

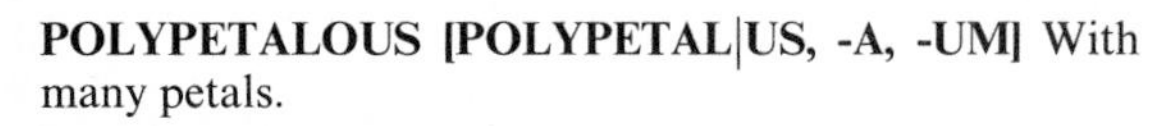

POLYPETALOUS [POLYPETAL|US, -A, -UM] With many petals.

POLYPLOID Plants which have more than the normal two sets of CHROMOSOMES in their cells; triploids have 3 sets, tetraploids, 4; pentaploids, 5; hexaploids, 6, etc. This condition may occur spontaneously but can also be induced by atomic radiation or drugs such as colchicine. Sometimes polyploid plants are more robust or hardier than the diploid sorts.

POLYPODY *(Polypodium)*. A genus of about seventy-five, RHIZOMATOUS ferns in the *Polypodiaceae* of cosmopolitan distribution. They have fronds borne singly on creeping or branching rhizomes and rounded or oblong SORI without INDUSIA; common polypody *(P. vulgare)* is a variable fern with deeply PINNATIFID, lanceolate-oblong fronds to 300mm or more long. In areas of high rainfall or humidity it sometimes grows as an EPIPHYTE.

● **POME** A fruit imbedded in a deep, fleshy cup-shaped RECEPTACLE, the seeds protected by a tough layer derived from the true, many celled fruit wall, e.g. apple, the core being the true fruit and the fleshy part the receptacle.

▲ **POMEGRANATE** *(Punica granatum)*. A deciduous shrub or small tree to 9.1m in the *Punicaceae* from S.E. Europe. It has somewhat angular, spiny twigs, 25–88mm long oblong to OBOVATE leaves and 25–38mm wide red flowers having 5–7 curiously crumpled petals. The 63–88mm wide globose, deep yellow, often red-flushed fruits have a leathery rind and numerous seeds each surrounded by a fleshy, juicy coat.

POMELO
see CITRUS

POMIFORM [POMIFORMIS, -E] With apple-like fruits.

POMOLOGY The study of fruits.

POMONA A book on fruits and their culture.

PONDWEED *(Potamogeton)*. A genus of 100 aquatic perennials in the *Potamogetonaceae* of cosmopolitan distribution. Some species have both floating and submerged leaves, others are entirely submerged, with translucent, LINEAR to OVATE leaves. The tiny green flowers have 4 PERIANTH segments and are carried in short spikes either under water or above: bog pondweed *(P. polygonifolius)* grows in the shallow water of bog pools and ditches with stems to 300mm or more and broadly elliptic, stalked floating leaves 19–57mm long; shining pondweed *(P. lucens)* is entirely submerged, with stems 45–180cm and oblong-lanceolate, 100–200mm long, translucent leaves; curled pondweed *(P. crispus)* has 30–120cm long, angled, submerged stems and 32–88mm long lanceolate leaves.

POOR MAN'S ORCHID *(Schizanthus pinnatus)*. An erect 600mm annual in the *Solanaceae* from Chile. It has PINNATIFID or bipinnatifid leaves and terminal RACEMES of ZYGOMORPHIC rose-purple, 5-petalled flowers of orchid-like appearance. There are several CULTIVARS in shades of red, pink, lilac and purple, often marked with yellow and contrastingly patterned or veined.

■

○

POOR MAN'S WEATHER GLASS
see PIMPERNEL

■ **POPLAR** *(Populus)*. A genus of 35 deciduous trees in the *Salicaceae* from the northern hemisphere. They have ovate to triangular-ovate leaves and tiny, petal-less DIOECIOUS flowers in pendulous catkins before the leaves. The tiny seeds are covered in white, cottony hairs, hence the American vernacular name cottonwood. Several species are grown as ornamentals and a number of man-made hybrids for forestry purposes. These latter are among the fastest growing of trees attaining 30m in about thirty years under ideal conditions. The timber is used for matches, boxes and packing cases of all kinds: black poplar *(P. nigra)* makes a broad-headed tree to 30m, the deeply fissured, black-brown trunk having characteristic large, rounded burrs when mature. The leaves are DELTOID-OVATE, the blades 50–100mm long and the male catkins red; Lombardy poplar *(P.n.* 'Italica') is a narrowly columnar form; balsam poplar covers several species, notably western *(P. tricocarpa)* and eastern *(P. tacamahaca)* with large leaves, silvery-white beneath and a strong sweet smell of balsam; aspen *(P. tremula)* grows to 15m or more with strongly flattened leaf stalks carrying grey-green, rounded to broadly-ovate leaves which move in the slightest breeze; white poplar *(P. alba)* grows to 22.5m or more with 3–5 lobed, broadly-ovate, 75–125mm leaves white woolly beneath.

○ **POPPY** *(Papaver)*. A genus of 100 annuals and perennials in the *Papaveraceae* from the north temperate zone, Australia and S. America. They have LANCEOLATE to OVATE, often PINNATIFID or bipinnatifid leaves and large, cup-shaped, 4-petalled flowers. The seed capsule is ovoid to cylindrical, often with a flat or conical lid and a row of holes around the top through which the seeds are jerked in windy weather (censer mechanism). All parts exude a milky latex (sometimes orange or yellowish) when damaged. Field or common poppy *(P. rhoeas)* is an annual to 600mm with bipinnatifid, hairy, toothed leaves and 50–88mm wide scarlet flowers. The Shirley poppies of gardens are derived from this species. Opium poppy *(P. somniferum)* is a 0.6–1.2m annual with grey-green, ovate-oblong, undulate and PINNATELY lobed leaves with white, lilac or red flowers. Double flowered CULTIVARS are grown as ornamentals. The latex yields opium and the seeds (maw) are used to garnish bread, etc. Oriental poppy *(P. orientale)* is a robust, hispid perennial to 900mm or more with pinnate, lobed and toothed leaves and 100–150mm wide red, pink or white flowers, usually with a black, basal blotch; Iceland poppy *(P. nudicaule)* is a short-lived perennial with basal pinnatifid leaves only, and 50–75mm wide yellow, orange or red, fragrant flowers. *See also* CALIFORNIAN, HORNED and MEXICAN POPPIES.

PORTLAND ARROWROOT
see LORDS AND LADIES

PORT ORFORD CEDAR
see CYPRESS (Lawson's)

PORTUGAL LAUREL
see CHERRY LAUREL

POSEIDON WEED *(Posidonia)*. A genus of 3 marine perennials in the *Posidoniaceae* from the Mediterranean and Australia. They rank among the few true flowering plants that live in the sea, having stout RHIZOMES and strap-shaped, dark-green leaves to 450mm. The tiny, greenish flowers have a PERIANTH of 3 scales, and are followed by 25mm long, plum-like drupes. The Mediterranean *P. oceanica* is abundant in shallow water and the leaves are washed up in enormous quantities. They are used as manure, or when dried as packing material for glass. Poseidon balls, such a feature of Mediterranean shores, are the leaf fibres woven into ovals and balls by the action of the waves, each one having a piece of rhizome as a nucleus.

POSTERIOR The side of a flower in a compound inflorescence which faces the stem, assuming always that it points upwards.

POTATO *(Solanum tuberosum)*. A perennial plant with stem TUBERS, in the *Solanaceae*, originally from Bolivia and Peru but now much grown elsewhere. It has 450–800mm, erect or semi-erect stems, interruptedly PINNATE leaves and CYMOSE CORYMBS of nodding white or purple flowers having 5 pointed COROLLA lobes and a cone of yellow stamens. The ovoid berries are greenish and poisonous. The wild and semi-wild potatoes of the Andes are very varied in appearance, in size of leaf, colour of flowers and of tubers. Some have very knobbly red or purple-fleshed tubers and many have coloured skins. Although being replaced by modern cultivars, they are still the staple food of the Altiplano Indians. Many hundreds of cultivars have been raised in Britain and the U.S.A. These vary mostly in tuber shape and size, though a few have red or purple skins or eye patterns (e.g. King Edward), and one of the salad potatoes or fir apples has purple flesh. Apart from being a staple item of diet in many temperate countries, potatoes are also a valuable source of starch for industry and are used to brew alcohol. *See* SWEET POTATO.

PRAECOX Precocious, appearing earlier than expected; used of late winter and early spring flowering plants.

PRAIRIE The natural grassland community that once covered the central plains of the USA and Canada. Areas still exist but much of it is now cultivated or extensively grazed. It comprises many species of grass, notably blue grama and other BOUTELOUA species, herbaceous perennials and annual flowering plants, e.g. larkspur, prairie clover, evening primrose, spiderwort, compass plant and many of the daisy family, plus some shrubs, e.g. prairie rose.

PRATAL [PRATENSIS, -E] Growing in meadows.

PRICKLY LETTUCE *(Lactuca serriola)*. An erect biennial in the *Compositae* from Europe, Asia and N. Africa, naturalised in N. America. It has narrowly OBOVATE, sometimes PINNATIFID, SESSILE leaves with SAGITTATE bases and prickly margins which are held vertically, usually aligned north and south. The cylindrical 13mm long flower heads of short, pale yellow, LIGULATE florets are carried in large terminal PANICLES.

PRICKLY PEAR
see CACTUS

PRICKLY POPPY
see MEXICAN POPPY

PRICKLY RHUBARB *(Gunnera)*. A genus of 50 perennials, some of massive dimensions, in the *Gunneraceae* from Malaysia, Tasmania, New Zealand, Hawaii and tropical America to Chile. They grow in damp or wet soils and have OVATE to OVATE-CORDATE leaves, often long stalked, and spikes of tiny green flowers followed by berry-like fruits: *G. chilensis* has leaves 1.2–1.5m wide on 0.9–1.2m prickly stalks and cone-shaped PANICLES of red fruits; *G. manicata* is the largest species and has leaves 1.5–3.0m wide, with a PELTATE base.

PRIMROSE
see PRIMULA

● **PRIMULA** A genus of 500 tufted perennials, mainly from the northern hemisphere. They have short RHIZOMES and basal leaves only, ranging from LANCEOLATE to OBOVATE, often toothed and sometimes lobed. The flowers are tubular or bell-shaped with 5 broad COROLLA lobes, often notched at the tip. They are DIMORPHIC, with pin-eyed and thrum-eyed blooms, the pin-eyed having the stigma showing at the mouth of the tube with stamens beneath, the thrum-eyed being on separate plants with the stamens showing and the stigma beneath. Pollen from one form is needed to pollinate the other. Best known in Britain is the primrose *(P. vulgaris)* with obovate, wrinkled leaves and 25–38mm wide yellow flowers; oxlip *(P. elatior)* has similar leaves and UMBELS of nodding, 13mm wide pale yellow flowers; bird's eye primrose *(P. farinosa)* has obovate-spathulate leaves to 10mm long and umbels of small rosy-lilac flowers; Chinese primrose *(P. sinensis)* has long-stalked, broadly ovate leaves with toothed lobes and umbels of rose-purple flowers to 38mm wide. Having similar flowers but in a variety of red and purple shades is the popular pot plant *P. obconica*, with ovate-cordate leaves; fairy primrose *(P. malacoides)* is another well-known pot plant, having candelabras of dainty, 13mm wide, fragrant flowers shades of rose, purple and white; Japanese primrose *(P. japonica)* is a robust bog plant to 450mm with candelabras of red-purple, 19–25mm wide flowers. It is often grown with the allied *P. pulverulenta* which has the stems covered with white farina; giant or Himalayan cowslip *(P. florindae)* can reach 400mm, with large umbels of fragrant, sulphur-yellow bells; auricula *(P. auricula)* is a dwarf, almost sub-shrubby plant with semi-erect rhizomes, obovate 50–75mm long leaves and umbels of fragrant, yellow, 19mm wide blooms covered with white farina; many cultivars are available in a wide range of colours. *See* COWSLIP

PRINCE'S FEATHER
see AMARANTH

▲ **PRIVET** Several species of *Ligustrum*, evergreen to deciduous shrubs and trees in the *Oleaceae* from Asia with one in Europe. They have lanceolate to ovate leaves in pairs, white tubular flowers with 4 spreading lobes and black to purple berries: garden or hedging privet *(L. ovalifolium)* from Japan grows to 4.5m and has semi-evergreen, elliptic leaves and heavily fragrant off-white flowers in 50–100mm terminal PANICLES. A golden variegated form is grown; wild privet *(L. vulgare)* is tardily deciduous to semi-evergreen, sometimes reaching 1.8m or more, with 50mm panicles of white flowers and glossy, black berries.

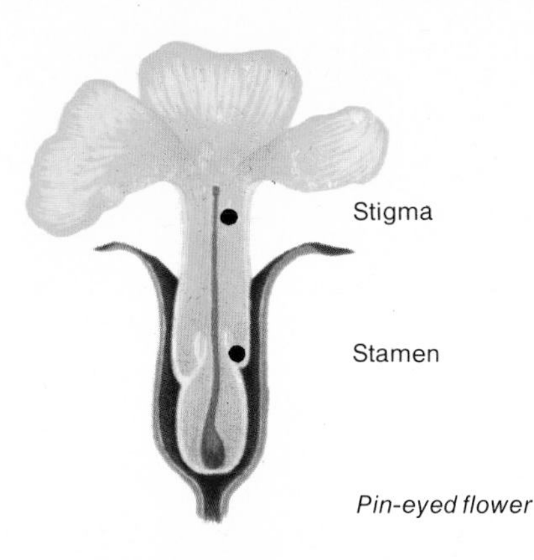

Pin-eyed flower ▲

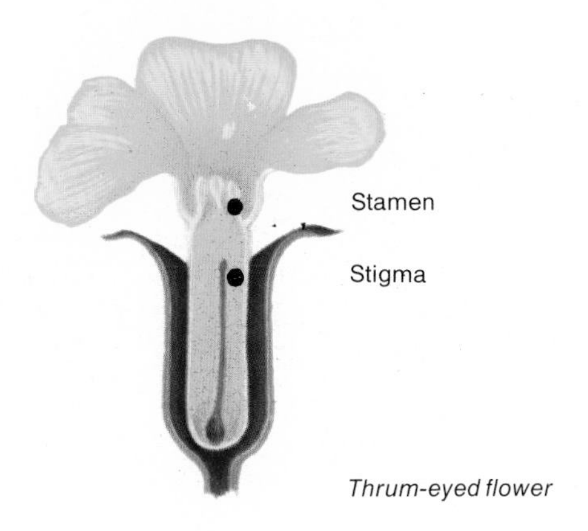

Thrum-eyed flower ▲

PROCAMBIAL STRANDS Columns of narrow, elongated cells that form in the growing tips of plants and which gradually mature into fully functional VASCULAR BUNDLES.

PROCER|US, -A, -UM Lofty, very tall.

PROCUMBENS Lying flat on the ground.

PROEMBRYO The first stage after fertilisation, when the egg cell in the EMBRYO SAC forms a cell wall and divides to produce a short chain of cells.

PROLIFER [PROLIFER|US, -A, -UM] Bearing many OFFSETS, or multiplying prolifically.

PROLIFERATION The production of abnormally excessive growth, usually where least expected, e.g. shoots from the middle of flowers or arising in the AXILS of sepals, petals or carpels, as with hen and chickens daisy. The twiggy masses in trees sometimes known as witches' brooms are also spoken of as proliferations.

PROPAGATION The means by which plants reproduce themselves. Seeds and spores are the most usual way in which plants perpetuate and spread themselves, but various so-called vegetative means are also displayed. These include STOLONS (blackberry), RUNNERS (strawberries) OFFSETS (houseleeks); in each case a small plantlet is produced at some distance from the parent. Gardeners use artificial ways to propagate plants: grafting desirable CULTIVARS on to the vigorous roots of the wild species, layering stems of shrubs (pegging them into the soil where roots form, the shoot then being severed), and taking cuttings (severed stems placed in a warm or humid atmosphere to induce independent rooting).

PROPHASE
see CELL DIVISION

PROPHET FLOWER *(Arnebia echioides)*. A perennial plant in the *Boraginaceae* from Armenia, USSR. It is clump-forming, with erect or semi-erect stems to 300mm, oblong-lanceolate roughly hairy leaves and terminal CYMES of tubular yellow flowers. The latter have 5 broad, spreading lobes, each with black-brown velvety spot which disappears as the flower ages.

PROSTRAT|US, -A, -UM Lying flat.

PROTANDROUS
see DICHOGAMY

PROTHALLUS
see ALTERNATION OF GENERATIONS

PROTOGYNOUS
see DICHOGAMY

PROTONEMA Branched, green threads produced when the moss spore germinates and from which the leafy moss shoots later arise.

PROTOPLASM The basic substance of all life on earth: a somewhat viscous, fluid, colourless material found in all plant and animal cells.

PROTOXYLEM Vertical columns of cells near the root and stem tip which form as vessels and later become proper xylem bundles.

PROXIMAL The side of a plant organ which faces the axis or stem that bears it.

PRUINAT|US, -A, -UM [PRUINOS|US, -A, -UM] Covered with a white waxy powder as though frosted.

PRUNE A sun-dried plum

PSEUDOBULB A swollen aerial stem, characteristic of many orchids.

PSEUDOCARP A fruit in which the ovary combines with some other structure, often the RECEPTACLE as in strawberry and apple.

PUBERUL|US, -A, -UM Minutely downy.

PUBESCENT [PUBESCENS] Covered with short, soft hairs.

PUCCOON [RED PUCCOON]
see BLOODROOT

PUFFBALLS Species of *Lycoperdon* and *Calvatia*, fungi in the *Gasteromycetes* from temperate regions. They are of globular or inverted pear-shaped, often whitish when young, brown later. When the puffballs are ripe, they become papery and rupture at the top. Thereafter any slight knock will send the internally-borne spores puffing out in clouds. Giant puffball *(Calvatia gigantea)* can be as large as a man's head and sometimes grows in rings in old pasture. When young and white it is edible.

PULCHELL|US, -A, -UM Beautiful, pretty, but small.

PULCH|ER, -RA, -RUM Beautiful.

PULSE Used for several members of the pea family *(Leguminosae)* grown for their ripe seeds (peas and beans), which have a high protein content.

PULVERULENT|US, -A, -UM As though dusted with powder or flour.

PULVINUS A swelling at the base of a leaf stalk, or at the top where it joins the blade. It is a feature of many plants that show strong leaf movements. *See also* NASTIC PLANTS.

PUMILIO [PUMIL|US, -A, -UM] Small or dwarf.

PUMMELO
see CITRUS

PUNCTAT|US, -A, -UM Dotted, marked with tiny spots or glands.

PUNGENS Ending in a sharp point, e.g. holly leaves.

PUNICE|US, -A, -UM Phoenician purple, a deep crimson.

PURGANS Purging or acting as an aperient.

▲ 

PURPLE HEART *(Setcreasea purpurea)*. A somewhat fleshy perennial in the *Commelinaceae* from Mexico. It has stems to 450mm or more, with purple, LANCEOLATE, stem-clasping 75–175mm long leaves and 25mm wide, 3-petalled rose-purple flowers from terminal boat-shaped bracts.

PURPLE MOOR-GRASS *(Molinia caerulea)*. A tufted deciduous perennial in the *Gramineae* from the north temperate zone. It is a grass of fens, bogs, damp heaths and moors, with flat, tapering leaves and 30–120cm stems bearing narrow, 75–300mm long PANICLES of tiny purplish or green flowering spikelets.

PURPURE|US, -A, -UM Purple; shades of colour between red and violet.

PURSLANE *(Portulaca oleracea)*. A prostrate or ascending fleshy leaved annual to 150mm tall in the *Portulacaceae,* of cosmopolitan distribution in warm temperate and sub-tropical climates. The oblong-ovate leaves are 9–19mm long and the 4–6-petalled AXILLARY flowers are up to 9mm wide. Much grown in the past as a leaf vegetable. Several other plants are known as purslane; *see* CLAYTONIA, SEA PURSLANE.

PUSILL|US, -A, -UM Very small; alternatively, weak or slender.

PUSTULE A raised, pimple-like spot, usually on leaves; sometimes natural, sometimes the results of the actions of parasites, disease or a physiological disorder, e.g. excessive humidity.

PYGMAE|US, -A, -UM Pigmy; dwarf or very small.

PYRAMIDAL ORCHID *(Anacamptis pyramidalis)*. A ground-dwelling perennial in the *Orchidaceae* from Europe, W. Asia and N. Africa. It is a tuberous-rooted plant with clusters of narrowly oblong-lanceolate, 75–150mm leaves, and 200–450mm stems bearing dense 19–50mm long pyramidal spikes of rose-purple flowers. Each blossom has a broad LABELLUM deeply cut into 3, narrowly oblong lobes.

● **PYRETHRUM** *(Chrysanthemum coccineum* syn. *Pyrethrum roseum)*. An herbaceous perennial in the *Compositae* from W. Asia and E. Europe, but much grown elsewhere both as an ornamental and for the production of the insecticide pyrethrum powder. It grows 300–600mm tall, with PINNATIFID leaves and 50–75mm wide pink, red or white daisy flower heads with yellow disks. Double-flowered CULTIVARS are grown.

QUADRANGULAR|IS, -E Four-angled, usually of stems.

QUAKING GRASS Several species of *Briza*, annuals and perennials in the *Gramineae* from Europe and Asia (others from Africa and S. America): common quaking or tottle grass *(B. media)* is a perennial to 500mm with flat, narrow leaves and spreading PANICLES of neat 4mm wide, ovoid, often purplish spikelets dangling from hair-like stalks; ▲ great quaking grass *(B. maxima)* is an annual with wide leaves (to 8mm) and 9.0–19mm long spikelets.

QUAMASH
see CAMASS

QUASSIA Two trees in the *Simarubiaceae*. (1) *Quassia amara*, a tree to 6m or more from tropical America with PINNATE leaves and tubular, red flowers in terminal clusters. The bitter bark and roots have been used against dysentry and as a tonic. (2) *Picrasma excelsa* (Quassia wood or chips). A larger tree with PINNATE leaves and small, green flowers from the W. Indies. An infusion of wood chips boiled in water was formerly used as an insecticide and can still be bought.

QUEENSLAND NUT *(Macadamia ternifolia)*. A densely leafy evergreen tree to 15m in the *Proteaceae* from N. E. Australia, but grown in Hawaii and elsewhere. It has 100–300mm long prickly-toothed, OBLANCEOLATE leaves in WHORLS of 3 or 4 and slender spikes to 200mm long of 25mm wide creamy flowers. These are followed by leathery fruits containing highly edible, globose nuts.

QUICKTHORN
see HAWTHORN

QUILLWORT *(Isoetes)*. A genus of 75, mainly aquatic perennials in the *Isoetaceae* from temperate and tropical regions. They are densely tufted plants with LINEAR leaves and SESSILE SPORANGIA imbedded in the leaf bases. Common quillwort *(I. lacustris)* is found mainly in mountain lakes and tarns. It has SUBULATE 75–200mm long leaves.

QUINCE *(Cydonia oblonga)*. A deciduous tree to 7.5m in the *Rosaceae* from Asia, but much cultivated in the Mediterranean countries and elsewhere. It has woolly young shoots and 50–100mm long OVATE leaves, grey-woolly beneath. The solitary, white or pink, 38–50mm wide bowl-shaped 5-petalled flowers are followed by fragrant, yellow, pear-shaped fruits to 100mm long. The latter are used to make jam and nurserymen use young plants as rootstocks upon which to graft pears. ● Japanese quince *(Chaenomeles speciosa* syn. *C. lagenaria* and *Cydonia japonica)*, sometimes known as japonica, is a shrub, sometimes spiny, with ovate-oblong, finely toothed, smooth leaves, scarlet to blood-red flowers in small clusters and yellow green ovoid fruits. It is much grown as an ornamental and has given rise to many CULTIVARS, several of hybrid origin.

QUININE [YELLOW or **PERUVIAN BARK]** *(Cinchona calisaya)*. An evergreen tree to 12.2m in the *Rubiaceae* from the Bolivian and Peruvian Andes. It has oval to oblong leaves and terminal PANICLES of small, tubular, fragrant pink flowers. The bark contains several medicinal substances, notably quinine, used for malaria and also in a cough mixture and as a tonic. Several other species also yield quinine.

QUINOA *(Chenopodium quinoa)*. An 0.9–1.5m annual in the *Chenopodiaceae* from S. W. America where it is grown as a grain crop. It has triangular-ovate leaves and tiny greenish flowers in dense terminal panicles. The comparatively large, rounded seeds can be reddish or white.

▲ **RACEME** An inflorescence composed of a usually erect stem bearing stalked flowers at intervals.

RACEMOS|US, -A, -UM Bearing flowers in RACEMES.

▲ Raceme

Raceme

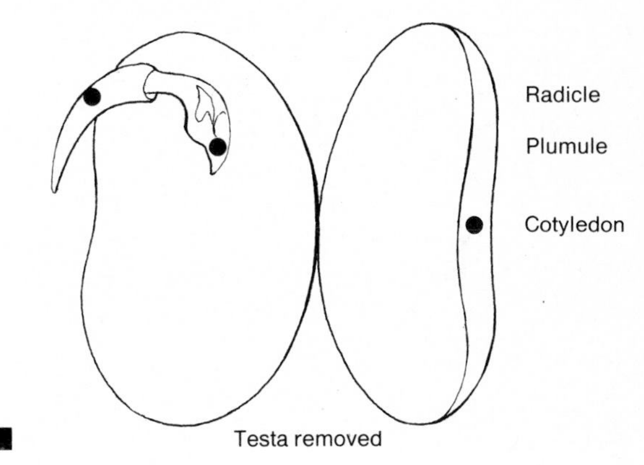

■ Testa removed

RACHIS The main stem or stalk of an inflorescence or compound leaf.

RADICAL Of leaves that apparently arise directly from the root, e.g. the basal leaves of dandelion.

RADICANS Having creeping or rooting stems.

■ **RADICLE** The first root to issue from the seed on germination.

RADISH *(Raphanus sativus)*. An annual or biennial plant in the *Cruciferae*, unknown in a truly wild state but certainly grown by the ancient Egyptians more than 2,000 years ago. There are several different sorts, all with rosettes of PINNATE leaves having large, terminal leaflets (LYRATE-PINNATIFID) and white, lilac, purple or yellow and purple 4-petalled flowers in terminal RACEMES. The cylindrical, pointed seed pods are somewhat inflated. The tuberous root may be white, red or blackish and turnip-shaped or cylindrical; some are hot to the taste, others sweet and mild.

RAFFLESIA Total parasitic plants in the *Rafflesiaceae* growing on the roots of tropical vines (members of the *Vitidaceae*) in Malaysia. *Rafflesia arnoldii* has the largest known flower of any plant, specimens of 900mm across being recorded. It has 5 yellowish petals and a brick-red, central bowl or nectarium and an odour of rotting meat to attract pollinating flies. The remainder of the plant is reduced to a MYCELIUM-like mass of tissue in the roots of the host.

RAGGED ROBIN *(Lychnis flos-cuculi)*. A perennial plant of damp or wet places in the *Caryophyllaceae* from Europe and Asia. It has erect stems to 750mm, opposite pairs of oblong-lanceolate leaves and DICHASIAL CYMES of rose-red, 25–38mm wide flowers from tubular CALYCES. The petals are deeply divided into 4 narrow segments, the centre two the longest.

RAGWORT *(Senecio jacobaea)*. A perennial plant to 1.2m in the *Compositae* from Europe, Asia and N. Africa and introduced to N. America and Australia where it is a serious weed being poisonous to stock if eaten in quantity. (A natural predator, the cinnabar moth has been used against it with some success in Australia). It has erect stems, LYRATE-PINNATIFID leaves and large compound terminal CORYMBS of 13–25mm wide, yellow daisy flowers.

RAIN TREE [SAMAN] *(Pithecellobium saman)*. A tree of 24.4m with a wide spreading head, in the *Leguminosae* from Central America. It has bi- to quadri pinnate leaves of ovate-oblong 38mm long leaflets, downy beneath and which fold together at the approach of rain. The small, yellow tubular flowers have red stamens and are borne in dense heads. The 150–200mm long, flattened pods contain a sweet pulp and are fed to cattle.

RAMBUTAN *(Nephelium lappaceum)*. An evergreen tree to 18.2m from S. E. Asia, very similar to LITCHI and in the same family. *Sapindaceae*. The fruits differ in being covered with soft red or yellow spines.

RAMENTA The flattened, scale-like hairs found on the stalks of many fern fronds.

RAMPION One bell-flower and several species of *Phyteuma*, perennials in the *Campanulaceae* from Europe and Asia. The phyteumas are distinct in having small flowers in dense spikes or heads which are tubular until the anthers have shed their pollen (pushed out of the top by the growing style), then open to 5 narrow petals: round-headed rampion *(P. tenerum*, syn. *P. orbiculare)* grows to 300mm or more, with long-stalked, LANCEOLATE, CRENATE leaves and 13–19mm wide, globose heads of purple-blue flowers; bell-flower rampion *(Campanula rapunculus)* is a biennial with oblong to ovate basal leaves and elongated PANICLES of erect. 19mm long purple bell-flowers. The white fleshy root and the leaves are edible.

● **RAMSONS** *(Allium ursinum)*. A cylindrically bulbous onion-ally in the *Alliaceae* from Europe to Turkey. It has 100–250mm long glossy, elliptic leaves and UMBELS of starry, white flowers having 6 pointed PERIANTH segments. All parts of the plant smell strongly of garlic when bruised.

RAPE [SWEDE, COLESEED] *(Brassica napus)*. A variable annual or biennial plant in the *Cruciferae* from S. Europe, but much cultivated elsewhere. It has LYRATE-PINNATIFID leaves and erect, branched RACEMES of 4-petalled, yellow flowers; rape or coleseed *(B. n. arvensis)* is non-tuberous rooted and is grown for its oil-rich seeds. Rape seed cake for cattle feeding is made from the seeds after the oil is extracted; Swede or Swedish turnip *(B. n. napobrassica)* is believed to be a rape-cabbage hybrid and forms a large, turnip-like root with a short, thick neck or trunk at the top. CULTIVARS may be purplish, whitish or yellowish with yellow or rarely whitish flesh, the common yellow-fleshed one is sometimes called rutabaga.

▲ **RASPBERRY** *(Rubus idaeus)*. A suckering, deciduous shrub to 1.8m in the *Rosaceae* from Europe to Asia and cultivated elsewhere. It has PINNATE leaves usually with 5 OVATE leaflets which are white tomentose beneath. The 5-petalled, white flowers are followed by rounded, 13–19mm long red or (rarely) pale yellow fruits composed of many tiny drupelets. Several CULTIVARS are grown including some which fruit in autumn.

RAT TAIL CACTUS *(Aporocactus flagelliformis)*. An EPIPHYTIC species in the *Cactaceae* from Mexico and Central America. It has weeping stems to 900mm long set with tiny, red-brown spines and bearing 50–75mm long trumpet-shaped crimson flowers.

■ **RAY FLORETS** The strap or tongue-shaped outer florets of a daisy flower, also known as ligulate florets. They are formed of 5 slender petal-lobes fused edge to edge. Dandelion and hawkweed are entirely composed of ligulate florets.

○ **RAY, JOHN (1627–1705)** Fine all-round naturalist; son of an Essex blacksmith and educated at Cambridge, where he became a lecturer. His local observations resulted in a record of the plants found near Cambridge. Published in 1660, and like all his works in Latin, it is the first of the County Floras of Britain and includes garden and crop plants listed alphabetically (no satisfactory method of classification had then been worked out, nor had the binomial system of naming). The names are often cumbersome, but the work is scientifically most accurate.

In 1662, with the Act of Uniformity, he found himself unable to remain at Cambridge and spent thereafter much

time travelling in Britain and on the continent. He retired to his old Essex home in 1672, and there wrote his *History of Plants*. Ray used the number of seed leaves as a basis for classification, creating the divisions MONOCOTYLEDONS and DICOTYLEDONS which have been followed ever since. The rest of his system was superseded by that of LINNAEUS, but that great man described Ray as an 'incomparable botanist'. The wideness of his interests is shown by the naming of Ray's bream, Ray's wagtail, as well as *Polygonum raii*, a shoreline knotgrass which he first recognised as a distinct species. The Ray Society was founded in 1844, its object to publish botanical papers which might otherwise remain unknown. Main books: *A Catalogue of Plants Growing Around Cambridge* (1660; modern English translation, London 1975); *Catalogue of the Plants of the British Isles* (1670); *History of Plants* (3 vols, 1686, 1688 and 1704).

RAZOR STROP FUNGUS [BIRCH POLYPORUS] *(Piptoporus betulinus)*. A PARASITIC or SAPROPHYTIC *Basidiomycetes* fungus in the *Polyporaceae* from the north temperate zone. Restricted to birch trees and forms on their trunks white-fleshed, very pale brown on greyish bracket-like fruiting bodies to 300mm across. When mature they are dry and corky and cut sections are used by entomologists for mounting small insect specimens; also sometimes recommended for stropping razors.

RECEPTACLE (1) The usually enlarged and modified stem tip which bears the floral whorls, i.e. sepals, petals, stamens and pistil.
(2) The broadly flattened stem tip which bears the florets and bracts of a composite (daisy) flower.

RECESSIVE GENE A gene which reproduces its character only when present in both parents. When present in one parent alone, its character does not appear, being suppressed by its opposite number which is then said to be the dominant gene. In MENDEL'S pea experiments (*see* HEREDITY), tall dominates short, the gene for shortness being recessive (heterozygous). The short character only reappears when in a later generation it is combined with another gene for shortness (homozygous).

RECLINAT|US, -A, -UM Reclining, bent over downwards, sometimes resting on the ground, with the tip turned upwards.

RECURV|US, -A, -UM Bent or curved back.

RED BUCKEYE
see HORSE-CHESTNUT

RED BUD *(Cercis canadensis)*. A tree very much like JUDAS' TREE, but with deep pink buds and pink flowers, from east central and N. America.

RED CEDAR
see JUNIPER (Pencil Cedar)

RED-HOT POKER *(Kniphofia)*. A genus of 25 tufted perennials in the *Liliaceae* from E. and S. Africa and Malagasy. They are of tufted growth with LINEAR, KEELED, tapered leaves and mainly dense, terminal spikes of stiffly drooping, tubular flowers. Common red-hot poker *(K. uvaria)* is a robust species with grey-green leaves 25mm wide and up to 900mm long. The coral-red flowers age to

orange and yellow and are carried in poker-like spikes on 0.9–1.8m stems. There are many hybrid CULTIVARS of red, yellow and cream.

RED INK PLANT
see POKEWEED

RED SNOW
see CHLAMYDOMONAS

● **RED VALERIAN** *(Centranthus ruber)*. A woody-based perennial in the *Valerianaceae* from southern Europe to Turkey and N. Africa, naturalised in Britain and elsewhere. It has opposite pairs of OVATE, LANCEOLATE leaves to 100mm long and terminal panicled CYMES of red, pink or white tubular flowers each about 9mm long and with a basal spur. The CALYX later forms a hairy PAPPUS on the ripe, seed-like fruits.

▲ **REDWOOD** *(Sequoia sempervirens)*. An evergreen tree in the *Taxodiaceae* from coastal ranges of California and Oregon. It is the tallest tree known, exceptional specimens topping 110.0m. It has thick, reddish, fibrous bark, very resistant to forest fires and LINEAR, 6–25mm long leaves in two ranks. The ovoid cones are about 25mm long. An important timber tree in its native country and grown in many temperate countries as an ornamental.

REED A name given to several grassy plants, particularly those growing by water, but strictly belonging to *Phragmites australis* syn. *P. communis*, a vigorous, RHIZOMATOUS plant in the *Gramineae* of cosmopolitan distribution, except for certain tropical areas. It can colonise vast areas of swamp and shallow water around ponds etc. The erect stems reach 1.8–3.0m tall with broad, flat, GLAUCOUS leaves and nodding terminal PANICLES of soft, dull purple spikelets. Much used for durable thatching.

REED GRASS *(Phalaris arundinacea)*. A RHIZOMATOUS plant in the *Gramineae* from the north temperate zone. It grows to 0.6–1.2m tall with flat leaves and 100–150mm long narrow panicles of small, purplish spikelets. *P.a.* 'Picta' has white striped leaves (GARDENER'S GARTERS). *See also* ARUNDO.

■ **REED-MACE** *(Typha)*. A genus of about twenty RHIZOMATOUS grass-like aquatic perennials in the *Typhaceae* from temperate and tropical regions. They have long, LINEAR leaves and tiny unisexual flowers tightly packed into terminal poker-like spikes. Common or great reed-mace or cats-tail (often wrongly called bulrush), *T. latifolia*, grows 1.2–2.4m tall with 150–300mm long dark velvety brown spikes.

REEDSWAMP Areas of marsh or fen or the edges of lakes and rivers, under water for most of the year, dominated by reeds *(Phragmites)* with reed grass, reed-mace and broad-leaved flowering plants such as purple LOOSESTRIFE and WATER DROPWORT.

REFLEXED [REFLEX|US, -A, -UM] Bent back abruptly.

REGAL|IS, -E Of outstanding merit or of regal appearance.

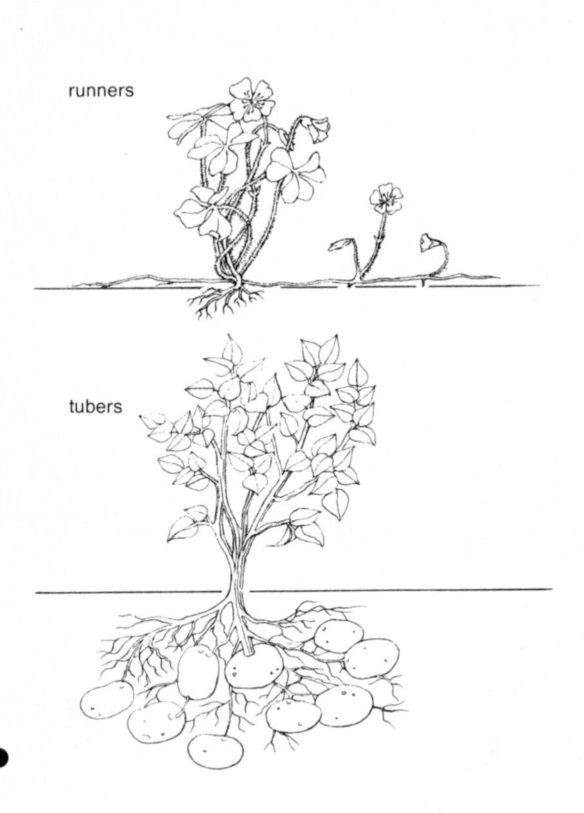

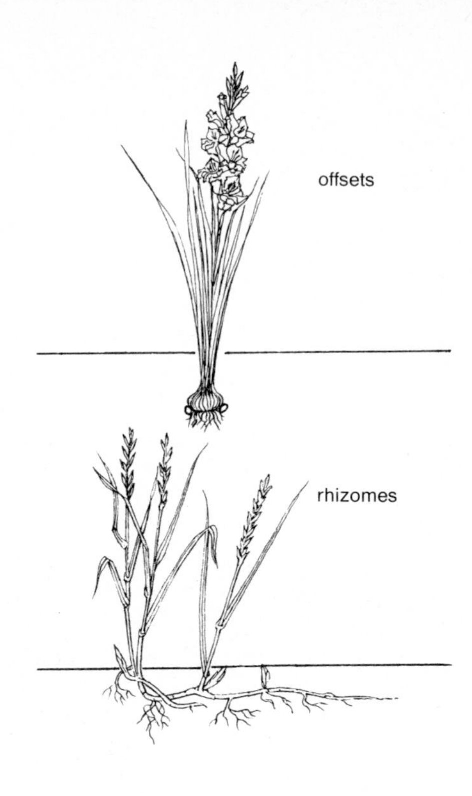

REINDEER MOSS Several species of *Cladonia*, whitish to grey-green LICHENS of circumpolar distribution. True reindeer moss *(C. rangifera)* is like a small, blue-grey leafless shrub with erect, branched stems. It is abundant only in arctic and sub-arctic regions and forms the staple diet of reindeer. *Cladonia impexa* is a common species of heaths and moors even at low altitudes, and forms elaborately branched, yellow-tinged cushions.

RENIFORM [RENIFORMIS, -E] Kidney-shaped, usually of leaves.

REPAND [REPAND|US, -A, -UM] Having an uneven and waved margin.

REPENS Prostrate; growing flat on the ground.

REPLUM
see SILIQUA

REPRODUCTION Plants reproduce themselves in two basic ways, asexually and sexually. Asexual covers all those ways a plant can spread without seeds, e.g. RHIZOMES, STOLONS, RUNNERS, OFFSETS, aerial BULBILS, etc. Sexual methods involve the formation of egg and sperm cells (male and female GAMETES), the fusion of which gives rise to spores or seeds.

REPTANS Creeping on the ground and often rooting.

RESIN A mixture of a volatile oil and a solid, gum-like substance which hardens when the oil evaporates. It is common among the coniferous trees, particularly pines, which have special resin ducts or channels in the leaves, stems and roots. It appears to be a waste product, but is of value to the plant in covering wounds with a protective layer. Resin yields turpentine and rosin on distillation.

RESPIRATION A basic life process common to animals and plants, taking in oxygen and giving out carbon dioxide. In plants it is known as internal or cell respiration, the oxygen being used in a chemical reaction to provide energy.

RESTHARROW Several species of *Ononis*, annuals, RHIZOMATOUS perennials or small shrubs in the *Leguminosae* from Europe, Asia and N. Africa. They have TRIFOLIATE leaves and short terminal RACEMES of pea-shaped flowers: common restharrow *(O. repens)* is a prostrate perennial with wiry rhizomes, glandular hairy leaves and 9–15mm long pink flowers; spiny restharrow *(O. spinosa)* is an erect or ascending spiny, shrubby plant to 300mm with similar flowers and leaves.

RESTING
see DORMANT

RESURRECTION PLANT *(Selaginella lepidophylla)*. A curious club-moss in the *Selaginellaceae* from southern USA to Peru. It forms a rosette of 100–200mm long frond like branches bearing tiny OVATE, overlapping leaves. The SPORANGIA are carried in 6–13mm long spikes. When dry the whole plant rolls inwards to form a tight ball, but soon unrolls when moist, even after several months of dessication. *See also* ROSE OF JERICHO.

▲

RETICULATE [RETICULAT|US, -A, -UM] Bearing an obvious network of veins, e.g. leaves and bracts.

RETINOSPORA An obsolete name for several coniferous trees, notably *Thuja* and *Chamaecyparis* which were originally named from juvenile material, the foliage of which often differs greatly from that of mature trees.

RETROFLEX|US, -A, -UM Bent or curved back.

RETUSE [RETUS|US, -A, -UM] Having a rounded, shallowly notched tip.

REVOLUTE [REVOLUT|US, -A, -UM] Rolled back, often referring to the leaf margins.

RHIZOIDS Colourless, unicellular hairs which act as roots and root hairs all in one, characteristic of mosses, liverworts and the PROTHALLUS of ferns.

RHIZOME Underground or partly above ground, horizontally growing stems. They may be slender and fast growing as in couch grass or slower and fleshy, acting as storage organs as in SOLOMON'S SEAL.

RHIZOMORPH Underground, cord-like strands of HYPHAE characteristic of certain fungi, notably HONEY FUNGUS. They enable the fungus to spread from tree to tree via the roots.

RHIZOPHORE Aerial root-like structures borne on the stems of many club mosses *(Selaginella)* and which produce true roots when they reach the soil.

▲ **RHODODENDRON** A genus of about 600 deciduous and evergreen trees and shrubs from the northern hemisphere and S. E. Asia to N. Australia. They have simple, LINEAR to OVATE-CORDATE leaves, bell or tubular flowers, usually slightly ZYGOMORPHIC, with 5–10 lobes and stamens, in terminal UMBEL-like clusters. The woody capsules contain very small, air-borne seeds often with wing or tail-like appendages. Several species are EPIPHYTIC. Rhododendrons can be divided into two large groups on the presence or absence of minute, scale-like hairs on the stems, leaves, etc. Species with scales are lepidote, those without, elepidote. *Rhododendron* includes several species known as azaleas. Originally the genus *Azalea* contained deciduous species with 5 stamens to each flower, but it was eventually realised that some plants did not fit well into either category and that there was insufficient reason for keeping them separate, and all azaleas became part of rhododendron. *R. ponticum* from Turkey, Spain and Portugal, which is well naturalised in Britain and elsewhere, can reach 6m with evergreen, oblong-lanceolate leaves and funnel-shaped purple flowers with 10 stamens; *R. luteum* is a typical deciduous azalea from E. Europe to W. Asia, with hairy, OBLANCEOLATE leaves and heavily fragrant yellow flowers with 5 stamens. It has given rise to many hybrid CULTIVARS popular in gardens; mountain laurel (in its native USA), *R. maximum*, is an evergreen tree to 12.2m with oblanceolate leaves to 250mm and bell-shaped, pink to rose-purple flowers, usually with 10 stamens; swamp honeysuckle covers several species of deciduous azaleas from N. America, notably *R. viscosum*, a 1.8–3.0m shrub with ovate to obovate, sticky-hairy leaves and 19–25mm long slender, trumpet-shaped white, pink-striped flowers with 5 stamens.

RHOMBOIDAL [RHOMBOIDE|US, -A, -UM] Roughly diamond-shaped.

RHUBARB *(Rheum rhaponticum)*. A long-lived perennial in the *Polygonaceae* from Siberia. It has long, red or green and red stalked, broadly ovate, deeply cordate, undulate leaves to 600mm or more in length and erect stems to 1.8m carrying large PANICLES of small whitish flowers with 6 PERIANTH segments. There are several CULTIVARS available, selected for the colour of the 'sticks' (leaf-stalks), sweetness and earliness. It seems fairly certain that all the garden rhubarbs are of hybrid origin.

RHYTIDOPHYLL|US, -A, -UM With wrinkled leaves.

RIB The main vein or veins of a leaf.

RIBBON GRASS
see GARDENER'S GARTERS; REED GRASS

RIB GRASS
see PLANTAIN (ribwort)

RICE *(Oryza sativa)*. An aquatic annual grain crop in the *Gramineae* from S. E. Asia, but grown in most tropical and sub-tropical countries and in many warm temperate regions. It provides the staple food of about half of the world's population. Rice is an erect, tufted grass from 45–150cm in height with 125–375mm long PANICLES of oblong-ovoid, single-flowered, spikelets with or without AWNS. The grains are usually off-white, but red, brown and blackish forms are known. There are hundreds of CULTIVARS, mainly differing in length of grain, time taken to mature and suitability for the climate. They can be grouped roughly into awned rice grown in warm temperate climates and awnless rice grown in the tropics. Rice is mainly grown in specially flooded fields called paddies, though some cultivars can be grown in ordinary, moist soil. In some areas rice is sown on flood plains just before the rains come and the land is inundated.

RIMU Several species of *Dacrydium*, evergreen, coniferous trees in the *Podocarpaceae* from New Zealand. (Others are from Tasmania, Australia, S. E. Asia and Chile.) Rimu or red pine, *(D. cupressinum)* grows to 54m, pyramidal when young with gracefully pendent branches. The leaves are awl-shaped and the 'cone' is red and fleshy with a single seed imbedded in the top. Mountain rimu *(D. laxifolium)* is a prostrate or semi-erect shrub with branches to 300mm and linear to linear-oblong leaves.

RINGENS Gaping, e.g. the mouth of a 2-lipped flower such as dead nettle.

RIPARI|US, -A, -UM Growing by rivers or streams.

RIVULAR|IS, -E [RIVAL|IS, -E] Growing by brooks or small streams.

ROBIN'S PINCUSHION The bedeguar gall, an irregularly rounded mass of small, woody galls covered with stiff red hairs on wild rose. It is caused by a small gall wasp. *See* GALL.

ROCAMBOLE [SAND LEEK] *(Allium scordoprasum)*. A perennial, bulbous-rooted plant in the *Alliaceae* from

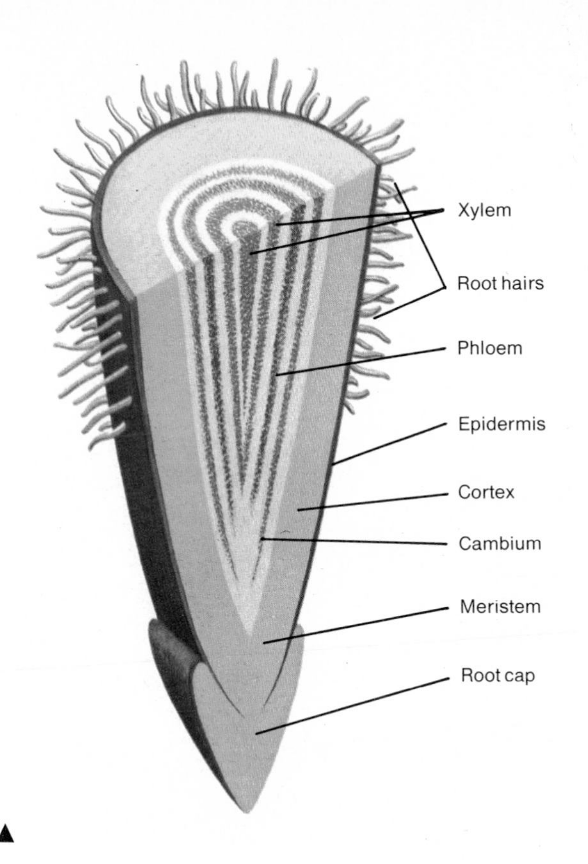

▲

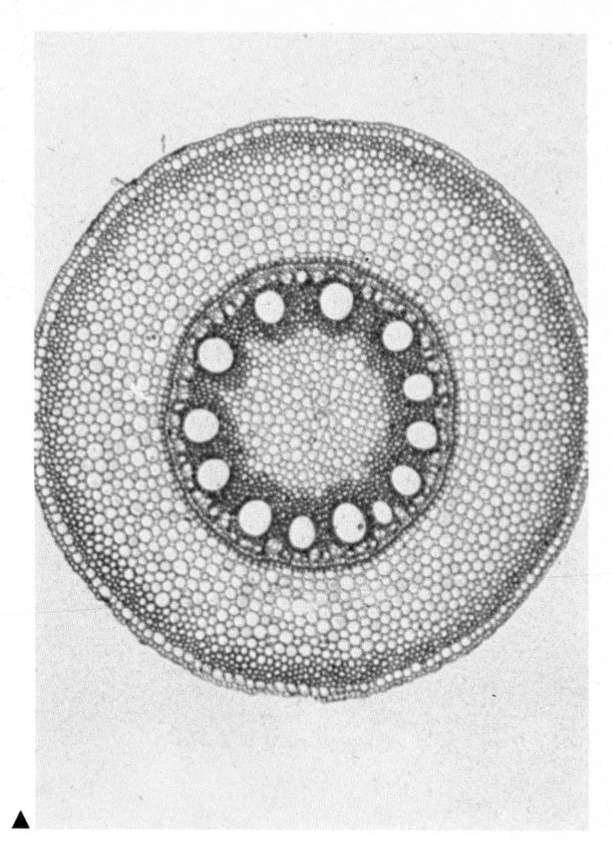

▲

Europe to Turkey and Syria. It has ovoid bulbs, LINEAR, KEELED leaves and globose heads of purple bulbils mixed with a few red-purple flowers, having 6 perianth segments. The bulbs and bulbils are used in much the same way as garlic.

ROCKET A name for several different plants in the cabbage family, the *Cruciferae,* but primarily to certain members of *Sisymbrium,* annuals and perennials from temperate regions: London rocket *(S. irio)* is an annual to 600mm with deeply PINNATELY lobed stalked leaves, the terminal lobe the largest. The yellow, 3mm wide 4-petalled flowers are borne in erect RACEMES. For yellow rocket *see* WINTER CRESS; for dame's rocket *see* DAME'S VIOLET. *See also* SEA ROCKET.

ROCKFOIL
see SAXIFRAGE

ROCKROSE *(Helianthemum).* A genus of 100 evergreen sub-shrubs in the *Cistaceae* mainly from the Mediterranean and from Europe and Asia. They have LANCEOLATE to OBLONG-OVATE leaves and 5-petalled, often showy yellow, red or white flowers that open out flat. Common rock rose *(H. chamaecistus* syn. *H. nummularium)* is almost prostrate except for the flowering stems. The leaves are oblong-oval and the 25mm wide flowers yellow. It has given rise to many fine hybrid CULTIVARS in shades of red, pink, yellow and white, some double-flowered.

▲ **ROOT** Develops from the radicle, serving to anchor the plant and to absorb water and mineral salts via its hairs, passing them upward through the xylem. Growth is by division of meristem cells. The primary radicle may persist as a tap root, may branch out into a fibrous root system, or die and give place to adventitious roots.

ROOT CAP A hood-shaped cap of mucilaginous cells which protects the actively growing root tip as it pushes through the soil. The outer cells of the cap are rubbed off against soil particles, but are renewed from within.

ROOT HAIRS Slender, elongated single cells which grow out from the EPIDERMIS of the root just behind the growing tip. They adhere closely to soil particles, extracting moisture from them by OSMOSIS.

ROOT NODULE Tiny galls which form on the roots of many leguminous and some other plants, caused by a bacterium *(Rhizobium)* which is able to fix atmospheric nitrogen. This is passed on to the plant in return for sugars, a good example of symbiosis (*see* MYCORRHIZA).

ROOT PARASITE PARASITIC plants which attach themselves to the roots of their host plant, e.g. BROOMRAPE.

ROSE *(Rosa).* A genus of 250 shrubs and scrambling or climbing plants in the *Rosaceae*, all but one from the northern hemisphere. They have PINNATE leaves, usually thorny stems with straight or hooked prickles, and wide open, 5-petalled, often fragrant flowers in shades of red, pink, purple, yellow or white. The red, orange or black-purple fruits known as hips or heps are fleshy, hollowed-out RECEPTACLES, containing nutlet-like ACHENES and often

bearing the numerous remains of sepals at the top. Roses have been popular for centuries and many hybrids and cultivars grace our gardens today, most of them with double or semi-double flowers: Austrian briar *(R. foetida)* has erect stems to 1.5m, straight prickles and 50–63mm wide, rich yellow flowers with a heavy odour; Banksian rose *(R. banksiae)* is a climber to 6m, almost without prickles and small white or yellow flowers in clusters; Bourbon *(R. X borboniana)* is a hybrid of the China rose with hooked prickles and semi-double, sweetly scented pink flowers; burnet of Scotch rose *(R. pimpinellifolia* syn. *R. spinosissima)* is a suckering, twiggy bush to 900mm with straight prickles and 25–38mm wide cream or pink flowers followed by black-purple hips; cabbage or Provence rose *(R. centifolia)* grows to 1.5m with double, fragrant pink flowers; China *(R. chinensis)* is a shrub or climber to 6m with stout, hooked prickles and PANICLES of, or solitary, crimson, pink or white 50mm wide flowers borne at intervals from summer to autumn (recurrent flowering). An important parent of many modern roses; Damask rose *(R. damascena)* grows 0.6–1.5m with red or white, double flowers; *R.d.* 'Versicolor', with red and white petals, is known as York and Lancaster rose; dog rose *(R. canina)* grows to 3m with hooked thorns and pink to white 38–50mm wide flowers – a common hedgerow rose in Britain; field rose *(R. arvensis)* is similar, but with slender, spreading green stems and white flowers; musk rose *(R. moschata* and *R. brunonii)* are climbers to 7.5m or more with sweet musk-scented 25–50mm wide flowers in clusters; prairie rose *(R. setigera)* is a spreading shrub or climber to 4.5m with 50mm wide pink flowers; Ramanas rose *(R. rugosa)* is a dense prickly suckering shrub to 2.4m with 50–75mm wide, fragrant red or white, single or double flowers and large rounded, orange-red hips; sweet briar *(R. eglanteria)* is much like dog rose, but with aromatic, glandular foliage; tea rose *(R. odorata)* is a tall shrub or climber to 9.1m with 50–75mm wide white, yellowish or pink double or semi-double flowers – an important parent of the modern hybrid tea roses; white rose of York *(R. X alba)* is a hybrid between dog and damask roses to 2.4m with white, single or semi-double, 63–75mm wide, fragrant blooms.

ROSELLE *(Hibiscus sabdariffa)*. A sub-shrubby plant in the *Malvaceae* from W. Africa but much planted in the tropics. It has PALMATE, deeply 3–5 lobed leaves to 150mm long and 5-petalled, somewhat funnel-shaped yellow flowers with a maroon centre. Fruit roselle has 13–19mm long CALYCES which become red and fleshy and are boiled with sugar to produce a refreshing drink; also made into sauces, jellies and chutneys. Fibre roselle *(H. s. altissima)* forms an erect, unbranched plant with spiny inedible calyces and is grown for its fibre.

ROSEMARY *(Rosmarinus officinalis)*. A sweetly aromatic, evergreen shrub in the *Labiateae* from S. Europe to Turkey. It has opposite pairs of LINEAR 19–50mm long leaves and pale purple-blue, tubular, 2-lipped flowers to 19mm long. The leaves are used as a culinary herb in meat dishes and in sachets and pot-pourri like lavender.

ROSE OF JERICHO *(Anastatica hierochuntica)*. Annual plant in the *Cruciferae* from Morocco to Iran, it is a spreading, branched plant to 150mm with OBOVATE leaves and spikes of small, white 4-petalled flowers. When mature, in the dry season, the leaves fall off and the

branches roll inwards forming a wickerwork ball which becomes detached from the soil and blows about. When the rains come, the plant expands and the seeds are released.

ROSE OF SHARON
see ST JOHN'S WORT

ROSEROOT
see STONECROP

▲ **ROSETTE** A crowded WHORL of leaves spread out flat or almost so, arising from the top of a root at ground level as in many biennial plants, e.g. Canterbury bell, or from the top of a stem as in some palms and cycads.

ROSE|US, -A, -UM Rose-coloured, pink.

ROSTELLUM A projecting or beak-like structure on the column of an orchid flower which represents a functionless STIGMA.

ROSTRAT|US, -A, -UM Bearing a beak-like point.

ROSULATE [ROSULAT|US, -A, -UM] Bearing a rosette.

ROTUND|US, -A, -UM Almost circular.

ROUGE
see SAFFLOWER

ROWAN [MOUNTAIN ASH] *(Sorbus aucuparia)*. A slender, deciduous tree to 15m or more in the *Rosaceae* from Europe to Turkey and the mountains of Morocco. It has PINNATE leaves usually of 13–15 oblong, toothed leaflets and flattened heads (compound corymbs) of creamy-white 5-petalled flowers. The berry-like fruits, technically POMES, are rounded, scarlet to 9mm wide. A yellow-fruited form is known. Several other *Sorbus* species are grown as ornamentals, some with white or pink fruits.

ROYAL FERN *(Osmunda regalis)*. A clump-forming herbaceous perennial fern in the *Osmundaceae*. Native to wet places, it has thick RHIZOMES and densely matted black root systems. The 0.6–3.0m high fronds are bipinnate with 19–63mm long leaflets. The upper leaflets of the fertile fronds have reduced blades and are entirely occupied by brown SPORANGIA. The chopped roots, mixed with moss are used by gardeners as a medium in which to grow exotic epiphytic orchids.

■ **RUBBER [PARA RUBBER]** The congealed and smoked latex of *Hevea braziliensis*, an evergreen or briefly deciduous tree to 30m or more in the *Euphorbiaceae* from Brazil, but grown elsewhere in tropical countries. The leaves are TRIFOLIATE, each leaflet about 150mm long, and the small yellow, bell-shaped, scented flowers are carried in PANICLES. The 3-lobed, 25–50mm wide seed capsules explode noisily when ripe. The rubber latex is obtained by making narrow incisions into the bark at an angle of 25–30°, spiralling half round the trunk. This severs the maximum number of latex vessels. The incisions are re-cut every other day to keep the latex flowing. Other species of *Hevea* and several trees in the *Moraceae* yield rubber, e.g. Panama rubber *(Castilla elastica)* and indiarubber tree (*Ficus elastica, see* FIG).

● 

■ 

RUBBER PLANT A gardener's name for young pot-grown specimens of *Ficus elastica, see* FIG.

RUBELL|US, -A, -UM [RUBENS] Reddish.

RUB|ER, -RA, -RUM Red.

RUBIGINOS|US, -A, -UM Brownish or rusty red.

RUDERAL|IS, -E Of rubbish heaps and waste places.

● **RUE** *(Ruta graveolens)*. An evergreen, acrid 300–900mm shrub in the *Rutaceae* from S. Europe. It has PINNATE leaves with OBOVATE, GLAUCOUS leaflets and terminal CORYMBS of mustard yellow 13–19mm wide flowers having 4 or 5 cupped petals.

RUFESCENS Becoming reddish.

RUGOS|US, -A, -UM Having a wrinkled appearance.

RUNCINATE [RUNCINAT|US, -A, -UM] PINNATIFID or boldly SERRATE, the teeth or lobes pointing towards the base.

RUNNER A slender, prostrate stem with long INTERNODES, rooting at the NODES and forming new plants, e.g. strawberry.

RUNNER BEAN [SCARLET RUNNER] *(Phaseolus coccineus)*. A tuberous-rooted perennial twining climber in the *Leguminosae* from Central America. It grows to 3m, with TRIFOLIATE leaves and AXILLARY RACEMES of scarlet (sometimes white) flowers. The slender, flattened pods may reach 500mm when mature, but are eaten while young and tender. The seeds (beans) are red, variously spotted or marbled purple-black. A dwarf, non-climbing cultivar is grown. Very similar in both climbing and non-climbing forms are French, kidney, waxpod and haricot beans *(P. vulgaris)*. They have largely white flowers and smaller, cylindrical pods which may be picked green or allowed to ripen and the dried beans harvested as haricots. Some CULTIVARS have yellowish, purple or mottled pods and the seeds may be white, red, brown or black.

RUPESTR|IS, -E Growing on rocks.

RUPICOL|US, -A, -UM Growing in rocky places.

▲ **RUPTUREWORT** Several species of *Herniaria,* prostrate perennial or annual plants in the *Caryophyllaceae* from Europe, Asia and N. Africa. Smooth rupturewort *(H. glabra)* is an annual or biennial with OVATE-LANCEOLATE, 3–6mm usually hairless leaves and minute, white petalled flowers in AXILLARY clusters forming leafy spikes; fringed or ciliate rupturewort *(H. ciliolata)* is similar, but perennial and mat-forming, with CILIATE leaves and axillary flower clusters only.

RUSH Several grassy-leaved plants are known by this name, but strictly it refers to *Juncus,* a genus of 300 annuals and perennials of cosmopolitan distribution from cool, moist climates. They are tufted in habit, with LINEAR or cylindrical tough evergreen leaves and small PANICLES of tiny greenish or brownish flowers with 6 PERIANTH segments. These often appear to be borne on the sides of

▲

the stems near the tips, but they are really terminal, the apparent stem tips being a stiff bract: hard rush *(J. inflexus)* is 300–600mm tall with grey-green, hard stems and loose panicles 38–50mm long; soft rush *(J. effusus)* is similar but with glossy, bright green stems; heath rush *(J. squarrosus)* has spreading tufts of wiry, 75–150mm leaves and 150–300mm flowering stems with obviously terminal panicles; toad rush *(J. bufonius)* is an annual, with FILIFORM, 13–50mm long leaves and sparsely branched, terminal panicles. *See also* FLOWERING RUSH, Dutch rush under HORSETAIL, REED-MACE (bulrush), WOODRUSH.

RUST FUNGI Small parasitic *Basidiomycetes* fungi producing flask or cup-shaped fruiting bodies which rupture the tissues of stem or leaf and show rust-coloured spore masses. Many of the different sorts of rust fungi have complicated life histories. Some live on one kind of host plant only (AUTOECIOUS), e.g. mint rust; others need two or more hosts (HETEROECIOUS), e.g. gooseberry rust, which also lives on sedges. At different times of the year, rusts produce different sorts of spores. Summer spores (uredospores) are thin-walled, usually pale brown or yellow. In autumn, dark brown, thick-walled resting spores (teleutospores) are produced. These germinate after a period of dormancy and form short, broad, tubular HYPHAE which produce BASIDIA bearing solitary basidiospores. The basidiospores attack a host plant, and in spring small angular AECIDIOSPORES form in special cup-shaped structures known as cluster cups or aecidial cups. These various spore stages may be produced on the one host plant, or on the various host plants of heteroecious rusts. Wheat rust for example, a serious disease of wheat and other cereals, forms its summer and autumn spores on wheat, but the spring aecidiospores on common barberry. To destroy the disease therefore one must remove the barberry where it lives in winter when the grain crop has gone.

■ **RUSTYBACK FERN** *(Ceterach officinarum)*. A small fern of rocks and walls in the *Aspleniaceae* from Europe and N. Africa to the Himalaya. It has 38–200mm PINNATE fronds, the backs entirely covered with rust-coloured scales and bearing narrowly oblong SORI.

RUTABAGA
see RAPE

RUTILANS [RUTIL|US, -A, -UM] Orange-red.

RYE *(Secale cereale)*. An annual cereal plant in the *Gramineae* probably from Turkey, but much cultivated since Iron Age times. It resembles barley in general appearance but the awned spikelets are usually 2-flowered and have very narrow GLUMES with stiffly hairy KEELS. Rye is an important crop in the colder northern areas of Europe and Asia where other grain crops fail. It is used to make bread, and in brewing for whisky and gin. The straw is used for paper-making, thatching and straw hats.

RYE GRASS Several species of *Lolium,* annual and perennials in the *Gramineae* from Europe, Asia and N. Africa, but grown and naturalised elsewhere. They are usually tufted plants with flat leaves and spikes of flattened spikelets arranged alternately: perennial rye grass *(L. perenne)* grows to 450mm with 6–13mm long spikelets of 8–11 narrow florets. It is much used for hard-wearing lawns

and as a fodder grass; Italian rye-grass *(L. multiflorum)* is similar, but usually only an annual or biennial and has awned spikelets.

SABAL
see PALM (palmetto)

SABULOS|US, -A, -UM Growing in sandy places.

SACCATE [SACCAT|US, -A, -UM] Pouched or bag-shaped.

SACCHARUM [SACCHARIN|US, -A, -UM] Sugary.

SAFFLOWER [SAFFRON THISTLE] *(Carthamus tinctoria)*. An annual plant to 900mm in the *Compositae*, probably from N. India to Turkey, but much cultivated in Afghanistan, Ethiopia and the Nile Valley since early times and in the USA since 1925. It has oblong to ovate, spine-toothed leaves and thistle-like, orange-yellow flower heads having spiny bracts, (a spineless cultivar is now grown). The 9mm long white or pale-grey ACHENES have a high oil content and the flowers yield the red dye safflower carmine, much used in the Middle East and India. It is not a fast colour and is now largely replaced by aniline substitutes for cloth dyeing, though still used for colouring rouge and confectionery.

SAFFRON CROCUS
see CROCUS

SAFFRON THISTLE
see SAFFLOWER

SAGE *(Salvia officinalis)*. An evergreen shrub to 600mm in the *Labiatae* from S. Europe. It has finely wrinkled, grey-green oblong-ovate leaves and erect RACEMES of 25mm long, purple or white, 2-lipped tubular flowers. There are cultivars with purple and variegated leaves. Sage has many culinary uses and was once used medicinally.

SAGITTAL|IS, -E, [SAGITTAT|US, -A, -UM] Shaped like an arrowhead, i.e. triangular-ovate with 2 barb-like, pointed lobes.

● **SAGO** Starchy food reserves in the trunks of several palms and cycads. The palms are most important commercially: *see* PALM (sugar, sago and fishtail). *Cycas circinalis* and *C. revoluta* from S.E. Asia are palm-like trees to 3m (*C. circinalis* sometimes to 12.2m) with rigid, leathery arching leaves and cone-like seed heads. In palms and cycads, the trunks are cut into sections and split, the pith is then scraped out. It is ground and washed and the starch grains settle, to be later dried to give sago flour. Pearl sago is made from a paste of the flour passed through a type of sieve and heat dried.

SAINFOIN *(Onobrychis viciifolia)*. An erect 300–600mm perennial in the *Leguminosae* from Europe and Asia, naturalised elsewhere. It has PINNATE leaves with 13–25 linear-oblong to obovate leaflets and dense terminal RACEMES of rich pink or red 9–13mm long pea-flowers. The rounded one-seeded pods are pubescent and tubercled.

ST DABEOC'S HEATH *(Daboecia cantabrica)*. A 300–450mm evergreen shrub in the *Ericaceae* from W. Europe (including Ireland). It has elliptic to linear 9mm

▲ 

■ 

long leaves which are dark green above, white tomentose beneath. The 6–9mm long nodding rose-purple or white urn-shaped flowers are carried in terminal clusters. Often cultivated as an ornamental but needs acid soil.

ST GEORGE'S MUSHROOM *(Tricholoma gambosum)*. A *Basidiomycete* fungus from temperate climates mainly in grassland. It is creamy to pale buff with flattish, slightly waved cap to 150mm wide and a stout stipe or stalk. It has a mealy smell when fresh and is very good to eat when cooked.

ST JOHN'S BREAD
see CAROB

■ **ST JOHN'S WORT** *(Hypericum)*. A genus of 400 annuals, perennials, shrubs and small trees in the *Hypericaceae* from the temperate regions of the world and tropical mountains. They have LINEAR to OVATE simple leaves in pairs or WHORLS and 5-petalled mainly yellow, sometimes red flowers with many slender stamens. Several species are grown in rock gardens and shrubberies: rose of Sharon or Aaron's beard *(H. calycinum)* is a vigorously RHIZOMATOUS, evergreen shrub to 600mm with elliptic leaves and yellow terminal flowers 50–75mm wide. It is much planted as ground cover and to stabilise banks. Also much grown is 'Hidcote', similar to rose of Sharon but non-suckering and forming a bush to 1.5m with bowl-shaped flowers: common St John's wort *(H. perforatum)* is an erect perennial with elliptic to oblong 9–19mm long leaves furnished with many translucent glandular dots. The 19mm wide yellow flowers are carried in large, terminal clusters.

ST PATRICK'S CABBAGE
see SAXIFRAGE

▲ **SALAD BURNET** *(Poterium sanguisorba)*. A tufted perennial smelling of newly cut cucumber, in the *Rosaceae* from Europe to Iran and N. Africa, naturalised in N. America. It has PINNATE, slightly GLAUCOUS leaves of 9–25 orbicular to oval, stalked, toothed leaflets and erect, branched stems to 450mm bearing dense globose heads of small green or purple tinged petal-less flowers. Somewhat similar is great burnet *(Sanguisorba officinalis)* with taller stems, larger oblong-ovate leaves and dull red ovoid flower heads.

SALICIFOLI|US, -A, -UM With willow-like leaves.

SALIGN|US, -A, -UM Willow-like.

SALIN|US, -A, -UM Growing in salt marshes.

SALLOW
see WILLOW

SALSIFY [VEGETABLE OYSTER] *(Tragopogon porrifolius)*. A biennial plant in the *Compositae* from the Mediterranean region. It has basal rosettes of long, LINEAR-LANCEOLATE leaves and erect stems 45–120cm tall. The purple flower heads are composed of LIGULATE florets and a ring of slender, INVOLUCRAL bracts which project beyond. The cylindrical white fleshy roots are cooked as a vegetable.

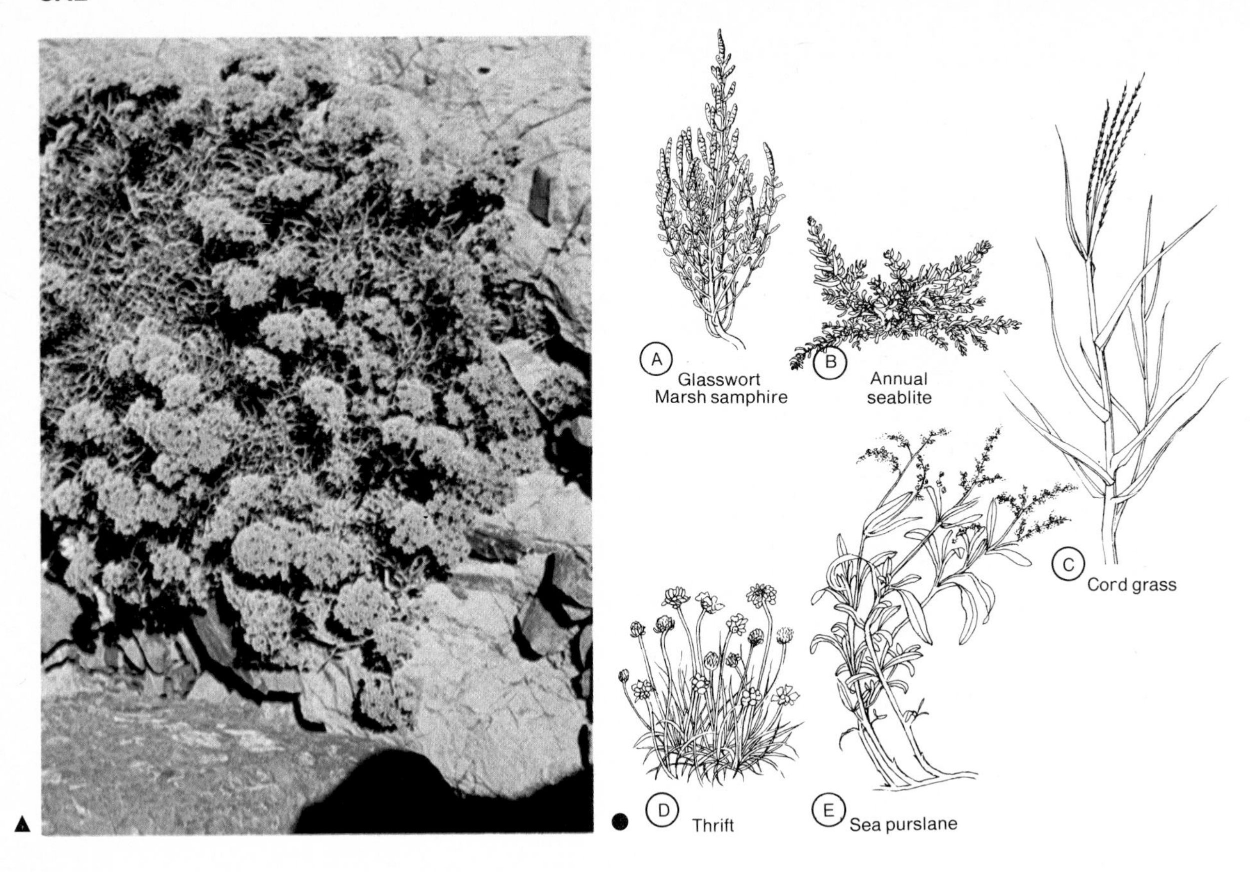

● **SALT MARSH** Areas of mud and sand bordering estuaries and bays. They are usually traversed by arms or channels of the sea and show a strong zonation of plant life, from the lower levels inundated at each tide to the higher levels that may be covered with sea water only at high spring tides. The first land plants to colonise the bare mud are the various kinds of glasswort or marsh samphire and cordgrass. Then come sea-blite, sea aster and other grasses. These catch and hold sandy mud and gradually raise the level above the higher tides. Scurvy grass, sea plantain, sea lavender, thrift, sea arrow grass, sea purslane and many others are characteristic of salt marshes.

SALTWORT *(Salsola kali)*. A prostrate or semi-prostrate, somewhat prickly annual in the *Chenopodiaceae* from sandy shores of Europe, Asia, N. Africa and N. America. It has reddish striped stems, almost cylindrical fleshy leaves 9–38mm long and tiny greenish, solitary AXILLARY flowers. The more erect and bushy inland form *S. k. tenuifolia* (Russian thistle) is more erect and bushy and is one of the most prominent TUMBLE WEEDS of N. America.

SAMAN
see RAIN TREE

SAMARA A dry fruit, part of the wall of which is modified to form a wing; e.g. maple and ash, whose winged fruits are also known as keys.

▲ **SAMPHIRE [ROCK SAMPHIRE]** *(Crithmum maritimum)*. A sea-side perennial in the *Umbelliferae* from Atlantic Europe, the Mediterranean region and around the Black Sea. It has PINNATE leaves, each leaflet divided into 3, almost fleshy, cylindrical lobes, and UMBELS of tiny, yellow-green flowers on 150–300mm stems. The leaves are pickled. Golden samphire *(Inula crithmoides)* is 150–900mm tall, with fleshy, OBLANCEOLATE, 3-toothed leaves and daisy-like flower heads with yellow RAY and orange DISK florets. *See also* GLASSWORT (marsh samphire).

SANDALWOOD *(Santalum album)*. A semi-parasitic tree to 12.2m in the *Santalaceae* from S. E. Asia. It has ovate-elliptic leaves and PANICLES of reddish, 4-lobed, bell-shaped flowers followed by black cherry-like fruits. The heartwood yields a valuable aromatic oil used in perfumery and the fragrant wood is made into ornamental boxes, fans, etc. Mountain sandalwood *(Exocarpus bidwillii)* from New Zealand is a leafless shrub with erect, flattened, grooved branches to approximately 300mm. It has minute, greenish flowers, followed by ovoid, nutlet-like fruits partially embedded in the top of swollen, red PEDICELS, like an egg in an egg-cup.

SAND SPURREY *(Spergularia rubra)*. A small annual or biennial in the *Caryophyllaceae* from sandy or gravelly places in Europe, N. Africa, Asia, N. America and naturalised in Australia. It has pairs of LINEAR leaves with AWNED tips and decumbent stems to 150mm or more. The 3–4mm wide flowers have 5 sepals and 5 smaller pink petals.

SANDWORT *(Arenaria)*. A genus of 250 annuals, perennials and dwarf shrubs in the *Caryophyllaceae* found world-wide, but mainly in the north temperate zone. Commonest in Britain is thyme-leaved sandwort *(A. serpyllifolia)*, an annual or biennial plant, usually decum-

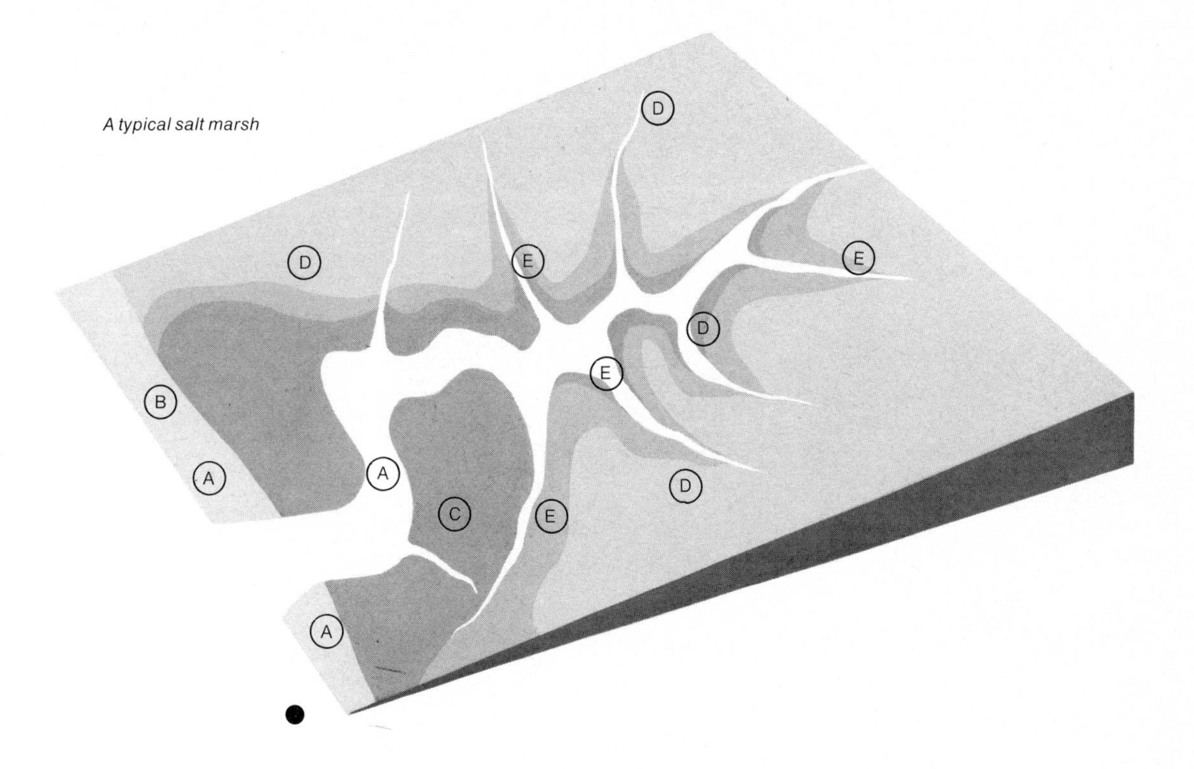

A typical salt marsh

bent, with 25–150mm long stems, OVATE leaves in pairs and terminal CYMES of white, 4–7mm wide, 5-petalled flowers with 5 ovate sepals. Very similar is lesser thyme-leaved sandwort *(A. leptoclados)*, distinguished by its smaller flowers, LANCEOLATE sepals and almost straight-sided capsule.

SANGUINE|US, -A, -UM [SANGUINOLENT|US, -A, -UM] Blood red.

SANICLE [WOOD SANICLE] *(Sanicula europaea)*. A woodland perennial in the *Umbelliferae* from Europe to Syria and Iran, and in N. Africa. It is an erect, tufted perennial to 600mm with long-stalked, deeply 3–5 lobed, PALMATE leaves and small groups of partial UMBELS on long stalks bearing tiny white or pinkish flowers. The fruits are covered with hooked hairs for animal dispersal.

SAP The life fluid of plants, mainly water with dissolved minerals and sugars that move in a steady stream from the roots to the leaves; rising in the XYLEM tissue, descending in the PHLOEM.

SAPID|US, -A, -UM Edible and pleasantly flavoured.

SAPIENT|US, -A, -UM Fit for wise men.

SAPODILLA [MARMALADE PLUM, CHICLE] *(Achras zapota* syn. *Manilkara achras)*. An evergreen tree to 18.2m in the *Sapotaceae* from Mexico and Central America but now widely grown in the tropics. It has elliptic to obovate, 50–150mm long leaves, tiny 6-petalled green and white flowers and globose to ovoid fruits. The latter are 50–100mm wide, greyish to rusty-brown with translucent yellow-brown flesh and black shining seeds. They are considered somewhat sickly to the European palate. The milky latex, obtained by tapping the trunk as for rubber, is made into chewing gum by boiling; it is also used in dentistry. The wood is hard and durable and was used by the Mayas to build their temples.

SAPONACE|US, -A, -UM Soap-like; forming a lather.

SAPROPHYTE A plant that relies on the dead remains of others for its food supplies. This it does in symbiosis with a fungus as endotrophic MYCORRHIZA. Saprophytes lack CHLOROPHYLL and are usually brownish or yellowish.

SAPWOOD The outer layer of wood in trunks and branches that still carry water and dissolved food to and from the roots and leaves.

SARMENTOSE [SARMENTOS|US, -A, -UM] Producing long runners as in strawberry.

SARSAPARILLA A drug formerly used for rheumatism, gout, syphilis and as a tonic, obtained from the RHIZOMES of several species of *Smilax*, perennials in the *Smilacaceae* from the tropics and subtropics. They are somewhat prickly climbing plants with LANCEOLATE to OVATE leaves and greenish, yellowish or whitish 6-TEPALLED flowers followed by black or red berries. Indian sarsaparilla is similar to the above, being used especially as a tonic. It comes from the root of *Hemidesmus indicus*, a small twining plant in the *Periplocaceae* from India and Malaysia, and is made into a sweet, fragrant syrup.

SASSAFRAS *(Sassafras varifolium)*. An aromatic large shrub or tree to 30m or more in the *Lauraceae* from east N. America. Has very variable leaves; elliptic to OVATE, bilobed or trilobed, 50–125mm long, the undersides being GLAUCOUS. The unisexual, petal-less flowers have a 6-lobed CALYX and are followed by deep slate-blue, berry-like DRUPES on red stalks. An aromatic oil is distilled from the bark, twigs and roots. Infusions of the bark were formerly used for many complaints, including rheumatism, gout and eye troubles.

SATIV|US, -A, -UM A plant deliberately sown or cultivated.

SAUERKRAUT Fermented cabbage.

SAUSAGE TREE *(Kigelia pinnata)*. A spreading tree about 15m tall in the *Bignoniaceae* from Central Africa. Has large PINNATE leaves and long, pendent RACEMES of purple, bell-shaped flowers direct from the branches (cauliflorus). Usually only one fruit develops at the end of the stem, which elongates to 0.9–1.5m. Each fruit is a grey, salami-sausage-shaped gourd 200–350mm long. It is used in native medicine but not eaten.

● **SAVANNA** Floristic region. Natural tropical grassland which occurs in areas of low rainfall between deserts and forest. Although various coarse, tall grasses form the dominant vegetation, in the higher rainfall areas well-spaced scrub or small trees occur, often *Acacia* in Africa and parts of S. America. Bulbous plants are also a feature growing and flowering after the seasonal rains.

SAVORY Two species of *Satureja*, members of the *Labiatae* from S. Europe, used for seasoning in cookery. Summer savory *(S. hortensis)* is an annual to about 225mm tall with 13mm long, aromatic, linear to oblong leaves and dense spikes of white to purplish tubular, 2-lipped flowers; winter savory *(S. montana)* is a semi-evergreen sub-shrub to 300mm with leaves to 25mm and pale purple flowers.

SAXATIL|IS, -E Growing among rocks.

SAXICOL|US, -A, -UM Growing on rocks.

▲ **SAXIFRAGE** *(Saxifraga)*. A varied genus of 370 annuals and perennials in the *Saxifragaceae*, mainly arctic and alpine in the northern hemisphere, but also in S. America. They range from dense low hummocks to larger leaved plants of loose, tufted growth. The 5-petalled flowers are borne in PANICLE-like CYMES, sometimes solitarily, and range through shades of white, red, yellow, pink and purple. Many species and hybrids are grown as ornamentals, particularly in rock gardens. Livelong *(S. aizoon* syn. *S. paniculata)* forms low hummocks of 25–75mm wide rosettes of OBOVATE or tongue-shaped leaves, the toothed margins white with chalk glands. The 50–300mm stems bear white, sometimes red dotted flowers: Pyrenean saxifrage *(S. longifolia)* forms a single MONOCARPIC rosette to 150mm wide with narrow, tongue-shaped leaves bearing chalk glands and 300–600mm long, pyramidal panicles of white flowers; rue-leaved saxifrage *(S. tridactylites)* is an erect 25–150 annual with TRIFID, SPATHULATE, glandular, hairy leaves and tiny white flowers; mossy saxifrage or Dovedale moss *(S. hypnoides)* forms mats of slender

▲

■

stems bearing LINEAR or 3–5 linear-lobed leaves and 75–150mm stems of white flowers; yellow mountain saxifrage *(S. aizoides)* forms loose mats of fleshy, oblong-linear leaves and 50–150mm stems of yellow or coppery to dark-red blooms; purple saxifrage *(S. oppositifolia)* has prostrate mats of thickened, oblong to obovate 2–6mm long leaves with chalk glands and 13–19mm wide rose-purple flowers; St Patrick's cabbage *(S. spathularis)* has clusters of rosettes formed of stalked, leathery, sub-orbicular, rounded-toothed leaves and 150–450mm tall panicles of starry white flowers, dotted with yellow and crimson; wood saxifrage *(S. umbrosa)* is similar with oval to obovate-oblong leaves; London pride *(S. X urbium)* is a hybrid between the two and combines their characters; mother-of-thousands *(S. stolonifera* syn. *S. sarmentosa)* is a popular house plant, having tufts of red-stalked, orbicular-cordate, marbled, 50–100mm wide leaves and 225–300mm panicles of white flowers with 2 of the 5 petals longer and drooping. Its vernacular name comes from the abundant branched, red runners which produce numerous tiny plantlets.

SCABIOUS Several annual and perennial plants in three genera belonging to the *Dipsacaceae* from Europe, Asia and N. Africa. They are erect plants with PINNATE or entire leaves in pairs and small, tubular, ZYGOMORPHIC flowers crowded together in shallowly, convex heads (CAPITULA), like those of the daisy family *(Compositae)*: devil's bit scabious *(Succisa pratensis)* grows 150–900mm tall with OBOVATE-LANCEOLATE leaves and mauve to blue-purple, ■ 19–25mm wide flower heads; field scabious *(Knautia arvensis)* has entire to PINNATIFID leaves, stems to 900mm and 25–38mm wide bluish-lilac flower heads; small scabious *(Scabiosa columbaria)* is a slender, wiry-stemmed plant 150–600mm tall, with deeply pinnatifid leaves and 19–32mm wide blue-lilac flower heads, the outer florets larger than the inner like the ray florets of a daisy; Caucasian scabious *(S. caucasica)* is similar, but much larger with 75mm flower heads in shades of purple, blue and white. It is much grown in gardens for cut flowers.

SCABRID [SCAB|ER, -RA, -RUM, SCABROS|US, -A, -UM] Rough.

SCALAR|IS, -E Ladder-like, as the arrangement of the pairs of leaflets in some pinnate leaves.

SCALE LEAVES Small rudimentary leaves often lacking chlorophyll. They may be protective as those enclosing the winter buds of trees; act as food stores, as in bulbs; have no special function as those on STOLONS or RHIZOMES, and on plants like broom and gorse where the stems perform the function of leaves.

SCALLION The slender, stem-like bulbs of chives and Welsh onions.

SCALY BULB Those bulbs which have separate leaf-like scales, e.g. those of lily *(Lilium)*.

SCANDENS Having climbing stems.

SCAPE Long, leafless flowering stems that arise directly from the base of a plant, e.g. daffodil, lily of the valley.

SCAPIGER|US, -A, -UM Bearing a scape.

▲

SCARIOUS [SCARIOS|US, -A, -UM] Thin, dry and membraneous.

SCHIZOCARP A dry fruit formed of two or more CARPELS which split into one-seeded portions when ripe, e.g. hollyhock, mallow, geranium, nasturtium *(Tropaeolum)*.

SCHIZOMYCETES [FISSION FUNGI] Obsolete terms for BACTERIA.

SCLEREIDE A single cell thickened with LIGNIN which forms the sclerenchyma, the tissue of FIBRES.

SCOPULOR|US, -A, -UM Growing on crags and cliffs.

SCORPIOID Coiled like a scorpion's tail, applied to monochasial CYMES which are coiled when young and at the tips, e.g. viper's bugloss, forget-me-not, comfrey.

SCORPION GRASS
see FORGET-ME-NOT

SCOTS [SCOTCH or COTTON] THISTLE *(Onopordum acanthium)*. A biennial, thistle-like plant to 1.8m or more in the *Compositae* from Europe to W. Asia, naturalised in N. America. It has 450–900mm long stems bearing several 32–50mm wide globular spiny, purple flower heads.

SCOURING RUSH
see HORSETAIL

● **SCREW PINE** *(Pandanus)*. A genus of 600 evergreen shrubs and trees, some prostrate, in the *Pandanaceae* from the old-world tropics but cultivated elsewhere. They have buttress roots and LINEAR, often spiny-toothed leaves, densely borne on the stems in spiral fashion. The tiny MONOECIOUS, petal-less flowers are carried in spikes or RACEMES protected by spathe-like bracts. The fruits are borne in head-like clusters. Several species with variegated leaves are grown as ornamentals and the leaves of others are used for thatching.

SCRUB Areas of shrubs or bushy trees, often transitional between grassland and woodland.

▲ **SCURVY GRASS** *(Cochlearia)*. A genus of 25 small annuals and perennials, often by the sea, in the *Cruciferae* from Europe, Asia and N. America: common scurvy grass *(C. officinalis)* is a variable, tufted species with long-stalked, orbicular-cordate basal leaves and procumbent to ascending, 50–300mm long stems bearing RACEMES of tiny, 4-petalled white flowers. It grows by the sea and on mountains, the seaside populations having fleshy, larger leaves. The latter are pleasantly acrid and contain vitamin C; formerly eaten by sailors to prevent scurvy.

SCUTAT|US, -A, -UM Shield-shaped.

SCUTCH
see COUCH GRASS; BERMUDA GRASS

SEA ASTER *(Aster tripolium)*. An erect, 150–900mm tall perennial of salt marshes and sea cliffs in the *Compositae* from coastal and inland saline areas of Europe to Central

●

●

▲

Asia. Has oblanceolate basal and narrowly-oblong to linear stem leaves and terminal CORYMBS of 13–19mm wide daisy flowers with blue-purple RAY and yellow DISK florets. RAYLESS forms occur *(A.t. discoideus)*.

● **SEA BEET** *(Beta vulgaris maritima)*. A maritime fleshy-leaved biennial or perennial in the *Chenopodiaceae* from Europe and N. Africa to S.E. Asia. It has stalked, RHOMBOID, 100mm long leaves and 300–600mm stems bearing spikes of rounded clusters of 2–3, tiny green flowers. Later, the single-seeded fruits adhere together in GLOMERULES. Sugar beet, spinach beet, chard, mangold and beetroot have all evolved via mutation from this plant.

SEA-BLITE Two species of *Suaeda,* plants of saline places in the *Chenopodiaceae* from the northern hemisphere and S.E. Asia, with linear, fleshy leaves and, in their AXILS, clusters of 2–3 tiny flowers with 5 fleshy, PERIANTH segments: annual seablite *(S. maritima)* is a variable species having prostrate to erect, 75–250mm long stems and GLAUCOUS or red-tinged leaves 3–25mm long; shrubby seablite *(S. fruticosa)* is an evergreen shrub with semi-erect stems to 1.2m and leaves rarely above 19mm long.

SEA BUCKTHORN *(Hippophaë rhamnoides)*. A suckering, somewhat spiny deciduous shrub or small tree to 9.1m in the *Elaeagnaceae* from Europe and Asia. It has LINEAR-LANCEOLATE, 9–75mm long leaves, covered with silvery scales, particularly beneath. The tiny, DIOECIOUS, petal-less flowers open before the leaves and are wind-pollinated. They are followed by 6mm long orange, berry-like fruits formed from the fleshy RECEPTACLES.

SEA DAFFODIL *(Pancratium maritimum)*. A bulbous-rooted perennial in the *Amaryllidaceae* from sandy beaches around the Mediterranean. It has grey-green, strap-shaped leaves to 600mm and UMBELS of slender white, daffodil-like flowers to 150mm long. Each bloom has narrow lanceolate petals and a funnel-shaped CORONA formed by the fusion of winged stamen filaments.

SEA HEATH *(Frankenia laevis)*. A mat-forming evergreen perennial in the *Frankeniaceae* inhabiting the margins of salt marshes from W. Europe and the Mediterranean to W. Asia. It has wiry stems and LINEAR 3mm long leaves crowded in short lateral shoots. The 4mm wide 5-petalled flowers are pink.

▲ **SEA HOLLY** *(Eryngium maritimum)*. A silvery, blue-green perennial in the *Umbelliferae* from the sandy and pebbly beaches of Europe. It has branched 300–600mm stems, spiny, holly-like leaves and dense, firm, ovoid heads of tiny, 5-petalled bluish flowers rising from a WHORL of spiny bracts.

SEA-KALE *(Crambe maritima)*. A perennial plant of sandy and shingly beaches and sea cliffs, in the *Cruciferae* from the Atlantic coast of Europe to the Baltic and Black Sea. It has GLAUCOUS, OVATE, long-stalked basal leaves to 300mm, usually PINNATELY lobed and wavy-margined. The erect 450–600mm stems terminate in CORYMBS of 4-petalled white flowers about 13mm wide. The ovoid seed pods float and are dispersed by the sea. Sea-kale is cultivated for its asparagus-flavoured shoots. These must be blanched by covering the crowns with soil in early spring or bringing the roots into a warm dark place in winter.

■ **SEA LAVENDER** Several species of *Limonium*, perennial plants in the *Plumbaginaceae* from the northern hemisphere (others from the southern hemisphere). Mainly tufted plants with obovate, oblanceolate or spathulate leaves and wiry-branched stems carrying CYMOSE spikelets of tubular to funnel-shaped flowers from the AXILS of broad, SCARIOUS bracts. The 5-lobed corollas are usually a shade of blue-purple or lavender. Common sea lavender *(L. vulgare)* is a salt marsh plant with oblong-lanceolate leaves to 100mm long and 100–300mm high stems of blue-purple blooms.

SEA LETTUCE *(Ulva lactuca)*. A semi-translucent green seaweed in the *Ulvaceae* from temperate regions. It forms large, wavy fronds to 300mm or more, mainly inhabiting the inter-tidal zone. It is sometimes eaten in Britain under the name green laver, and regularly so in China and Japan where it is also used as a medicine. *See also* LAVER.

SEA MILKWORT *(Glaux maritima)*. A succulent, procumbent perennial in the *Primulaceae* from the north temperate zone. It has elliptic-oblong to obovate 3–13mm long leaves and petal-less flowers with tiny, bell-shaped CALYCES.

SEA PEA *(Lathyrus japonicus* syn. *L. maritimus)*. A perennial plant of shingle beaches in the *Leguminosae* from the north temperate zone. Has PINNATE leaves with 6–8 obovate 19–38mm long leaflets and small terminal tendrils. The 13–16mm long purple to blue pea-flowers are carried in short-stemmed AXILLARY RACEMES.

SEA PINK
see THRIFT

SEA PURSLANE *(Halimione portulacoides)*. A semi-prostrate, mealy shrub in the *Chenopodiaceae* from Europe to Turkey and N. and S. Africa. It has ascending branches to 0.6m with elliptic to linear leaves and spikes of tiny unisexual flowers, the female ones without a PERIANTH but enclosed by two, 3-lobed triangular BRACTEOLES.

○ **SEA ROCKET** *(Cakile maritima)*. A drift-line annual in the *Cruciferae* from W. Europe, the Mediterranean and Black Sea coasts (an allied species is in N. America). It is a prostrate or ascending plant to 150–450mm tall, with OBOVATE to OBLANCEOLATE, PINNATELY lobed leaves (but sometimes entire), and terminal RACEMES of lilac, purple or white 4-petalled flowers about 13mm wide. The small pods are dispersed by the sea.

SEA SANDWORT *(Honkenya peploides)*. A fleshy-leaved, STOLONIFEROUS perennial in the *Caryophyllaceae* from north temperate arctic zone sandy shores. It is prostrate with shortly ascending shoots bearing pairs of SESSILE, OVATE leaves 6–19mm long, and small greenish-white, 5-petalled flowers in terminal leafy CYMES.

SEA SPLEENWORT *(Asplenium marinum)*. A small, evergreen fern in the *Aspleniaceae* from sea-cliffs of Atlantic Europe and the W. Mediterranean. It has PINNATE, leathery fronds 150–300mm long and linear SORI along the secondary veins of the PINNAE.

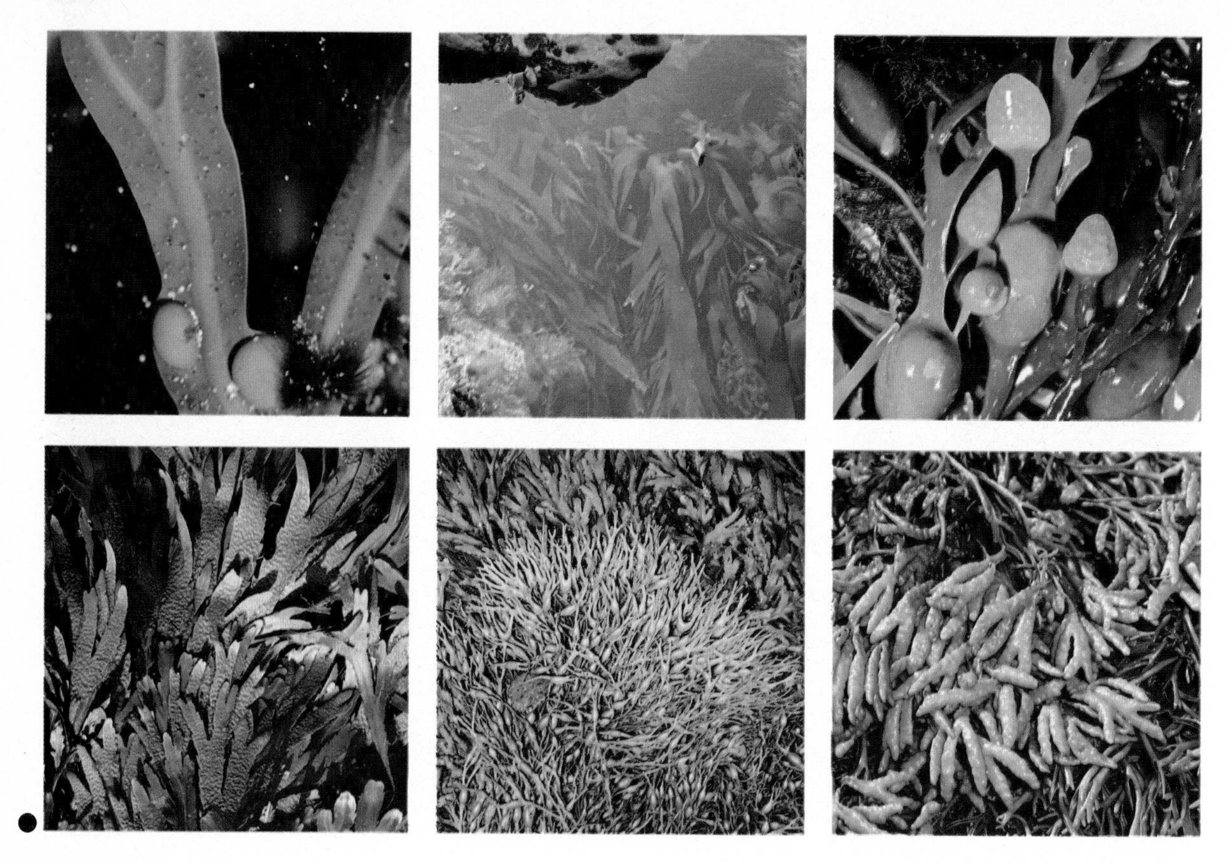

SEAWEED A vast group of marine algae in the THALLOPHYTA found in all the seas of the world. About 800 different kinds have been recorded for the area around the British Isles alone. They do not have true roots, but sucker-like structures called holdfasts. Some species are free-floating like gulfweed *(Sargassum)* which covers vast areas east of the W. Indies. Many of them have thickened, stalk-like bases and leaf-like fronds or thalli (singular: thallus). There are 3 main groups of marine algae: red, green and brown, the thalli of each being a shade of these colours. They vary much in form, from the delicate tracery of sea cockscomb, a rose-red species, to the massive 15–30m ribbons of the giant kelps. Many seaweeds bear air bladders, particularly the commonly washed-up knotted WRACK. Reproduction can be by simple fragmentation of the thallus or by the fusion of male and female GAMETES (sex cells). In the simple seaweeds, these are motile and similar in appearance, but in the more advanced kinds the female cell is larger and non-motile and known as an egg, while the smaller male is an ANTHEROZOID. Fusion of these produces a ZYGOTE or spore which later grows into a new plant. Agar is made from several red seaweeds; it is an important substance in laboratory work for culturing bacteria and is still superior to gelatin.

SECONDARY THICKENING The increase in girth of woody stems by the repeated lateral division of the CAMBIUM cells. In for example a one-year old lime twig, there is, beneath the bark, a sheath of VASCULAR TISSUE known as xylem and phloem, in two distinct layers. Between these two layers is the actively growing cambium. Each year it forms a new layer of phloem to the outside and xylem to the inside. This cambial activity is most vigorous in spring and the xylem formed then is of large vessels (water-conveying tubes). Later, as growth slows down, the vessels are smaller. This annual formation of large and small-vesselled xylem is easily recognisable as dark and light rings when a stem or trunk is cut across. They are a fairly reliable indicator of the age of a tree and also give an idea of seasonal differences. Bad weather such as cold or dry springs will result in smaller vessels in narrower bands and vice-versa. The study of annual rings from an old tree often reveals a sequence of wet and dry periods and indicates weather patterns of the past.

SECUND [SECUND|US, -A, -UM] Directed or facing one way only, usually of flowers in an inflorescence.

SEDGE A name often loosely applied to many members of the *Cyperaceae*, but primarily referring to *Carex*, a genus of more than one thousand five hundred species of cosmopolitan distribution, especially in temperate climates and wet soils. They are perennial RHIZOMATOUS plants with LINEAR, KEELED, tapering leaves. The tiny unisexual flowers lack a PERIANTH and are protected by GLUMES and borne in simple branched spikes. The solitary OVARY of each female flower is completely sheathed in a membraneous covering (the PERIGYNUM) which tapers to a beak through which the stigmas protrude. The fruit is a triangular nutlet enclosed within the perigynium which develops a distinctive shape characteristic of the species and used for identification purposes: common sedge *(C. nigra)* is a creeping or tufted plant usually 150–450mm tall, the dying leaf bases becoming a black, fibrous mass. There are usually three spikes a stem, the lower one female, the upper two males; pendulous sedge *(C. pendula)* is probably

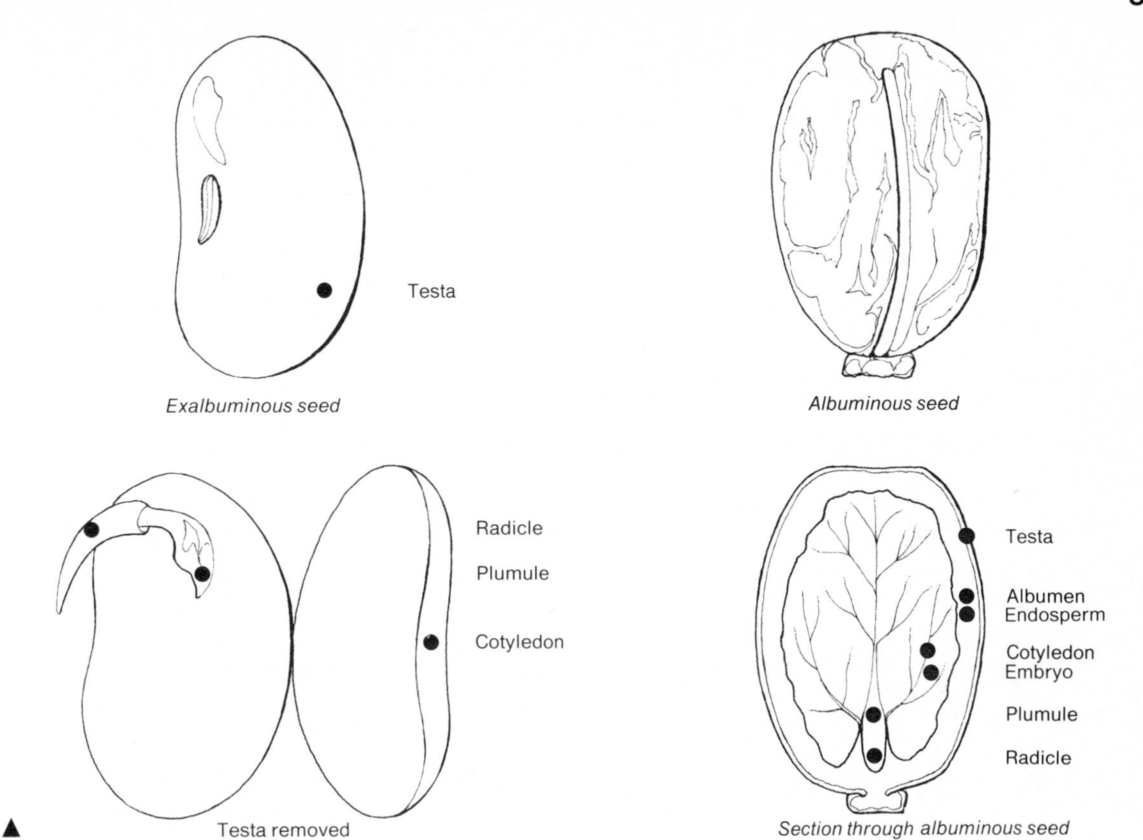

Exalbuminous seed

Albuminous seed

▲ Testa removed

Section through albuminous seed

the best known and most distinctive sedge in Britain. It forms dense tufts of broad, ribbed leaves and 0.9–1.5m stems arching at the top and bearing several slender, pendulous male and female spikes to 75mm long. It is sometimes cultivated; sand sedge *(C. arenaria)* grows on dunes by the sea with long, straight RHIZOMES which produce rows of plantlets looking as though planted with a rule. About 5–12 short, brownish male and female spikes are borne close together at the top of 100–300mm stems; great fen sedge *(Cladium mariscus)* is a vigorous 75–300cm tall aquatic plant of fens and reed swamps. It has keeled, sharply saw-edged leaves and narrow PANICLES of small, stalked rounded spikes.

▲ **SEED** The mature OVULE of a flowering plant after fertilisation and which usually contains an EMBRYO plant and a store of food. The food store may be in large fleshy seed leaves (non-endospermous or exalbuminous) or separate with the embryo plant imbedded in it (endospermous or albuminous). The seeds of orchids are the exception to this, being minute and containing a group of undifferentiated cells. Seed may vary enormously in size from the dust-like begonia seeds to the vast, woody globes of the coconut. Each seed has an outer skin or testa, often handsomely patterned, e.g. castor oil *(Ricinus)* or with an intricate pattern of tubercles (campion). A minute pore in the testa (the micropyle) allows water to have access to the embryo and a small scar shows where the seed was attached to the wall of the ovary. *See also* GERMINATION; DORMANT.

SEED DISPERSAL Most plants produce vast numbers of seeds, a single specimen of shepherd's purse can shed 60,000, an orchid as many as 74,000,000. It is thus important that seeds or the fruit that contains them are dispersed as far away from the parent plant as possible so as to stand a chance of growing to maturity. Many plants have evolved fruit and seeds able to be dispersed by wind, water and animals (including birds). A fourth group of plants uses mechanical means of dispersing their seeds: the pods of many legumes, e.g. broom and gorse, explode violently flinging the seeds many feet away. Pansy capsules open into three sections or valves upon which are arranged many glossy, tapered seeds. As the valve dried it folds upwards, squeezing the seeds away like a cherry stone between thumb and forefinger, *see also* OXALIS; CENSER MECHANISM. Wind-dispersed seeds and fruits have plumes of hair, e.g. willow-herb and dandelion, or wings, e.g. maple, ash and pine. Water-dispersed fruits have woody or corky or otherwise waterproofed coats, e.g. sea-kale, coconut; or, as in the case of water lilies, the seeds are surrounded by mucilaginous tissue (an ARIL) containing air bubbles. Animal-dispersed seeds and fruits are characterised by hooks and barbs which hold on to fur and feathers, e.g. burdock and goosegrass; or they may be eaten and travel in the gut of a bird until voided, perhaps many miles away. The seed of berried fruits are usually very hard and resistant to digestive juices, though passage through the gut may well soften the testa and make GERMINATION quicker and more certain. Ordinary seeds, particularly many garden weeds, can also travel in the mud on the feet of animals (including man) and on the wheels of farm and other vehicles. Man unwittingly spreads many seeds around in packing materials, animal feeds and ship's ballast.

SEEDLING The very young plant soon after the germination of the seed. *See also* COTYLEDON.

SEGETAL|IS, -E [SEGET|US] Growing in cornfields.

SEGMENT One of the divisions into which a divided leaf or petal is cut.

SELF-HEAL [ALL-HEAL] *(Prunella vulgaris)*. A tufted, spreading perennial in the *Labiatae* from the north temperate zone, naturalised in Australia. Has OVATE, 19–50mm long leaves and 50–250mm tall stems bearing purple-tinged bracts and clusters of tubular, 2-lipped purple, pink or white flowers forming short, oblong spikes.

SELF STERILITY A common situation: where self-pollination fails to produce viable seeds. It is a provision to prevent in-breeding which may weaken the vigour of a species. Plants which exhibit self-sterility will only set seed with pollen from another individual of the same species.

SEMPERVIRENS Evergreen.

SENESCENS Growing old or appearing aged.

SENIL|IS, -E Aged. In plants, usually of those covered with long, white hair.

SENNA The pods and leaves of several species of *Cassia*, shrubs and trees in the *Leguminosae* from Africa and Arabia, grown in India and elsewhere. The commonest seen is *C. acutifolia*, a bushy plant to 900mm or so with PINNATE leaves and 5-petalled, yellow, red-veined flowers in racemes. Somewhat similar is 'tinnevelly' senna *(C. angustifolia)* and the N. African *C. obovata*. False or purging senna *(C. fistula)* is also much grown as an ornamental under the names golden shower and pudding-pipe tree. It is a small tree with 300mm pinnate leaves and PENDULOUS RACEMES of yellow flowers. The 300–600mm long pods are laterally partitioned and contain a substance with laxative properties. *See also* BLADDER SENNA.

SENSITIVE FERN *(Onoclea sensibilis)*. A 450mm tall herbaceous perennial RHIZOMATOUS fern in the *Aspidiaceae* from N. America and Asia, naturalised elsewhere. It has triangular-ovate, pinnatifid to bipinnate sterile fronds and pinnate fertile fronds with rolled pinnules holding globose SORI.

SENSITIVE PLANT
see MIMOSA

SEPAL
see CALYX

SEPALOID The PERIANTH of a flower resembling sepals only.

SEPARATION LAYER
see ABSCISS LAYER

SEPIUM Growing in hedges.

SEPTAL Of hedgerows.

SEPTATE Divided by a partition or septum, as are the pods of many members of the *Cruciferae*, e.g. honesty.

SEPTENTRIONAL|IS, -E North or northern.

SEPULCHRAL|IS, -E Growing around burial places or tombs.

SERICE|US, -A, -UM Silky hairy.

SERIES A division of a genus between sub-genus and species. Very large genera, e.g. *Rhododendron*, are split up into series (or groups) which closely resemble each other.

SEROTIN|US, -A, -UM Late maturing or late-flowering.

SERPENS Creeping flat on the ground.

SERPYLLIFOLI|US, -A, -UM Having thyme-like leaves.

SERRATE [SERRAT|US, -A, -UM] Edged with saw-like teeth.

SERVICE TREE Two species of *Sorbus*, deciduous trees in the *Rosaceae* from Europe to N. Syria and N. Africa. The service tree *S. domestica* much resembles ROWAN, but is a broader, taller tree with either apple or pear-shaped 25mm long fruits which are edible when bletted (allowed to rot, as with medlar): wild service *(S. torminalis)* grows to 15m with a head of spreading branches and 63–100mm OVATE leaves having deeply pointed lobes, the lowest pair longer and spreading. The white, 5-petalled 10mm wide white flowers are grouped into loose CYMOSE clusters. The brownish, finely dotted ovoid, 13mm long fruits are sometimes called chequer berries and were formerly made into jams, jellies and a cider-like drink.

SESAME [SIMSIM] *(Sesamum indicum)*. A 300–600mm tall annual in the *Pedaliaceae* from Africa, but much grown in India and tropical Asia. It has 75–125mm long oblong-ovate to lanceolate leaves and tubular, 5-lobed white, pink or mauve flowers in the upper leaf AXILS. The 32mm long oblong-ovoid capsules yield white seeds very rich in oil which is used for making cooking and salad oil and margarine. The whole seeds are used in confectionery. Alternative names are beniseed, gingelly and til.

SESQUIPEDAL|IS, -E A foot and a half (450mm) long.

● **SESSILE [SESSIL|IS, -E]** Without a stalk, or apparently so.

SETA A bristle or bristle-like hair.

SETACE|US, -A, -UM Bearing bristles or slender prickles.

SETOS|US, -A, -UM Covered with stiff hairs.

SHADBUSH
see AMELANCHIER

SHADDOCK
see CITRUS

SHAG BARK
see HICKORY

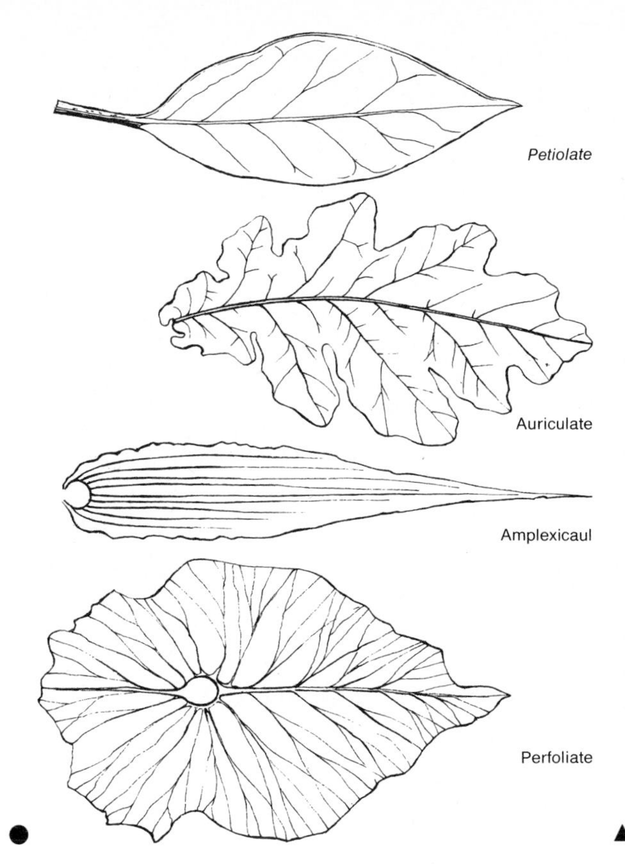

SHALLOT *(Allium ascalonicum)*. A perennial clump-forming onion in the *Alliaceae* from Syria. The ovoid, reddish or yellowish bulbs grow in clusters on or just below the surface of the ground and bear cylindrical leaves and 225m stems with UMBELS of 6-TEPALLED lilac flowers. Shallots are much cultivated for pickling and flavouring. Although still known under the above Latin name, shallot is now classified under *Allium cepa aggregatum*. *See* ONION.

SHAMROCK Several plants with TRIFOLIATE leaves go under this name, mainly white clover, hop trefoil, black medick and wood sorrel; but lesser clover *(Trifolium dubium)* is most favoured in Ireland.

▲ **SHEEP'S-BIT** *(Jasione montana)*. A tufted biennial or perennial in the *Campanulaceae* from Europe and the Mediterranean. It has 50–300mm long decumbent stems, linear-oblong, undulate or crenate leaves and rounded SCABIOUS-like heads of purple-blue or blue florets which open to 5 slender petals.

SHELL FLOWER [BELLS OF IRELAND] *(Molucella laevis)*. A 300–600mm annual in the *Labiatae* from Syria. It has broadly OVATE, coarsely CRENATE long-stalked leaves and spikes of small tubular, 2-lipped flowers, each one sitting in a large, bell-shaped, pale green CALYX which enlarges after flowering. It is grown for dried flower arrangements.

SHE-OAK *(Casuarina stricta)*. A curious evergreen tree to 18.2m in the *Casuarinaceae* from S.E. Australia and Tasmania. It is apparently leafless, bearing at each joint WHORLS of 9–12 pointed scales, joined at the base like a collar around the twigs. The tiny petal-less flowers are unisexual, the male in cylindrical spikes, the females in round heads. The 25–50mm long globular to ovoid fruit clusters look like cones.

SHEPHERD'S CLOCK
see GOAT'S-BEARD

SHEPHERD'S NEEDLE *(Scandix pecten-veneris)*. A slender 150–450mm tall annual in the *Umbelliferae* from Europe to the Himalaya, naturalised elsewhere. It has bi- or tri-pinnate leaves, small UMBELS of tiny, white flowers surrounded by whorls of linear bracts and slender, cylindrical fruits 32–88mm long. Once common as an arable weed but now declining.

SHEPHERD'S PURSE *(Capsella bursa-pastoris)*. A common annual or biennial weed in the *Cruciferae* of cosmopolitan distribution due to man's activities. It has a neat rosette of OBLANCEOLATE leaves which may be ENTIRE or deeply PINNATIFID, and erect RACEMES of tiny, 4-petalled white flowers. The distinctive seed pods (siliculae) are inverted heart-shaped.

SHIELD FERN Several species of *Polystichum*, perennial ferns in the *Aspidiaceae* from the N. temperate zone. Soft shield fern *(P. setiferum)* has soft, arching, bipinnate fronds, lanceolate in outline and up to 900mm or more long, arranged in tufts. The tiny, round SORI are arranged in two rows down the middle of each pinnule; hard or prickly shield fern *(P. aculeatum)* is similar, but with smaller, firm, leathery fronds, each pinnule being prickle-pointed.

SHOO FLY PLANT *(Nicandra physaloides)*. A 0.6–1.2m annual in the *Solanaceae* from Peru. It has OVATE, wavy-toothed, or slightly lobed leaves, and circular blue flowers with 5 broad lobes emerging from an inflated lantern-like CALYX. The almost dry, berry-like fruits are enclosed in enlarged calyces.

SHOOT A young leafy stem, sometimes with the leaves still in an immature form.

SHORT DAY PLANT
see LIGHT

SHRIMP PLANT *(Beloperone guttata* syn. *Drejerella guttata)*. A popular, soft shrub to 900mm in the *Acanthaceae* from Mexico, often grown as a house plant. It has ovate to elliptic leaves and down-curving flower spikes composed of brownish-pink, overlapping bracts and tongue-like 2-lipped white flowers.

SHRUB A woody stemmed plant with lateral branches coming from just below or near ground level and under 7.5m in height.

SHRUBBY CINQUEFOIL *(Potentilla fruticosa)*. A deciduous shrub in the *Rosaceae* from the N. temperate zone. It varies greatly in stature and hairiness but is usually about 900mm tall, having PINNATE leaves of 5–7 OBLONG-LANCEOLATE leaflets and 5-petalled, yellow flowers about 19mm wide. Several forms and varieties are grown as ornamentals, with white, yellow, orange and red flowers and grey-hairy to bright green leaves.

SICKENER *(Russula emetica)*. A *Basidiomycetes* fungus from the temperate zone, usually associated with coniferous trees. It has a flattened, scarlet cap to 88mm, white gills and stipe (stalk); the flesh has a faint honey scent but is acrid and may cause vomiting.

SIDE OATS
see BOUTELOUA

SIDE SADDLE FLOWER
see PITCHER PLANTS

SIEVE ELEMENTS
see VASCULAR BUNDLES

SIGNAL ARM GRASS
see BOUTELOUA

SILICULA A siliqua which is at least as broad as long, e.g. shepherd's purse.

SILIQUA The pod-like fruits of members of the cabbage family *(Cruciferae)*. It is formed of two elongated carpels divided by a sheet of tissue known as a false septum, e.g. the silvery membrane in honesty.

SILK COTTON TREE
see KAPOK

● **SILKY OAK** *(Grevillea robusta)*. An evergreen tree to 30m or more in the *Proteaceae* from east Australia, but naturalised elsewhere. It has silvery downy shoots, 150–225mm long PINNATE leaves and one-sided RACEMES

▲ 

■ 

of golden-yellow, tubular flowers brimming with nectar. Young plants have ferny, bipinnate leaves to 450mm long and are frequently grown as short-term pot plants.

SILVATIC|US, -A, -UM Growing in woods; *see* SYLVATICUS.

SILVER CORD MOSS *(Bryum argenteum)*. A tiny congested or cushion-forming silvery-green moss from the temperate regions. The individual stems bear closely overlapping leaves and have a cord-like appearance. It is one of the few mosses found in cities and is seen in cracks in pavements, etc.

SILVER LEAF FUNGUS *(Stereum purpureum)*. A PARASITIC *Basidiomycete* fungus from temperate regions. It attacks plum, cherry and other trees, giving the leaves a silvery cast due to a layer of air under the epidermis. The fruiting bodies (which only appear on dead wood) are bracket-like, often borne in tiers, with undulate margins and a purplish hue when moist.

SILVER TREE *(Leucadendron argenteum)*. An erect tree to 9m in the *Proteaceae* from S. Africa, but often grown in other warm, dry climates. It has OBLONG-LANCEOLATE 63–125mm SESSILE leaves covered with long, silky, silver-grey hairs and small, terminal cone-like heads of tiny, tubular yellowish flowers.

▲ **SILVERWEED** *(Potentilla anserina)*. A tufted, vigorously STOLONIFEROUS perennial in the *Rosaceae* from the northern hemisphere, naturalised in the southern. Has PINNATE, 75–150mm long leaves of 15–25 silvery, hairy, oblong, deeply SERRATE leaflets, with much smaller ones in between each pair. The 5-petalled yellow, 19mm wide flowers are solitary on 150–300mm stems.

SIMIL|IS, -E Similar; resembling another species.

SIMPLE Undivided, of leaves such as privet.

SIMPLEX Simple, undivided, unbranched.

SINUATE [SINUAT|US, -A, -UM] Having a waved or undulate margin.

SISAL HEMP
see AGAVE

SKULL CAP *(Scutellaria galericulata)*. A tall, RHIZOMATOUS perennial in the *Labiatae* from the north temperate zone, with slender stems, pairs of sessile, oblong, lanceolate leaves and leafy RACEMES of curved, tubular, 13–19mm long blue-violet flowers with expanded, 2-lipped mouths. The CALYX closes after the flower dies.

■ **SKUNK CABBAGE** Two marsh-dwelling members of the *Araceae* from N. America. (1) *Symplocarpus foetidus* has 100–150mm barrel-shaped, purple blotched and streaked leathery spathes with a pointed tip that curls down. It shelters a thick SPADIX of tiny fleshy flowers with a foetid carrion smell to attract pollinating flies. The 300–600mm cabbage-like leaves expand as the flowers fade. (2) *Lysichiton americanum* has 150–200mm long yellow boat-shaped spathes and oblong-lanceolate leaves to 1.5m. *L. camtschatcense* from N. E. Asia has white spathes.

SLEEP MOVEMENT
see NASTIC

SLIME FUNGI
see MYXOMYCETES

SLIPPER FLOWER
see LADY'S SLIPPER

SLOE
see BLACKTHORN

SMILAX A florist's name for *Asparagus medeoloides* (*see* ASPARAGUS) but also a genus in its own right: *see* SARSAPARILLA.

SMOKE TREE *(Cotinus coggygria* syn. *Rhus cotinus).* A bushy shrub to 2.4m or more in the *Anacardiaceae* from S. Europe to the Caucasus. It has slender-stalked, OBOVATE 38–75mm long leaves and terminal, much branched panicles of fine hairy, often pinkish filaments, some of them bearing tiny 4–6-petalled flowers. The frothy panicles create a smoky appearance at a distance. There are several purple-leaved cultivars grown as ornamentals.

SNAIL FLOWER (1) *(Phaseolus caracalla).* A perennial twining climber in the *Leguminosae* from tropical America. It has TRIFOLIATE leaves with ovoid-rhomboid leaflets and AXILLARY RACEMES of purple and yellow pea-flowers, the KEELS of which are spirally twisted like a snail shell. (2) Species of ARISAEMA. (3) Those flowers that are pollinated by slugs and snails, e.g. *Aspidistra*.

SNAKE BEAN
see COW PEA

SNAKE GOURD [VIPER GOURD, SERPENT CUCUMBER] *(Trichosantha cucumerina* syn. *anguina).* A climbing plant in the *Cucurbitaceae* from S. E. Asia, but grown elsewhere in the tropics, particularly India. It has rounded 3–7 angled or lobed leaves with solitary female and short racemes of male flowers in the AXILS. Each white blossom has 5 lobes fringed with long filaments. The cylindrical green and white striped cucumber-like fruits (orange when ripe) can reach 1.8m, but are usually less. They are sliced and eaten boiled.

SNAKEROOT A name applied to several plants reputed to be an antidote to snakebite, including species of DUTCHMAN'S PIPE, SANICLE and BANEBERRY. The following are best known: black snakeroot *(Cimicifuga racemosa)* a perennial similar to baneberry in the *Ranunculaceae* from east N. America. It grows to 1.8m or more with long, terminal RACEMES of petal-less flowers having prominent white stamens and PETALOIDS: button snakeroot *(Liatris scariosa)* is 0.3–1.2m erect perennial in the *Compositae* from east N. America with a rounded, corm-like tuber and LANCEOLATE to LINEAR leaves. The rounded 25mm wide, purple flower heads are borne in loose racemes, the top flowers opening first. Several other similar liatris are grown as ornamentals.

SNAPDRAGON
see ANTIRRHINUM

▲

■

● **SNEEZEWORT** *(Achillea ptarmica)*. A clump-forming perennial in the *Compositae* from Europe and Asia, naturalised in N. America. It has slender, angular stems to 600mm, linear-lanceolate, finely toothed leaves and loose terminal CORYMBS of 19mm wide, daisy-flowers having almost rounded ray florets.

SNOWBALL TREE (*Viburnum opulus* 'Sterile'). A mutant form of GUELDER ROSE with ball-like clusters of sterile florets only; much grown as an ornamental.

SNOWBERRY Several species of *Symphoricarpus*, suckering deciduous shrubs in the *Caprifoliaceae*, mainly from N. America (one in China). They have rounded to oval leaves in pairs, terminal clusters of small pink to white bell-shaped flowers, and globular white, pink or purple fruits: common snowberry *(S. rivularis* syn. *S. racemosa)* grows to 1.8m or so, with 19–38mm long oval leaves (sometimes lobed on vigorous stems) pink flowers and glistening white berries to 19mm wide; coral berry *(S. orbiculatus)* is similar but the berries are rose-purple and smaller.

■ **SNOWDROP** *(Galanthus)*. A genus of 20 bulbous species in the *Amaryllidaceae* from Europe, particularly southern and eastern. They have linear, often grey-green leaves and solitary, pendulous flowers having 3 white spreading outer tepals and 3 smaller, notched inner tepals, marked with green: common snowdrop *(G. nivalis)* grows to 150mm or more with 25mm wide blooms; *G. n. reginae-olgae* flowers before the leaves in autumn; *G. elwesii* grows to 250mm or more with larger flowers and broader, grey leaves.

SNOWFLAKE *(Leucojum)*. Several bulbous species similar to snowdrop in the *Amaryllidaceae* from Europe ▲ and the Mediterranean region: spring snowflake *(L. vernum)* is like a robust snowdrop, but all the 6 tepals are alike, white, tipped green; summer snowflake *(L. aestivum)* flowers in spring, the 300–600mm stems bearing UMBELS of 3–7 flowers.

SNOW-IN-SUMMER *(Cerastium tomentosum)*. A fast growing, mat-forming RHIZOMATOUS perennial in the *Caryophyllaceae* from S. E. Europe to the Caucasus, naturalised elsewhere. It has opposite pairs of LINEAR-LANCEOLATE leaves and dichasial CYMES of 13mm wide, white flowers with 5 deeply lobed petals. The whole plant has dense white hairs. It is much confused and hybridised with the allied *C. biebersteinii*, a generally more silvery plant.

SNOWY MESPILUS *(Amelanchier ovalis)*. An erect, bushy, deciduous shrub to 2.4m in the *Rosaceae* from southern Europe. It has 25–50mm long oval to obovate toothed leaves and RACEMES of white, 5-petalled flowers followed by blue-black, currant-like fruits. Other species of AMELANCHIER are sometimes known as snowy mespilus.

SOAPBERRY Several species of *Sapindus*, deciduous and evergreen trees in the *Sapindaceae* from the tropics and sub-tropics. They have PINNATE leaves, tiny, often greenish 4 or 5-petalled flowers and berry-like DRUPES, the flesh of which contain saponine and can be used as a soap substitute. *S. saponaria*, a small Jamaican tree, is the best known.

SOAPWORT [BOUNCING BET] *(Saponaria officinalis)*. A vigorous, RHIZOMATOUS perennial in the *Caryophyllaceae* from Europe and Asia. It has erect, 300–900mm stems, pairs of ovate to elliptic 50–100mm long leaves and terminal CYMES of 5-petalled, pink flowers. A double-flowered form is often naturalised in Britain.

SOBOLE [SOBOLIFER|US, -A, -UM] Bearing vigorous shoots, particularly from ground level.

SOFT GRASS *(Holcus mollis)*. A RHIZOMATOUS perennial in the *Gramineae* from Europe. It has sparsely pubescent flat leaves and 300–600mm erect stems, hairy only at the nodes. The tiny whitish florets have a small AWN and are carried in loose PANICLES. Yorkshire fog *(H. lanatus)* is similar, but softly hairy all over and the tiny awn protrudes from the spikelet.

● **SOLDANELLA** A genus of 11 small tufted alpine perennials in the *Primulaceae* from the mountains of Europe. They have orbicular to kidney-shaped leaves and nodding bell-shaped flowers with fringed lobes. They are sometimes known as snowbells because they bloom as the snow melts and can actually melt a hole in thin snow by the heat of their respiration. Common or alpine snowbell *(S. alpina)* has 25–75mm wide, kidney-shaped leaves and 50–150mm stems bearing 2 or 3 violet-blue bells.

SOLITARY Used of flowers when they are borne one to each leaf AXIL.

▲ **SOLOMON'S SEAL** *(Polygonatum)*. A genus of 50 herbaceous perennials in the *Liliaceae* from the northern hemisphere, with thick, fleshy RHIZOMES bearing seal-like scars from previous years' stems. There are two main groups, those with erect stems and WHORLS of narrow leaves and those with arching stems and broader, alternate leaves: common Solomon's seal *(P. X hybridum* in gardens, often grown as *P. multiflorum)*, grows to 750mm with alternate OVATE leaves and pendulous, tubular white and green flowers and blue-black berries; whorled Solomon's seal *(P. verticillatum)*, also growing to 750mm, has whorls of 3–6 LINEAR-LANCEOLATE leaves and smaller flowers followed by red berries.

SOMATIC Of the soma or body, all those cells that make up the body of a plant, but excluding the sex cells.

SOMNIFER|US, -A, -UM Sleep-inducing.

SOREDIA An asexual form of reproduction in LICHENS. Each soredia comprises a few algal cells surrounded by fungal threads. Together they form minute green granules which can grow into a new lichen plant.

SORIFEROUS Bearing SORI. *See* FERNS.

SOROSIS A fleshy multiple fruit, e.g. pineapple.

SORREL Several species of *Rumex*, perennial plants in the *Polygonaceae* from the north temperate zone, naturalised elsewhere. They have long-stalked, narrow leaves and branched PANICLES of tiny flowers with 6 PERIANTH segments followed by triangular fruits: sheep's sorrel *(R. acetosella)* is a variable, usually reddish, slender plant with linear-lanceolate, often HASTATE leaves with spreading lobes and stems to 250mm. It spreads widely by adventitious shoots from the roots and is a serious weed in sandy soils: common sorrel *(R. acetosa)* grows to 900mm with oblong lanceolate, HASTATE leaves to 75mm long, the basal lobes small and down-pointing. It is a clump-forming plant with acid leaves sometimes used as a vegetable, or raw in salads.

SOR|US, -I Groups of SPORANGIA on the leaf of a fern, sometimes covered with a membrane, the INDUSIUM.

SOUR SOP *(Annona muricata)*. An evergreen tree to about 4.5m in the *Annonaceae*, from C. America but widely grown in the tropics. It has PINNATE leaves, fragrant fleshy, green and yellow, 6-petalled flowers and heart-shaped fruits. The latter are green, to 250mm long, bearing rows of fleshy spines. The juicy white aromatic flesh is much used for drinks and for flavouring ice cream.

SOUTHERN BEECH *(Nothofagus)*. A genus of 35 deciduous and evergreen trees in the *Fagaceae* from the south temperate zone. Much like the northern hemisphere beeches, but the leaves, flowers and nutlets are usually smaller: Australian beech *(N. moorei)* grows to 45m with glossy evergreen, OVATE-LANCEOLATE leaves 38–75mm long; roble beech *(N. obliqua)* from Chile reaches 30m, with deciduous, irregularly toothed and shallowly lobed ovate to oblong leaves; black beech *(N. solandri)* grows to 24.4m or more with ovate to elliptic-oblong 9–13mm long evergreen leaves; silver beech *(N. menziesii)* is an evergreen tree to 27.3m with a silvery trunk and broadly ovate, double-toothed leaves to 13mm long. Both it and black beech are from New Zealand.

SOUTHERNWOOD
see ARTEMISIA

SOWBREAD
see CYCLAMEN

SOW THISTLE Several species of *Sonchus*, annuals and perennials in the *Compositae* from Europe, Asia and Africa. They exude a milky sap when broken, have stem-clasping often prickly-margined leaves and flower heads of yellow LIGULATE florets; perennial sow or field milk thistle *(S. arvensis)* grows 0.6–1.5m with RUNCINATE-PINNATIFID leaves and CORYMBS of 38mm wide flower heads, the INVOLUCRES densely covered with yellowish glandular hairs; common or smooth sow thistle *(S. oleraceus)* is a 30–120cm annual with angular stems and runcinate-pinnatifid leaves with broad terminal lobes and glabrous 19-25mm flower heads; prickly sow thistle *(S. asper)* is similar, but has spiny-toothed leaves.

SOY BEAN *(Glycine max* syn. *G. soja)*. An erect, often bushy, hairy annual to 600mm in the *Leguminosae* from E. Asia, but much grown elsewhere. It grows 30–180cm tall with TRIFOLIATE leaves, AXILLARY clusters of white or purplish pea-flowers and 50–75mm long hairy pods bearing 50–100mm beans which may be green, brown, black or yellowish. The beans have a high protein content and are used to make flour, sauces, and increasingly in artificial meat. They also are rich in oil which is used for margarine and in cooking oils, also for soap, paint and plastics. The beans also provide bean-sprouts as in MUNG BEAN.

SPADIX [SPADICES] A fleshy flower spike with the flowers imbedded in pits or sitting flush on the surface: characteristic of the arum family *(Araceae)*, e.g. lords and ladies and flamingo flower.

SPANISH BROOM *(Spartium junceum)*. An apparently leafless, green-stemmed shrub to 3m in the *Leguminosae* from the Mediterranean region. It has reed-like shoots with sparse, oblong-linear leaves which soon fall and terminal RACEMES of 25mm wide, shining yellow, fragrant pea-flowers.

SPANISH MOSS [OLD MAN'S BEARD] *(Tillandsia usneoides)*. An epiphytic plant in the *Bromeliaceae* from southern USA to S. America. It forms pendulous masses of slender, often spirally twisted stems to 900mm or so, bearing filiform, silvery scurfy leaves and small, 3-petalled yellow-green flowers.

SPATHE A green or coloured bract enclosing one to many flowers. It may be large and fleshy (arum lily), membranous (narcissus) or almost woody (certain palms).

SPATHULATE [SPATHULAT|US, -A, -UM] Spatula-shaped, broad at the tip and tapering to the stalk, especially used of leaves.

SPATTERDOCK *(Nuphar advena)*. A North American water lily, rather like a large Yellow WATER LILY.

SPAWN Mainly a gardener's term for a mass of HYPHAE of the cultivated mushroom grown on sterilised manure and used to start new mushroom beds.

SPEARWORT Several species of *Ranunculus*, marsh, fen and waterside plants in the *Ranunculaceae* from Europe and Asia. They are perennials with ovate to elliptic leaves and yellow buttercup-like flowers: greater spearwort *(R. lingua)* grows to 1.2m with half-clasping, oblong-lanceolate leaves to 150mm or so and 25–50mm wide flowers; lesser spearwort *(R. flammula)* is 300–450mm tall with 13–19mm wide flowers.

SPECIES A group of individuals which breed together and have the same constant and distinctive characters, though small differences may occur, e.g. in flower colour and hairiness.

SPECIOS|US, -A, -UM Showy, splendid.

SPECTABIL|IS, -E Spectacular, showy.

SPEEDWELL *(Veronica)*. A genus of 300 annual and perennial plants in the *Scrophulariaceae* from temperate regions, mainly the northern hemisphere. They may be prostrate or erect, with LINEAR-LANCEOLATE to OVATE leaves, usually toothed. The rotate flowers appear to have 4 petals only, but the larger upper petal, represents the fusion of two. They are usually blue or blue and white and are borne in AXILLARY or terminal RACEMES or solitarily: common field speedwell *(V. persica)* is an annual weed of farms and gardens, having prostrate stems, triangular-ovate 13–25mm leaves and solitary bright blue and white flowers to 13mm wide; germander speedwell *(V. chamaedrys)* is a prostrate or ascending perennial to 300mm with the axillary 10–12 flowered racemes of similar blooms; spiked speedwell *(V. spicata)* is a variable

perennial plant usually having erect, 150–450mm stems, oblong-lanceolate leaves and dense terminal spikes of bright blue flowers. There are several CULTIVARS with white or pink blooms. *See also* BROOKLIME.

SPELT
see WHEAT

SPERM [SPERMATOZOID] The usually motile male cells or GAMETES of primitive plants from seaweeds to cycads.

SPERMATOPHYTE All the seed-producing plants which form one of the great divisions of the plant kingdom. It is divided into *Gymnosperms* (including the conifers and cycads) and *Angiosperms* (all the true flowering plants).

SPHAEROCEPHAL|US, -A, -UM Round-headed; e.g. the flower heads of globe thistle.

SPHAGNUM [BOG MOSS] Several mosses characteristic of boggy areas in the temperate zones. They are usually tufted or hummock-forming and may be the dominant plants over vast areas. Their partially decayed remains form moss peat. The individual stems are erect, bearing groups or WHORLS of arching or pendulous branches clad in tiny, overlapping leaves. The tip is often bunched, the congested branches forming a head in palm-like fashion. They vary in colour from deep red to very pale green, depending upon the species. Stalked, ovoid capsules can sometimes be found which open explosively to release the spores. Sphagnum moss has a great water-holding capacity and is remarkably sterile. It has been used to dress wounds since ancient times. Nowadays it is still in demand by gardeners as an ingredient for growing EPIPHYTIC orchids and by florists for making up wreaths.

SPICATE [SPICAT|US, -A, -UM] Bearing spikes.

SPICE The fruits, seeds, bark or roots of several plants used for flavouring food and drink. *See* ALLSPICE, CINNAMON, CLOVES, GINGER, NUTMEG, PEPPER.

SPICULE [SPICULAT|US, -A, -UM] A short point; the sharp needle-like cells in some fruits and roots.

SPIDER ORCHID Two species of *Ophrys*, tuberous-rooted, ground-dwelling plants in the *Orchidaceae* from Europe and N. Africa: late spider orchid *(O. fuciflora)* has elliptical-oblong leaves to 100mm and 100–300mm stems bearing 2–7 flowers. These have spreading pink sepals and an obovoid 9mm velvety-maroon LABELLUM with a pattern of yellow-green lines. Early spider orchid *(O. sphegodes)* is similar, but the sepals are yellow-green and the labellum purple-brown.

SPIDER PLANT *(Chlorophytum comosum)*. A tufted, fleshy rooted perennial in the *Liliaceae* from S. Africa. It has arching, keeled, linear leaves to 450mm and slender, branched racemes to 600mm or more, bearing 6-petalled white flowers and small plantlets which root when they touch the soil. A cream-striped cultivar is much grown as a pot plant.

▲

SPIDERWORT *(Tradescantia virginiana)*. A tufted perennial in the *Commelinaceae* from N. America. It has erect, rather fleshy stems to 750mm, LINEAR-LANCEOLATE leaves and terminal UMBELS of 3-petalled, 25mm wide blue-purple flowers with bearded stamen filaments. Several CULTIVARS of hybrid origin *(T. X andersoniana)* are grown in gardens in shades of purple, red, pink and white.

SPIKE An unbranched inflorescence comprising a stem or rachis bearing stalkless flowers.

SPIKELET A small or secondary spike, *see* GRASS.

SPINACH *(Spinacea oleracea)*. A fleshy leaved annual in the *Chenopodiaceae* from the E. Mediterranean region. It has triangular-ovate leaves and DIOECIOUS green flowers in branched terminal spikes to 600mm. It is much grown for its edible leaves, eaten as a vegetable.

SPINACH BEET [SWISS CHARD] *(Beta vulgaris cicla)*. A form of wild sea beet much like BEETROOT, but larger, with green leaves and lacking the swollen root. Sea-kale beet is a form with wide, white leaf-stalks and midribs which are cut out and eaten like sea-kale, the leaf blades like spinach. Rhubarb beet has red stalks.

▲ **SPINDLE** *(Euonymus europaeus)*. A deciduous shrub or small tree to 6m in the *Celastraceae* from Europe and W. Asia. It has 4-angled twigs, ovate-lanceolate to elliptic CRENATE leaves and small, AXILLARY CYMES of 9mm wide green flowers with 4 petals. The 4-lobed pink fruits open to show the seeds covered with orange ARILS. Also a term in CELL DIVISION.

SPINE A straight, stiff, sharp-pointed structure derived from a modified twig, branchlet, leaf stalk, flower stalk or stipule.

SPINOS|US, -A, -UM Bearing spines.

SPIRAEA A genus of 100 deciduous shrubs in the *Rosaceae* from the north temperate zone, of varied appearance, with LINEAR to broadly OVATE leaves which can be lobed or toothed, smooth or hairy. The small 5-petalled, white or pink flowers are borne in UMBELS, CORYMBS or terminal PANICLES. Several species and hybrids are grown as ornamentals: bridal wreath, foam of May *(S. X arguta)* is a garden hybrid to 1.8m with OBLANCEOLATE, 10–38mm long leaves and numerous clusters of glistening white flowers; bride-wort *(S. salicifolia)* is a vigorous suckering bush to 1.8m or more with 50–75mm long LANCEOLATE leaves and dense terminal panicles of pink blossoms.

SPIRAL|IS, -E Twisted or wound round like a watch spring.

SPIROGYRA Filamentous green algae, often common in ponds forming extensive woolly green masses known as blanket-weed.

SPLEENWORT Several species of *Asplenium*, perennial ferns in the *Aspleniaceae* mainly from the north temperate zone. They are tufted in habit with PINNATE or bipinnate fronds and linear to oblong SORI: black spleenwort *(A. adiantum-nigrum)* has somewhat leathery bipinnate fronds 100–300mm or more long and triangular-lanceolate

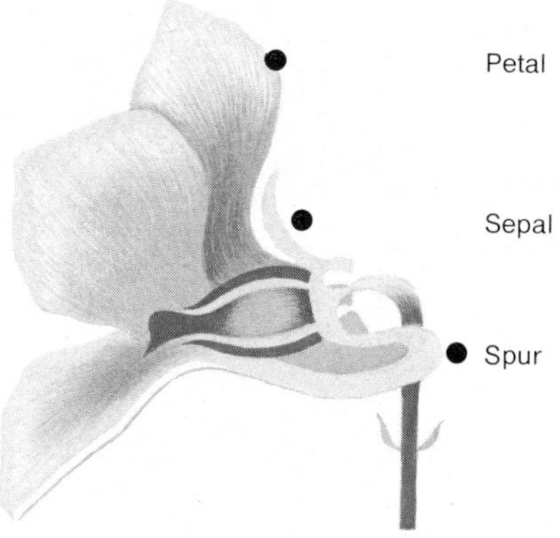

Section through spurred flower

in outline; maidenhair spleenwort *(A. trichomanes)* has slender, pinnate fronds to 200mm or more with 6mm long oblong PINNAE. *See also* WALL RUE; SEA SPLEENWORT.

SPORADIC Of plants that are widely spread in their native countries, but occur as isolated colonies elsewhere.

SPORANGI|UM, -A An asexually formed spore, produced by certain fungi, e.g. potato blight, and ferns. The tip of the HYPHAE which bears the sporangium is a sporangiophore.

SPORE Minute asexual reproductive bodies of one or a few cells together which give rise to a new individual either directly or indirectly, e.g. in ferns, mosses, and fungi. *See* ALTERNATION OF GENERATIONS.

SPOROGONIUM The spore capsule and its stalk in mosses and liverworts.

SPOROPHYLL The fertile frond of a fern that produces SORI and spores.

SPOROPHYTE The spore-bearing phase in the life cycle of such plants as ferns and some seaweeds. *See* ALTERNATION OF GENERATIONS.

SPRING BELL *(Forsythia)*. A genus of 7 deciduous shrubs in the *Oleaceae* mainly from Asia (one in Europe). They have LANCEOLATE, ELLIPTIC or OVATE leaves and yellow bell-shaped flowers with 4 spreading lobes which open before the leaves; golden spring bell *(F. suspensa)* has an arching or untidily pendulous habit with sometimes TRIFOLIATE leaves on the strong shoots. Commonest in the gardens of Britain is *F. X intermedia*, a more erect and floriferous hybrid from *F. suspensa*.

SPROUT (1) The tight leafy buds of BRUSSELS SPROUTS. (2) Any fast-growing leafy shoot, particularly from the base of the plant.

SPRUCE [SPRUCE FIR] *(Picea)*. A genus of 50 pyramidal, evergreen trees in the *Pinaceae* from the north temperate zone, usually alpine. They are distinguished from true firs *(Abies)* by having the needle-like leaves borne on peg-like projections from the twigs and cigar-shaped pendulous cones which do not break up when ripe. The seedlings have 4–15 COTYLEDONS. They are important timber trees, the wood having a wide range of uses from pit props to musical instruments. Several species yield resin, and a beer is made from the leaves and twigs of common or Norway spruce *(P. abies)* which is also much in demand in Britain as a Christmas tree. It grows to 61m with 13–25mm horny, pointed leaves and 100–150mm cones; Sitka spruce *(P. sitchensis)* can reach 54m or more with pointed leaves to 19mm which are aromatic when bruised, and 50–100mm cones.

SPUR A hollow projection of a sepal or petal containing nectaries, e.g. violet, nasturtium. *See also* DWARF SHOOT.

SPURGE Several species of *Euphorbia*, a genus of 2,000 species of annuals, perennials, trees, shrubs and succulents from subtropical and temperate regions. Although widely different in form, all are typified by having small

inflorescences called CYATHIA that look like single flowers. Most species have bracts around the cyathia, sometimes coloured as in the popular pot plant poinsettia *(E. pulcherrima)*, a deciduous shrub from Mexico with terminal heads of vermilion, LANCEOLATE bracts: Mexican spurge *(E. fulgens)* is also a shrub to approximately 1.8m with wand-like arching stems and small, AXILLARY scarlet-bracted cyathia; caper spurge *(E. lathyrus)* is a 30–120cm unbranched biennial with narrow long leaves to 200mm long and terminal UMBELS of cyathia and yellow-green bracts; Dalmatian spurge *(E. characias wulfenii)* is a shrub to 1.8m with narrow, grey leaves and large heads of chrome yellow bracts; wood spurge *(E. amygdaloides)* is sub-shrubby to 750mm with OBLANCEOLATE 450–900mm long leaves and greenish-yellow bracts; sea spurge *(E. paralias)* is a 200–400mm GLAUCOUS perennial, with dense, fleshy OVATE leaves and yellowish bracts; cypress spurge *(E. cyparissias)* is a perennial 100–250mm tall with LINEAR leaves and yellowish bracts that age reddish; sun spurge *(E. helioscopa)* is an annual weed of cultivation with one to several unbranched stems, obovate-serrulate leaves and yellow-green bracts; petty spurge *(E. peplus)*, also an annual weed, is smaller and bushier with greener bracts.

SPURGE LAUREL
see DAPHNE

SPURREY [CORN SPURREY] *(Spergula arvensis)*. A slender, 75–300mm tall annual weed in the *Caryophyllaceae* from most of the temperate world. It has WHORLS of linear, fleshy, convex leaves and dichasial CYMES of tiny, 5-petalled white flowers.

SQUALID|US, -A, -UM Squalid, dingy, neglected.

SQUAMAT|US, -A, -UM Scaly.

SQUARROS|US, -A, -UM Having overlapping leaves or bracts, the tips of which spread or curve outwards.

SQUASH
see CUCURBITS

○ **SQUILL** *(Scilla)*. A genus of 80 bulbous-rooted plants from temperate Europe, Asia and S. Africa. They have LINEAR to OBLONG-LANCEOLATE leaves and bell-shaped, or starry flowers with 6, usually blue or purple TEPALS: Siberian squill or Spring beauty *(S. sibirica)* has nodding, electric-blue bells on a 75–150mm stem; spring squill *(S. verna)* has starry blue to 13mm wide flowers on 50–150mm stems; autumn squill *(S. autumnalis)* is similar, but flowers in autumn before the leaves.

SQUIRREL TAIL GRASS
see BARLEY GRASS

SQUIRTING CUCUMBER *(Ecballium elaterium)*. A prostrate perennial in the *Cucurbitaceae* from the E. Mediterranean. It has OVATE-CORDATE, somewhat 3-lobed, rough textured leaves, solitary female and RACEMES of yellow male flowers in the AXILS and ovoid, 25–50mm long hairy fruits. The latter drop off when ripe and squirt a jet of seeds and liquid through the stalk hole.

STAGHORN MOSS
see CLUB MOSS

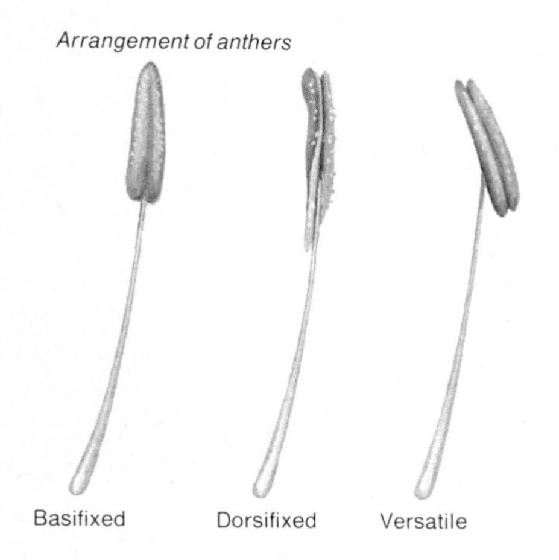

Arrangement of anthers

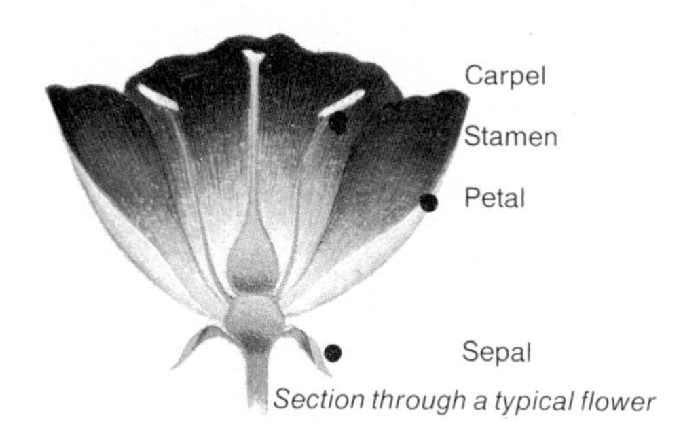

Section through a typical flower

STAGNAL|IS, -E Growing in the still water of ponds and ditches.

● **STAMEN** The male unit of a flower comprising two ANTHER lobes joined together at the top of a stalk or filament.

STAMINATE Unisexual male flowers.

STAMINODE A rudimentary stamen, sometimes functioning as a petal or nectary but producing no pollen.

STANDARD (1) The upper or top petal of a pea-flower.
(2) The 3 inner, often upstanding petals of an iris flower.
(3) A gardening term for a tree or plant with an unbranched leg, main stem or trunk and a head of branches.

STARCH SHEATH
see ENDODERMIS

STAR OF BETHLEHEM *(Ornithogalum umbellatum)*. A bulbous perennial in the *Liliaceae* from the Mediterranean region, but naturalised elsewhere. It has linear leaves with whitish midribs and CORYMBOSE RACEMES of 25–38mm glistening starry, white flowers having 6 tepals.

STARWORT
see ASTER

STELLATE [STELLAT|US, -A, -UM] Star-like, with radiating petals, bracts or branches.

STEPHANOTIS *(Stephanotis floribunda)*. Also called wax flower and Malagasy jasmine, this is an evergreen climber in the *Asclepiadiaceae* from Malagasy. It climbs to 3m or more with thick, leathery, ovate-elliptic leaves and AXILLARY clusters of sweetly scented, tubular white flowers having 5 petal-like lobes.

STERILE (1) Of flowers that do not produce seeds, e.g. the male flowers of a unisexual plant.
(2) The fronds of ferns that do not bear spores.
(3) Of a solitary plant that is self-sterile and produces no fruit or seeds.

STIGMA
see PISTIL

STINKHORN *(Phallus impudicus)*. A SAPROPHYTIC *Basidiomycete* fungus from temperate climates in woods and thickets. The fruiting body starts as a white jelly-filled globe the size of a golf-ball. This splits and a stout stalk and ovoid conical cap grow up rapidly, the cap covered with a layer of spores containing a greenish jelly. There is a smell of rotting meat, which attracts flies that feed on the jelly and distribute the spores.

STIPE An alternative name for the leaf stalks of ferns or the stems of *Basidiomycetes* fruiting bodies, e.g. mushroom.

STIPULAT|US, -A, -UM Having stipules.

STIPULE Outgrowths from the base of a leaf stalk, usually leaf-like and one either side, e.g. garden pea.

▲

Sometimes they are scale-like and protect the young leaves, soon falling, e.g. oak, beech, while in other cases they are hard and spine-like, e.g. false acacia.

STITCHWORT Several species of *Stellaria*, perennial plants in the *Caryophyllaceae* from Europe, Asia and N. Africa. They have pairs of narrow leaves and dichasial CYMES of white flowers with deeply bilobed petals: greater stitchwort *(S. holostea)* has slender decumbent stems 150–600mm long, LINEAR-LANCEOLATE leaves and showy, 19mm wide flowers.

STOCK Several species of *Matthiola*, annuals and perennials in the *Cruciferae* from the E. Mediterranean: common or garden stock *(M. incana)* is a 300–600mm sub-shrub, though often grown as an annual. It has grey, hairy lanceolate leaves and terminal RACEMES of 4-petalled 25–38mm wide purple, red or white flowers. There are several double cultivars; night scented stock *(M. bicornis)* is a 300mm annual with oblong, LANCEOLATE-PINNATIFID leaves and 19mm lilac flowers which open at night.

STOLON Usually aerial stems which root when they touch the soil; e.g. blackberry.

STOLONIFEROUS [STOLONIFER|US, -A, -UM] Bearing stolons.

STOMA [STOMATA] Small openings or pores in the leaf EPIDERMIS, which lead into the MESOPHYLL layer allowing air into the leaf tissue. Usually more of them occur on the undersides of the leaves, except in floating aquatics such as water lilies. Each stoma is guarded by two sausage-like cells (guard cells). When distended with water they are open, but when pressure is low they become flaccid and close together. Stomata are usually wide open by day and closed by night. Osmotic pressure is high by day when sugars are being formed in the guard cells, but low once the light fades and PHOTOSYNTHESIS ceases.

STONE CELLS Solitary or small groups of cells almost entirely filled with LIGNIN. Single cells are sclereides, but differ from those that form FIBRES by being roughly rounded in shape and not forming continuous tissue. The grittiness in the flesh of some pears is due to stone cells.

▲ **STONECROP** *(Sedum)*. A genus of 600 succulent plants in the *Crassulaceae* from the north temperate zone. Most of them are perennials, having fleshy, linear to orbicular leaves and CYMES of small, starry usually 5-petalled flowers in shades of white, red, pink, yellow, rarely blue: biting stone crop or wall pepper *(S. acre)* forms mats of interlacing stems overlapping, triangular ovoid leaves to 4mm long and yellow flowers; white stonecrop *(S. album)* is similar but the 6–13mm long leaves are oblong and the flowers white; orpine or livelong *(S. telephium)* is erect, to 600mm, with semi-GLAUCOUS, OBLANCEOLATE 19–75mm long leaves and dense, flattened heads of red-purple flowers; rose-root *(S. rosea)* is erect to 300mm having glaucous, densely borne OBOVATE leaves and DIOECIOUS yellow-green flowers with 4 petals.

STONE PINE
see PINE

● 

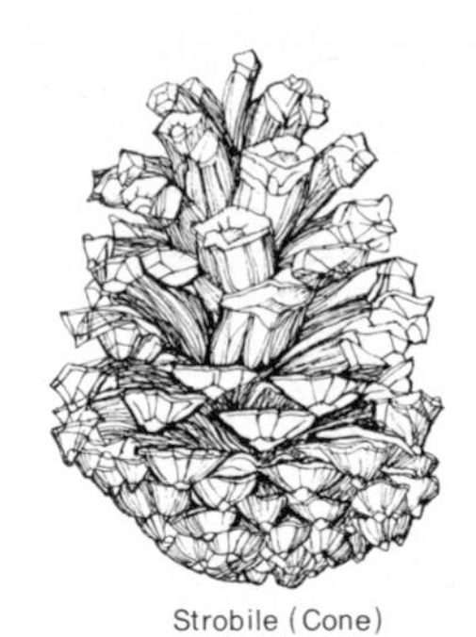

Strobile (Cone)

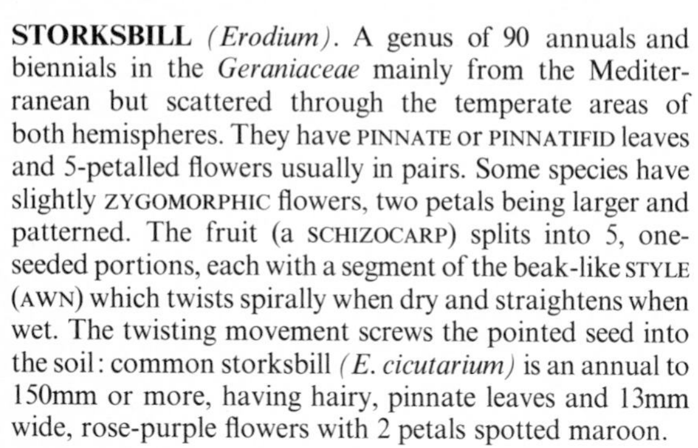

▲ *Section through cone* *Scale*

STORKSBILL *(Erodium)*. A genus of 90 annuals and biennials in the *Geraniaceae* mainly from the Mediterranean but scattered through the temperate areas of both hemispheres. They have PINNATE or PINNATIFID leaves and 5-petalled flowers usually in pairs. Some species have slightly ZYGOMORPHIC flowers, two petals being larger and patterned. The fruit (a SCHIZOCARP) splits into 5, one-seeded portions, each with a segment of the beak-like STYLE (AWN) which twists spirally when dry and straightens when wet. The twisting movement screws the pointed seed into the soil: common storksbill *(E. cicutarium)* is an annual to 150mm or more, having hairy, pinnate leaves and 13mm wide, rose-purple flowers with 2 petals spotted maroon.

STRAMINE|US, -A, -UM Straw-coloured.

● **STRAWBERRY** *(Fragaria)*. A genus of 15 perennials in the *Rosaceae* from N. and S. America, Europe and Asia. They have TRIFOLIATE leaves, 5-petalled white flowers in small CYMES and fruits comprising greatly enlarged fleshy RECEPTACLES studded with ACHENES or 'seeds'. They spread widely by runners. The garden strawberry *(F. X ananassa)* is a hybrid between N. American *F. virginiana* and the S. American *F. chiloensis* raised in France in the 18th century. There are now many cultivars some of which fruit in summer and autumn. The wild strawberry in Europe *(F. vesca)*, has coarsely toothed, 25–50mm OBOVATE leaflets and erect stems to 150mm or more with 13–19mm wide flowers and ovoid fruits to 19mm long; alpine strawberry *(F. v. semperflorens)* differs in having few or no runners and flowering from spring to late autumn; barren strawberry *(Potentilla sterilis)* is often mistaken for wild strawberry, but the plant is smaller and lacks runners.

STRAWBERRY GERANIUM
see SAXIFRAGE

STRAWBERRY TOMATO
see CAPE GOOSEBERRY

STRAWBERRY TREE *(Arbutus)*. A genus of 20 evergreen trees and shrubs in the *Ericaceae* from W. Asia, W. Europe, the Mediterranean region, North and Central America. They have OVATE to LANCEOLATE, leathery leaves, nodding terminal PANICLES of urn-shaped flowers and globular, red fruits covered with small, soft tubercles. The latter are edible but somewhat tasteless: common strawberry tree *(A. unedo)* grows to 12.2m, though often shrubby, with elliptic to obovate leaves, white or pinkish 6mm long flowers and bright red fruits to 19mm wide. Often cultivated is a hybrid of this, *A. X andrachnoides* with handsome reddish bark and white flowers.

STRIAT|US, -A, -UM Marked with fine, parallel lines or grooves, usually on leaves and bracts.

STRICT|US, -A, -UM Erect, or of upright growth.

STRIGOSE [STRIGOS|US, -A, -UM] Bearing straight stiff hairs or bristles lying close to the leaf or stem.

▲ **STROBILE [STROBILUS]** A cone; the spore-bearing structures of club mosses and the flowering and fruiting spikes of conifers. The latter are often woody when mature, composed of a central stem or RACHIS and overlapping woody scales which bear the seeds. Juniper and yew are exceptions in having fleshy berry-like cones.

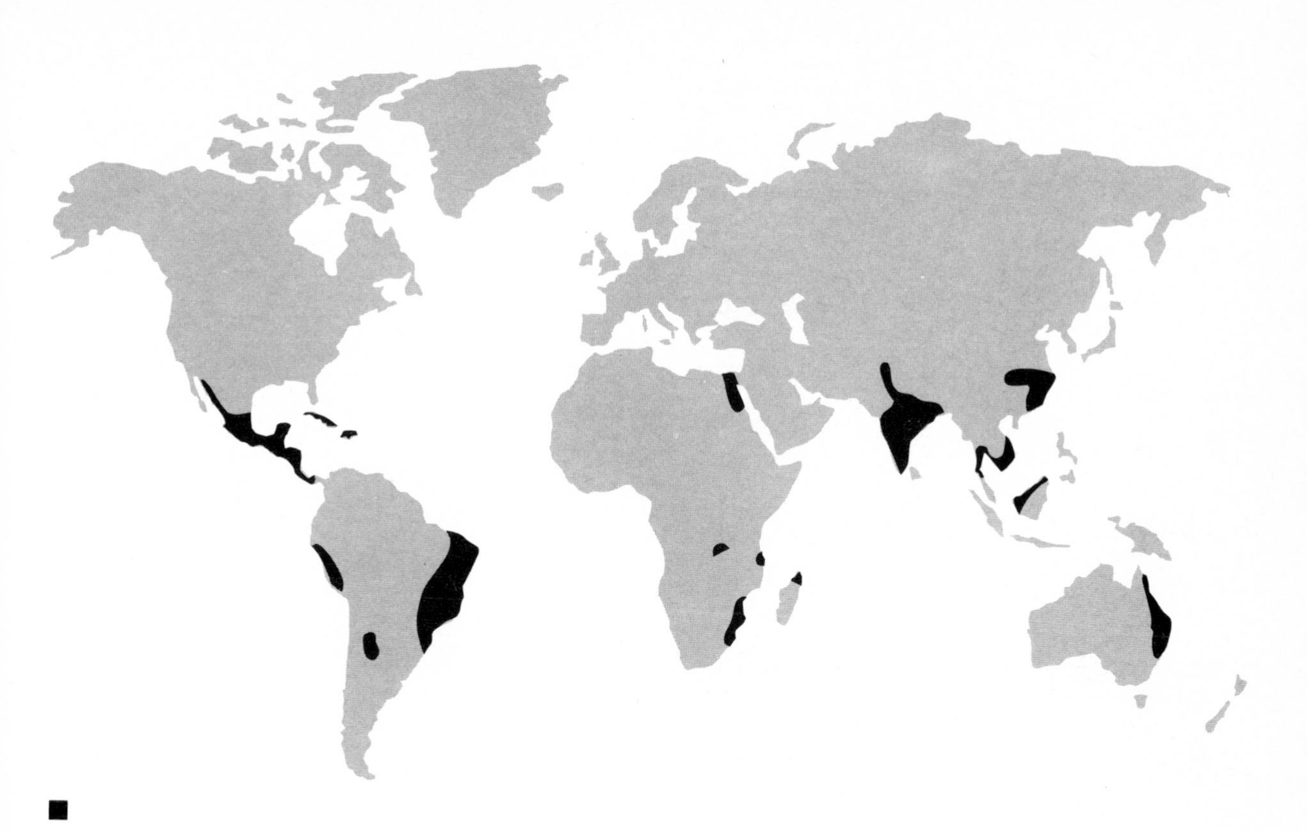

■

STYLE The stalk joining the ovary to the stigma. It can be long and slender, e.g. foxglove, short and thick as in tulip, or non-existent as in field poppy.

STYLOS|US, -A, -UM Having a prominent style.

SUAVEOLENS [SUAV|IS, -E] Sweet, pleasant.

SUBER Cork or corky.

SUBSHRUB A low shrub which is woody at the base only, the upper stems dying back each winter. In popular parlance it also refers to any small, soft-stemmed shrub.

SUBSPECIES A distinct, true breeding form of a species, often isolated geographically from the species itself, and differing more significantly than a variety.

SUBULATE [SUBULAT|US, -A, -UM] Awl-shaped, usually of leaves and bracts.

SUCCIS|US, -A, -UM Ending abruptly as though cut or broken off, e.g. the root stock of devil's bit scabious.

SUCCULENT PLANT A plant with either swollen leaves (as houseleek), or stems (as cacti), containing water storage tissue. They are often found in arid regions, in climates with long, dry summers, or on dry soils.

SUCKER (1) A shoot that arises below ground level, usually on a shrub or tree.
(2) Another name for HAUSTORIA.

SUFFRUTICOSE [SUFFRUTICOS|US, -A, -UM] Somewhat shrubby; a SUBSHRUB.

SUGAR APPLE
see CUSTARD APPLE

SUGAR BEET *(Beta vulgaris cicla)*. A biennial form of sea beet with a swollen root. It is virtually the same as large-rooted beetroot, but lacks the red pigment. Sugar is extracted by shredding the roots and heating in running water, the liquid obtained being concentrated by boiling until crystallisation takes place. Useful by-products are cattle feed from the leaves and cake from the root after sugar extraction, also molasses which is used for making industrial alcohol. MANGEL-WURZEL or mangold is an extra-large beet grown for cattle fodder only.

■ **SUGAR CANE [NOBLE CANE]** *(Saccharum officinarum)*. A clump-forming bamboo-like grass in the *Gramineae,* probably from New Guinea but much grown in the tropics and not known truly wild. It is of ancient hybrid origin and has been known in the S. W. Pacific since ancient times. Native peoples still chew pieces of cane for the sweet juice. Sugar cane grows 2.4–6.0m tall depending upon CULTIVAR, climate and soil. It has broad, linear, tapering leaves with a white midrib and large terminal PANICLES of small spikelets.

SULCATE [SULCAT|US, -A, -UM] Grooved or furrowed.

SULPHURE|US, -A, -UM Pale yellow.

SUMAC Several species of *Rhus*, a genus of 250 evergreen and deciduous shrubs, trees and climbers in the *Anacardiaceae* from temperate and sub-tropical regions. The leaves are mainly PINNATE or TRIFOLIATE and the tiny, 4–6 petalled flowers are carried in PANICLES. Japanese lacquer is obtained from the resin of *R. verniciflua*; and the berries of *R. succedana*, also from China and Japan, yield wax. Several species have highly irritant properties to people who are allergic, causing painful blistering that can lead to ulceration, e.g. the American trifoliate climber, poison ivy *(R. radicans)* and the shrubby poison oak *(R. toxicodendron)*. A number of species are grown for the bright colour of their autumn foliage, e.g. stag's horn sumac *(R. typhina)*, a suckering shrub with velvety, antler-like twigs and long, pinnate leaves.

SUMMER CYPRESS
see BURNING BUSH (Kochia)

● **SUNDEW** *(Drosera)*. A genus of 100, mainly perennial, carnivorous plants in the *Droseraceae* of world-wide distribution. The LINEAR to orbicular leaves bear large, sticky, glandular hairs which trap small creatures and absorb food substances from their bodies. Like BUTTERWORT and PITCHER PLANT, sundews often grow in acid, boggy soils where nitrogen and other minerals are low. Many species are rosette-forming, a few erect and even climbing. The 4–8-petalled flowers may be white, pink, red or purple small and CLEISTOGAMIC or large and showy. Common or round-leaved sundew *(D. rotundifolia)* is a rosette-forming species with round, stalked, leaves 9mm across and small, 6-petalled white flowers.

SUNFLOWER *(Helianthus)*. A genus of 110 species of annuals and perennials in the *Compositae* from N. and S. America. They have LANCEOLATE to broadly OVATE leaves and daisy flower heads usually with yellow RAY florets. Annual sunflower *(H. annuus)* is a giant, unbranched annual to 3.6m with flower heads 300mm or more across. CULTIVARS with chestnut-red florets and double heads are known. It is much grown as a crop plant for the oil-rich seeds. *See also* ARTICHOKE (Jerusalem).

SUN ROSE *(Cistus)*. A genus of 20 small to medium sized evergreen shrubs in the *Cistaceae* mainly from the Mediterranean region and the Canary Is. They have pairs of linear to oblong-ovate leaves and terminal clusters of saucer-shaped, 5-petalled flowers in shades of white, pink, purple and sometimes spotted with crimson: *Cistus ladanifer* syn. *C. ladaniferus*, grows to 1.8m with dark, narrowly lanceolate leaves and 75–100mm wide white flowers spotted with crimson in the centre. It yields the medicinal ladanum resin.

SUPERIOR OVARY
see HYPOGYNOUS

SUPIN|US, -A, -UM Lying flat, prostrate.

SUSPENSOR The chain of cells that attaches the embryo plant to the wall of the embryo sac after fertilisation.

SWAMP CYPRESS *(Taxodium distichum)*. A deciduous coniferous tree to 36m in the *Taxodiaceae* from the swamps of S.E. North America. It is pyramidal when young, broadening at maturity, with the 8–19mm long linear leaves arranged in two ranks on secondary branchlets which fall complete in autumn. The 13–25mm cones are globular. Growing in swamps or by water, this tree produces 'knees' – roots which grow vertically out of the ground, apparently for the purposes of aeration.

SWEDE
see RAPE

SWEET BAY
see BAY LAUREL

SWEET BRIAR
see ROSE

SWEET CHESTNUT *(Castanea sativa)*. A deciduous tree to 35m in the *Fagaceae* from S. Europe, W. Asia, N. Africa and planted elsewhere. It has grey bark often with a spiral fissuring on the trunks of old trees. The 150–200mm long boldly veined, oblong-lanceolate leaves are sharply toothed. Tiny, petal-less, whitish-yellow catkins are followed by densely spiny globular fruits to 40mm wide, each containing one to three globose to flattened, polished brown nuts.

SWEET FLAG *(Acorus calamus)*. An aromatic, iris-like perennial in the *Araceae* from central Asia and western N. America, and grown elsewhere. It has sword-shaped leaves to 900mm and tiny green flowers crowded on a 75mm spike which grows at right angles to the stem.

SWEET GALE
see BOG MYRTLE

SWEET GUM
see LIQUIDAMBAR

SWEET PEA *(Lathyrus odoratus)*. An annual climber in the *Leguminosae* from S. Europe. It has winged stems to 3m or so and leaves with 2 OVATE leaflets and branched terminal TENDRILS. Short RACEMES of fragrant pea-flowers having broad, erect STANDARD petals are borne in the upper leaf AXILS. The wild plants have 25mm red-purple and blue flowers. As the result of several MUTATIONS and the work of plant breeders, there are now cultivars in a wide range of colours, some with frilled standards up to 50mm wide.

SWEET POTATO *(Ipomoea batatas)*. A tuberous-rooted perennial in the *Convolvulaceae*, probably from S. America, but not known wild and much grown in tropics and warm temperate zones. It has trailing or climbing stems and leaves that can be OVATE-CORDATE or shallowly to deeply DIGITATELY lobed. The 25–50mm long funnel-shaped flowers are purple. The tubers are globular to spindle-shaped, up to 225mm long. Depending on the cultivar, they may have white, yellow, orange, reddish or purple flesh and the same range of skin colours plus brown.

SWEET SULTAN *(Centaurea moschata)*. An erect annual to 600mm in the *Compositae* from the E. Mediterranean. It has LYRATE or PINNATIFID leaves and smooth, knapweed-like flower heads with fragrant, slender, tubular florets in white, yellow and purple. *C.m. imperialis* is the larger flowered form grown in gardens.

SWEET VERNAL GRASS *(Anthoxanthum odoratum)*. A tufted perennial plant in the *Gramineae* from Europe, Asia, N. Africa and naturalised in N. America and Australia. It has short flat pointed leaves that smell of coumarin (new mown hay) when cut and 38–56mm long oblong, compact PANICLES of slender spikelets on stems to 450mm.

SWEET WILLIAM *(Dianthus barbatus)*. A tufted, short-lived perennial in the *Caryophyllaceae* from S. and E. Europe. It has pairs of oblong-lanceolate leaves and 309–600mm stems bearing flattened CORYMBS of spicily fragrant, 5-petalled flowers from tubular CALYCES. They can be various shades of red, splashed or zoned with white and other colours.

SYCAMORE
see MAPLE

SYLVATIC|US, -A, -UM [SYLVESTRIS, -E] From woods or forests, or growing wild.

SYMBIOSIS
see MYCORRHIZA

SYMPETALOUS An alternative for GAMOPETALOUS.

SYMPODIUM [SYMPODIAL] A mode of branching in trees and shrubs where the leading stem is formed each year from a lateral bud. Most broad leaved trees grow in this way, e.g. beech, elm, lime, etc.

SYNCARPOUS An ovary formed by the fusion of several CARPELS, e.g. lily, poppy.

SYNCONIUM A fruit formed of a hollowed RECEPTACLE, e.g. fig.

SYNONYM Abbreviated to 'syn.', this is an alternative but usually invalid name for a plant. When botanists are reclassifying plants, they may come across an older, overlooked name which according to international rules must take precedence; or fresh study may make it necessary for a plant to be moved from one genus to another. In either case the old name is quoted as a synonym.

SYSTEMATIC BOTANY The study of the relationships of plants.

TABULAR|IS, -E Flattened horizontally, plate-like.

TAIL FLOWER
see FLAMINGO FLOWER

● **TAMARISK** *(Tamarix)*. A genus of 54 deciduous trees and shrubs in the *Tamaricaceae* from Europe, Asia and N. Africa, and grown elsewhere. They tend to grow near to the sea or in salty soils and have tiny, scale-like leaves on small twigs which fall with the leaves. The tiny pink or white 4–5-petalled flowers are arranged in dense, spike-like RACEMES, Common tamarisk *(T. anglica* syn. *T. gallica)* is a feathery shrub to 2.4m or more with sea-green foliage and usually pink flowers.

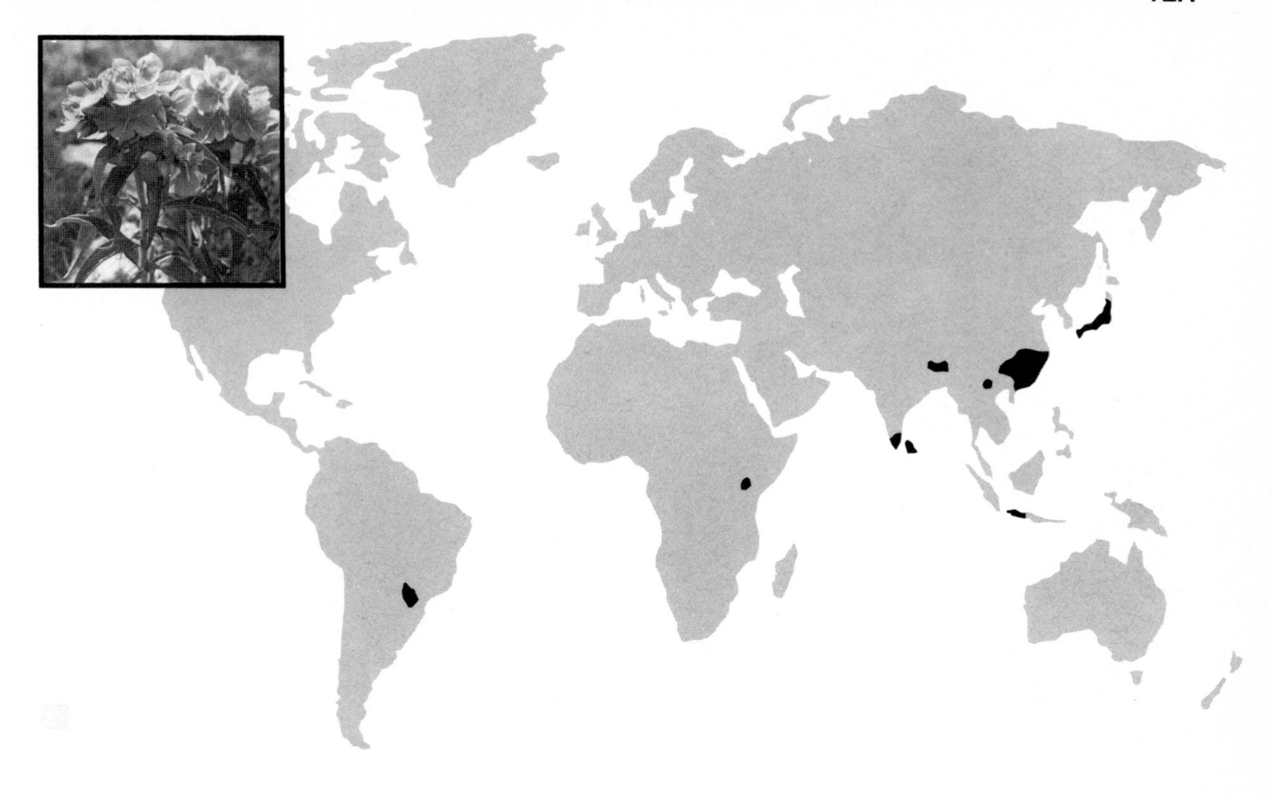

▲

TANGLE
see OARWEED

TANSY *(Tanacetum vulgare)*. An erect perennial to 1.8m in the *Compositae* from Europe to Siberia, but naturalised elsewhere. Formerly cultivated as a medicinal plant and for flavouring sweet and savoury foods. It has PINNATE leaves with PINNATIFID leaflets and yellow, button-like flower heads of DISK florets in terminal compound CORYMBS.

TAPA CLOTH
see PAPER MULBERRY

TAPE GRASS [EEL GRASS] *(Vallisneria spiralis)*. A submerged aquatic in the *Hydrocharitaceae* from Europe. It is a tufted plant spreading by runners, with semi-translucent tape-like leaves to 600mm and DIOECIOUS green flowers. The female flowers are solitary, borne on stems that reach the surface where they open. The male flowers are carried in short, ovoid spikes which break off when mature and float to the surface where they open.

TAP ROOT The vertically growing main root of a plant, which in trees provides anchorage. Sometimes it is the natural extension of the first seedling root (RADICLE).

TARE
see VETCH

TARO [DASHEEN] *(Colocasia antiquorum)*. A cormous rooted perennial in the *Araceae* from S.E. Asia, but much cultivated in the tropics and subtropics. It has OVATE-PELTATE leaves about 450mm long on 450–900mm stalks, and rarely produced, 150–300mm long, pale yellow, arum-like SPATHES. The ribbed, rounded, tuber-like corms contain a fine-grained starch.

TARRAGON
see ARTEMISIA

TAXON [TAXA] Any generally applied unit within classification. It stands for any grouping, for example: a variety, a species hybrid group, a genus, or family.

TAXONOMY The CLASSIFICATION of plants according to their mutual affinities. *See also* NOMENCLATURE.

▲**TEA** Used in a general way for many plants, the leaves of which can be infused with hot water to make a drink. Of primary importance is *Camellia sinensis* syn. *Thea sinensis*, an evergreen tree to 3.7m or so in the *Theaceae*, from N. Burma and N.E. Assam, but cultivated in China since early times, and then in Japan and elsewhere. Tea planting in India only started in 1818. Under cultivation, tea is regularly trimmed for its leaves and young shoots and rarely exceeds 1.8m. It has toothed, leathery, OBOVATE-LANCEOLATE, slender, pointed leaves to 150mm or more long and fragrant, 5–7 petalled white flowers. Black or Indian tea is manufactured by first allowing the leaves to wither, then breaking them on rollers after which the broken leaf ferments for about 2 hours. It is then dried with a forced draught of hot air, graded, sorted and packed into chests. Green tea is made by drying before fermentation. *See also* BERGAMOT (oswego tea), DUKE OF ARGYLL'S TEA TREE, LABRADOR TEA, MATÉ (Brazilian tea).

TEAK *(Tectona grandis)*. An evergreen tree to 45m in the *Verbenaceae*, from S. India to Malaysia. It has 200–300mm long OVATE leaves and PANICLES of small, bell-shaped, white flowers with 5–6 lobes. It is grown under forestry conditions and felled when 40–60 years old at 15–21m.

TEASEL *(Dipsacus fullonum)*. An erect biennial to 1.8m or more in the *Dipsacaceae* from Europe, the Near East and N. Africa. It has angled stems with small prickles and LANCEOLATE pairs of stem leaves which unite (CONNATE) to form a water-holding cup. The tiny, rose-purple tubular flowers are borne in a cone-like head among straight, spine-tipped BRACTS. Fullers' teasel *(D.f.* ssp. *sativus)* has the bracts ending in stiff, curved spines. It is still sometimes used to raise the nap on cloth.

TELEGRAPH PLANT *(Desmodium gyrans)*. An erect, 300–900mm annual in the *Leguminosae* from S.E. Asia. It has TRIFOLIATE leaves each composed of a large terminal, elliptic-oblong leaflet and two very much smaller ones which move in an elliptical orbit when temperatures exceed 75°F. The small, purple pea-flowers are carried in branched PANICLES.

TELEUTOSPORES
see RUST FUNGI

TELOPHASE
see CELL DIVISION

TEMPLE TREE
see FRANGIPANI

TENDRIL Stems, leaves or leaf stalks of climbing plants modified to thread-like organs which twist round any suitable support. Stem tendrils are found on cucumber, white bryony and leaf (leaflet) tendrils on sweet peas.

TENELL|US, -A, -UM Delicate, dainty.

TENU|IS, -E Slender, thin.

TEPAL The usually petal-like organs of a flower when the sepals and petals look exactly alike, as with e.g. tulip, bluebell, lily.

TERETE A stem which is smooth and rounded in cross section.

TERMINAL Usually of flowers and inflorescences that terminate a stem, as distinct from those which are AXILLARY.

TERNARY Borne in groups of three.

TERNAT|US, -A, -UM In WHORLS or bunches of three; usually of leaves or leaflets.

TERRESTRIAL Growing on the ground, used of orchids to distinguish these from tree-dwelling (EPIPHYTIC) species.

TESSELAT|US, -A, -UM Chequered or with roughly square spots or marks, e.g. the flowers of snake's head FRITILLARY.

TESTA The outer skin or coat of a true seed.

TESTACE|US, -A, -UM Terracotta or brownish-red.

TETRARCH
see DIARCH

TETRAMEROUS Used of flowers which have sepals and petals in WHORLS of 4; e.g. wallflower.

TETRAPLOID
see POLYPLOID

TETRAPTEROUS With 4 wings, usually used of fruits.

THALLOID Shaped like a THALLUS.

THALLOPHYTA The division of the plant kingdom which includes the most primitive plants, ranging from unicellular ALGAE to the more complex FUNGI. *See also* LICHENS.

THALLUS A usually flattened, leaf-like stem, not differentiated into true stems and leaves, e.g. LIVERWORT, *Marchantia*, and such seaweeds as OARWEED.

THECA The spore-bearing capsule of a moss.

THEROPHYTE Plants which pass the unfavourable season as seeds. The seventh of a 7-category system of classification of life forms.

● **THISTLE** Several prickly or spiny plants are called thistles, most of them in the daisy family *(Compositae)*. The main candidates are in the genera *Carduus* and *Cirsium*, and are annuals, biennials and perennials from the northern hemisphere. They are erect plants with spiny margined leaves and sometimes spiny stems and flower heads. The florets are tubular, often red or purple and the fruits have a pappus of long hairs (thistledown): musk thistle *(Carduus nutans)* is biennial with a spiny, winged stem to 900mm and nodding, 32–50mm wide, hemispherical, spiny, red-purple heads; spear thistle *(Cirsium vulgare)* is a biennial to 1.5m with erect, ovoid heads; creeping thistle *(C. arvense)* is a DIOECIOUS perennial with deep, fleshy roots which produce ADVENTITIOUS shoots, a serious weed of farms and gardens. The stems are spineless and the 13–25mm wide heads are pale purple; stemless thistle *(C. acaulon)* is a rosette-forming perennial of short grassland, with sessile or shortly stalked 25–38mm wide red-purple heads. *See also* CARLINE, GLOBE, MILK, SALTWORT (Russian), SCOTCH (Cotton), SOW THISTLE.

THORN-APPLE *(Datura stramonium)*. A branched annual to 900mm in the *Solanaceae* from the northern hemisphere. It has OVATE, coarsely toothed, UNDULATE leaves to 200mm long and slender, funnel-shaped flowers 56–75mm. The whole plant is narcotic and poisonous.

THRIFT *(Armeria maritima)*. A mat or hummock-forming perennial in the *Plumbaginaceae* from coasts and mountains in the northern hemisphere. Has LINEAR leaves 19–75mm long (sometimes reaching 150mm) and wiry leafless stems rarely to 300mm bearing dense globular heads of small, 5-petalled pink flowers.

THYME *(Thymus)*. A genus of about three hundred, small, wiry-stemmed shrublets in the *Labiatae* from Europe and Asia. They have linear to ovate, aromatic leaves in pairs and WHORLS of mainly small, tubular, 2-lipped flowers, sometimes in the AXILS of coloured bracts: common thyme *(T. vulgaris)* is a greyish, hairy shrub to 200mm with oblong-lanceolate leaves and lilac flowers. It is much used as a culinary herb, as is lemon thyme *(T. X citriodora)*, a 100–300mm shrublet with GLABROUS leaves; variegated forms are grown as ornamentals; wild thyme *(T. arcticus* syn. *T. drucei)* is a vigorous prostrate plant with broadly-ovate to lanceolate leaves, ranging from glabrous to densely hairy, with rose to red-purple flowers.

THYRSE A PANICLE which is broadest in the middle and tapers to the base and apex.

THYRSIFLOR|US, -A, -UM Bearing flowers in a THYRSE.

TIMOTHY
see CAT'S TAIL GRASS

TOADFLAX *(Linaria)*. A genus of 150 annual and perennial plants in the *Scrophulariaceae* from the northern hemisphere, mainly the Mediterranean. They have LINEAR to OVATE leaves and flowers like those of a snapdragon (see ANTIRRHINUM), but with a basal spur: common toadflax *(L. vulgaris)* is a 300–600m perennial that spreads by ADVENTITIOUS shoots from the roots. It has GLAUCOUS, linear-lanceolate leaves and yellow and orange flowers 13–25mm long.

TOADSTOOL
see BASIDIOMYCETES; FUNGI.

TOBACCO [VIRGINIAN TOBACCO] *(Nicotiana tabacum)*. A 0.6–2.4m annual or biennial in the *Solanaceae* from tropical S. America. It has clammy, glandular, hairy, OVATE to LANCEOLATE basal leaves and terminal PANICLES of tubular, pinkish flowers with 5 pointed lobes. Tobacco is manufactured first by cutting and partially drying mature leaves, then bulking them together to allow slight fermentation. Turkish tobacco *(N. rustica)* is a smaller, yellow-flowered plant with a higher nicotine content which is grown in USSR and N. India.

TOMATO *(Lycopersicon esculentum)*. A variable annual in the *Solanaceae,* originally from Peru and Ecuador but much grown and naturalised elsewhere. It has prostrate and ascending stems to 1.8m long and PINNATE leaves with 7–9 major PINNAE, often lobed, and smaller pairs in between. The yellow, nodding 13–19mm wide flowers have 6 pointed lobes and a cone of anthers. The red, orange or yellow globular to flattened-ovoid fruits are technically berries and may have 2–9 LOCULI and a thick, fleshy core. There are many CULTIVARS including early maturing dwarf, bush types suitable for cool, temperate climates. *See also* CAPE GOOSEBERRY (Strawberry tomato); TREE TOMATO.

TOMENTUM [TOMENTOSE, TOMENTOS|US, -A, -UM] Having a dense covering of short, firm, matted hairs.

▲ **TOOTHWORT** *(Lathraea squamaria)*. A perennial total parasite in the *Orobanchaceae* from Europe and Asia. It lives on the roots of trees and shubs, particularly hazel and

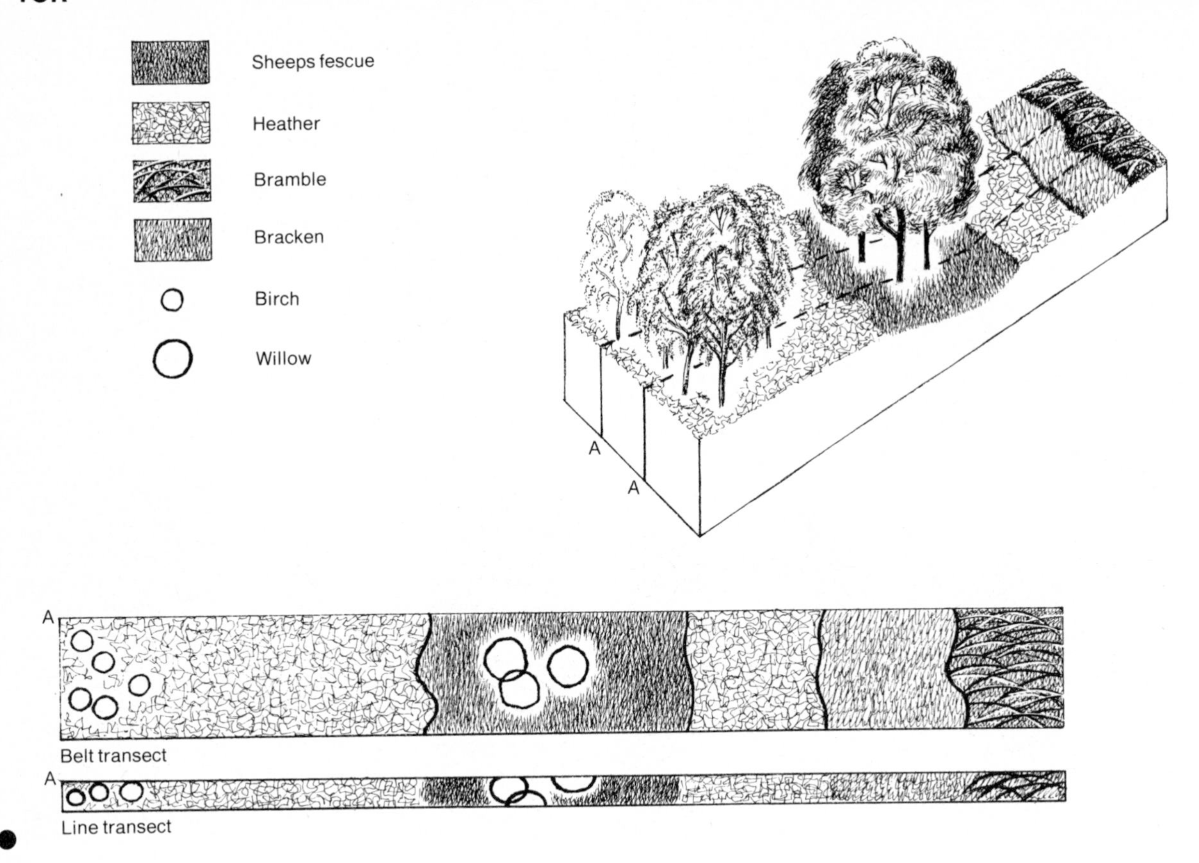

elm, having fleshy white RHIZOMES covered with scale leaves. The erect, 75–300mm aerial flowering stems are whitish or pinkish with a few scale leaves and a one-sided RACEME of tubular, dull purple-flushed flowers.

TORMENTIL *(Potentilla erecta)*. A tufted woody-rooted perennial in the *Rosaceae* from Europe, Asia and N. Africa. It has decumbent to erect slender stems 100–300mm long, TRIFOLIATE leaves with large, lobed STIPULES and 9mm wide, 4-petalled yellow flowers. A common plant of heaths and moors.

TOUCH-ME-NOT An alternative name for BALSAM.

TOWNHALL CLOCK
see MOSCHATEL

TRACHEIDS Elongated, cylindrical, water-conducting cells arranged end to end like a pipe. They are non-living cells, strengthened by bands or bars of LIGNIN.

● **TRANSECT** A method of sampling the frequency of species in a given environment. A straight line of tape or string is pegged out across the area and all the plants that touch are listed or represented diagrammatically on squared paper. The results are related to topography, soil and moisture. This is strictly a line transect. A belt transect is similar, but two lines are laid out, parallel to each other and all the plants between are recorded.

TRANSPIRATION The loss of water from a plant, particularly the leaves, in the form of vapour, equivalent to perspiration in animals. Water can evaporate from the entire aerial surfaces of a plant where the CUTICLE is not completely water-proof, but most is lost through the STOMATA or leaf pores. This continuous evaporation, particularly on hot or windy days, helps to cool the leaf surface and prevent heat damage to the delicate tissues. Transpiration also keeps the water flowing from roots to leaves (the transpiration stream) and indirectly assists in the absorption of mineral salts and their rapid movement up the plant. As water is lost from the outer leaf cells, suction pressure is set up, causing water to be drawn from adjacent cells. It is this pressure or pull in combination with OSMOSIS that gets water to the tops of tall trees.

TRAVELLER'S JOY [OLD MAN'S BEARD] *(Clematis vitalba)*. A woody, perennial, deciduous climber in the *Ranunculaceae* from Europe to the Caucasus and N. Africa. It has stems to 27.3m and PINNATE leaves, usually with 5 leaflets, the midrib acting as a tendril. The greenish-white flowers have 4-sepals and are borne in AXILLARY and terminal PANICLES. The ripe ACHENES have long, feather plumes formed from the STYLES.

TRAVELLER'S TREE *(Ravenala madagascariensis)*. An evergreen tree to 27m in the *Musaceae* from Malagasy. It has leathery, oblong-ovate leaves to 3m long, the overlapping, sheathing stalks of which hold drinkable water. The white flowers are carried in 150–200mm long boat-shaped SPATHES.

▲ **TREE** A woody plant 7.5m or more in height, usually with a single trunk at least 1.2m to the first branch. However, smaller plants with single trunks may also with some justification be called trees, and there are also dwarf trees

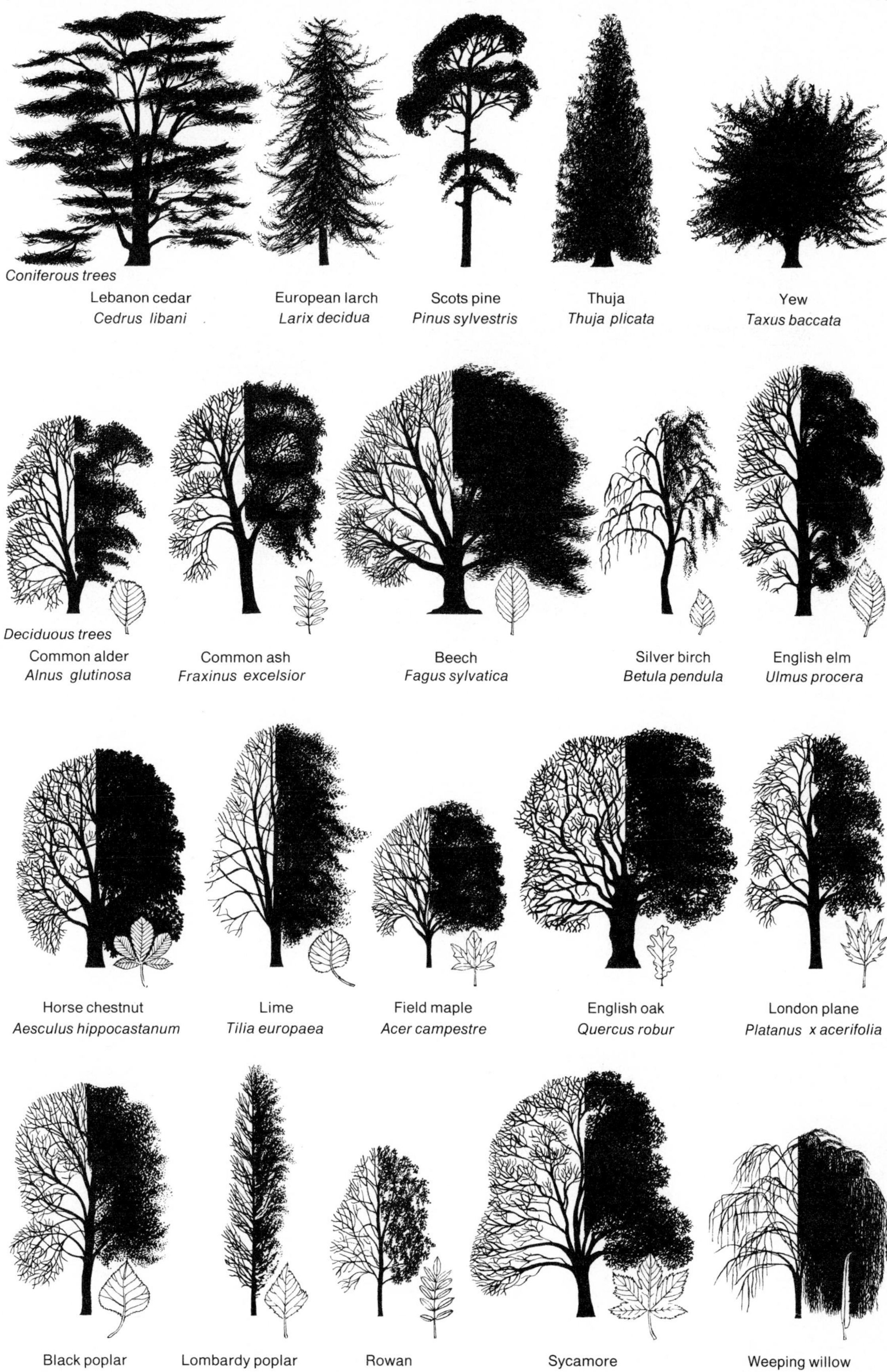
Coniferous trees
Lebanon cedar
Cedrus libani
European larch
Larix decidua
Scots pine
Pinus sylvestris
Thuja
Thuja plicata
Yew
Taxus baccata
Deciduous trees
Common alder
Alnus glutinosa
Common ash
Fraxinus excelsior
Beech
Fagus sylvatica
Silver birch
Betula pendula
English elm
Ulmus procera
Horse chestnut
Aesculus hippocastanum
Lime
Tilia europaea
Field maple
Acer campestre
English oak
Quercus robur
London plane
Platanus x acerifolia
Black poplar
Populus nigra
Lombardy poplar
Populus nigra 'Italica'
Rowan
Sorbus aucuparia
Sycamore
Acer pseudaplatanus
Weeping willow
Salix alba 'Tristis'

where form is more important that overall height. Trees are the largest and longest lived of all plants. The tallest broad-leaved tree is the Australian mountain ash *(Eucalyptus regnans)*, exceptional specimens of which exceed 105m. Generally, however, conifers far outstrip the broad-leaves and redwood holds the record at 111.7m. Bristle cone pine (another conifer) holds the longevity record, some specimens being estimated at almost 5,000 years. Despite the growing acreage of timber planted by man, their timber still represents one of the world's most important living, wild resources. Wood is still widely used in building construction, for furniture, and when pulped for paper and producing plastics.

TREE LINE
see LIMIT OF TREES

TREE OF HEAVEN *(Ailanthus altissima)*. A deciduous tree to 24.4m in the *Simaroubaceae* from China, but much grown elsewhere. It has 300–600mm long PINNATE leaves with 13–25, coarsely toothed leaflets and large terminal PANICLES of small, greenish flowers with 5 or 6 petals. The oblong, winged fruits are usually strongly red-flushed before they ripen.

TREE RINGS
see SECONDARY THICKENING

TREE TOMATO *(Cyphomandra betacea)*. A robust, rather fleshy-stemmed shrub or small tree to 4.5m in the *Solanaceae* from Peru. It has CORDATE-OVATE leaves to 300mm long, and pendulous PANICLES of whitish to greenish-pink fragrant flowers with 5-pointed lobes and a cone of yellow anthers. The egg-shaped red fruits are about 50mm long and have a flavour similar to the tomato.

TREFOIL
see BIRD'S FOOT TREFOIL; CLOVER

TRIANDROUS [TRIANDR|US, -A, -UM] Flowers which have 3 stamens.

TRIARCH
see DIARCH

TRICARPELLARY Flowers with 3 carpels.

TRICHOTOMOUS Branching or divided into threes.

TRICOLOR Having 3 colours or shades.

TRIDENS [TRIDENTAT|US, -A, -UM] Having 3 teeth, usually at the leaf apex.

TRIFID|US, -A, -UM With 3 deep lobes, most frequently used of leaves.

■ **TRIFOLIATE** WHORLS or groups of 3 leaves, or more commonly, leaves with 3 leaflets, the central one on a long stalk. Leaflets which have all 3 leaflets arising from the same point should correctly be called trifoliolate.

TRIGONOUS Three-angled, e.g. of beech nuts.

TRILOB|US, -A, -UM Three-lobed.

▲

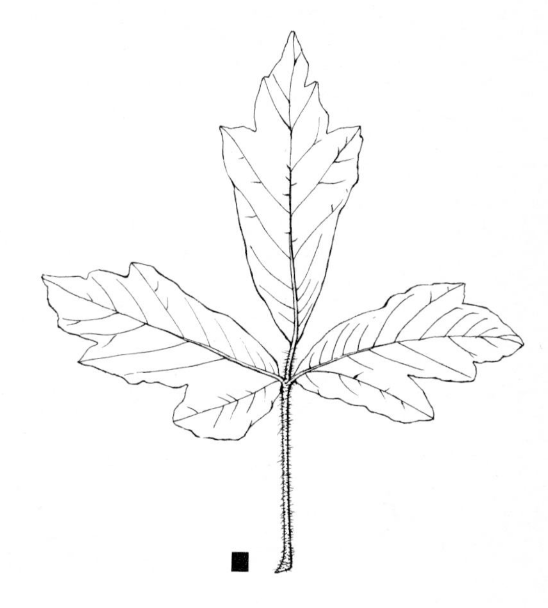

■

TRILOCULAR Three-celled, an ovary of 3, fused carpels.

TRIMEROUS Having the parts in WHORLS of 3, e.g. sepals and petals.

TRIMORPHIC Three-forms. For example, the three different kinds of flowers of purple loosestrife with short, medium and long styles.

TRIPLOID
see POLYPLOID

TRIPTER|US, -A, -UM Having 3 wings, e.g. of seeds or fruits.

TRIQUETR|US, -A, -UM Three-cornered, e.g. the stems of sedges.

TRIST|IS, -E Dull coloured, sad looking.

TRITERNATE Two sets of three; usually of leaves which divide into 3 stalks, each one bearing 3 leaflets, e.g. some of the leaves of ground elder and baneberry.

TRIVIAL|IS, -E Ordinary or commonplace.

TROPOPHYTE Plants of arid regions which are more or less dormant until the seasonal rains come, then grow vigorously, e.g. many bulbous-rooted and annual plants.

TRUFFLE Several species of *Tuber*, *Ascomycetes* fungi with underground fruiting bodies, and found in temperate regions. The French truffle *(T. melanosporum)* is best known, being an ingredient of paté de fois gras. It is irregular, brownish and tuber-like to 4in or more wide. Pigs are used to find them. English truffle *(T. aestivum)* is rounded, 25–75mm wide with fissured or warted skin.

TRUMPET LEAF
see PITCHER PLANTS *(Sarracenia)*

TRUNCAT|US, -A, -UM Ending abruptly or bluntly, as though broken off.

TRUNK The main stem of a tree.

TUBE The united, tube-like part of a GAMOSEPALOUS or GAMOPETALOUS flower, or the tube formed by stamen filaments, e.g. those of mallow, hibiscus and hollyhock.

TUBER Underground storage organs taking the form of swollen stems or roots. True tubers, e.g. of dahlia and lesser celandine, either do not bear buds at all, or only from the point where they are attached to the stem. True stem ▲ tubers, e.g. potato and Jerusalem artichoke, bear buds at intervals (the 'eyes' of potatoes) which produce aerial stems the following season.

TUBERCULAT|US, -A, -UM Bearing wart- or knob-like projections.

TUBIFLORAE A sub-family of the *Compositae* (daisy family), in which at least some of the florets in each flower head are tubular. There are two groups: the thistles, knapweeds, etc, with all tubular florets and the daisies with tubular disc florets and strap-shaped RAY florets. In the

▲

latter group, the disc florets are hermaphrodite, the ray florets female (gynomonoecious). *See also* LIGULIFLORAE.

● **TULIP** *(Tulipa)*. A genus of 100 bulbous perennials in the *Liliaceae* from Europe, Asia and N. Africa. They have erect stems, LINEAR to OVATE, sometimes channeled or waved leaves, and mainly solitary, 6-tepalled, showy flowers: some species have branched stems with 2–6 flowers. Tulips have been popular plants for centuries and hundreds of hybrids and cultivars are grown, varying greatly in shape and colour, from white and yellow through shades of pink and orange to red and maroon. These garden tulips have been classified into 14 divisions, some of the more familiar being the Dwarf Single, and Double Early Flowering, the tall late Darwins with deep, cup-shaped flowers, the Lily-flowered with tapered pointed petals and the Parrot with bizarrely crimped and fringed petals. Division 15 covers all wild species. *See also* MARIPOSA LILY (Californian tulip).

TULIP TREE *(Liriodendron tulipifera)*. A deciduous tree to 60m in the *Magnoliaceae* from eastern USA where it is also known as tulip or yellow poplar. The 100–150mm long broadly-OVATE, 4-lobed leaves are roughly saddle-shaped and the greenish-yellow, tulip-shaped flowers have 6 petals spotted at the base with orange. The fruit is a spike of narrow, flattened, wing-like fruitlets containing one or two seeds. It is an important timber tree in its homeland, producing American white wood.

TUMBLEWEEDS Certain bushy, annual plants from the prairies of the USA, the steppe country of Asia and other open, often semi-arid regions. When the plants have ripened their seeds, they break off at ground level and with the first strong wind bowl away, often travelling several miles, scattering seeds as they go. Saltwort (Russian thistle), several amaranths, and allies of hedge mustard are among the most spectacular of tumbleweeds forming globe-shaped plants 0.6–1.2m tall and across. On a miniature scale are the globular 50–75mm seed heads of the sand-binding grass *Spinifex hirsutus* which bounds along on the tips of its AWNS.

TUNDRA Floristic region. The extensive lowland areas which border the Arctic Ocean in Europe, N. America and Asia. The temperature remains below freezing for much of the year, except for 2–3 months in summer. There is permanently frozen soil deeper than 300–600mm below the surface, even in summer, and the vegetation is restricted to lichens, mosses, small shrubs and herbaceous plants. Typical plants are dwarf willows, dwarf birch, mountain avens, Labrador tea, Lapland rhododendron, alpine bearberry, clubmosses and sedges.

TURION (1) The dormant bud-like shoots of certain water plants, e.g. frogbit, which become detached and spend the winter at the bottom of a pond or lake.
(2) The usually scaly sucker shoots from below ground of such plants as asparagus, blackberry and raspberry.

TURMERIC *(Curcuma longa)*. A RHIZOMATOUS perennial in the *Zingiberaceae* from India but much cultivated in other parts of tropical Asia. It grows to 900mm or more with long stalked, oblong to elliptic leaves, 300–450mm long. The tubular flowers are yellow lipped and borne in spikes with white, pink-tipped bracts. The fleshy, tuber-

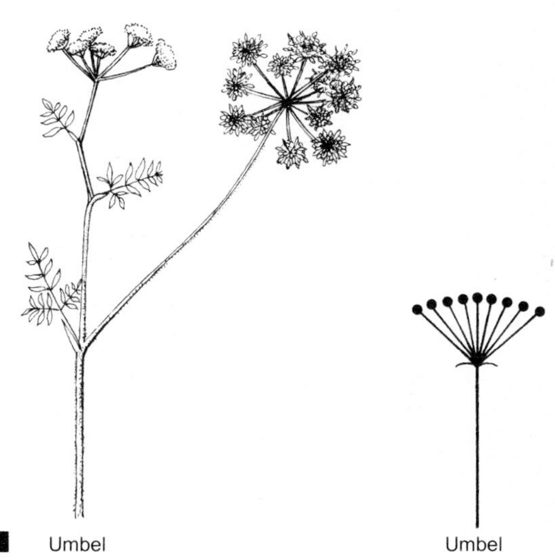

Umbel

Umbel

like RHIZOMES yield a yellow dye, and when dried and ground provide the turmeric used in curries.

TURNIP *(Brassica rapa)*. An annual or biennial with a swollen root in the *Cruciferae*, perhaps from central Europe, but not known truly wild and much cultivated throughout Europe and Asia since early times. It has LYRATE-PINNATIFID leaves and branched stems to 900mm bearing RACEMES of 4-petalled, bright yellow flowers. There are white and orange fleshed CULTIVARS both of which contain appreciable amounts of vitamin C, pectase (used to set jam), but no starch. The leaves are also high in vitamin C and are often eaten as a green vegetable. Wild turnip *(B.r.* ssp. *campestris)* is a similar plant but lacks the swollen roots. Some forms of it are grown as a substitute for RAPE for the oil-rich seeds. *See also* ARISAEMA (Indian turnip).

TURPENTINE
see RESIN; PINE.

▲ **TUTSAN** *(Hypericum androsaemum)*. An evergreen or semi-evergreen shrub to 600mm in the *Hypericaceae* from Europe from the Caucasus to Algeria. It has 50–100mm long, OVATE, slightly aromatic leaves and small CYMES of 5-petalled, 19mm wide yellow flowers. The ovoid fruits are berry-like, fleshy, turning first red then purple-black.

TWAYBLADE *(Listera ovata)*. A perennial plant in the *Orchidaceae* from Europe and Asia. It has an erect stem with two strongly nerved, broadly ovate leaves and a slender, 75–250mm RACEME of 13mm long, yellow-green flowers. Each blossom has 3 ovate sepals, 2 narrower spreading petals and a slender, forked LABELLUM.

TWIN FLOWER *(Linnaea borealis)*. A prostrate, evergreen perennial in the *Caprifoliaceae* from the cooler parts of the north temperate zone. Has pairs of broadly ovate to orbicular, 13mm long, CRENATE leaves and slender, erect stems bearing pairs of bell-shaped, 5-lobed pink flowers to 8mm long.

TWITCH GRASS
see COUCH GRASS

TYPE SPECIMEN When a plant new to science is found and described, that specimen is preserved (usually dried) and becomes the type specimen for reference.

UGLI FRUIT
see CITRUS

ULIGINOS|US, -A, -UM Growing in marshes or wet ground.

■ **UMBEL** An inflorescence of stalked flowers, all of which arise from the same point, e.g. cow parsley and other members of the *Umbelliferae*.

UMBO [UMBONAT|US, -A, -UM] Raised like the central boss on a shield.

UMBRACULIFORM|IS, -E Umbrella-shaped.

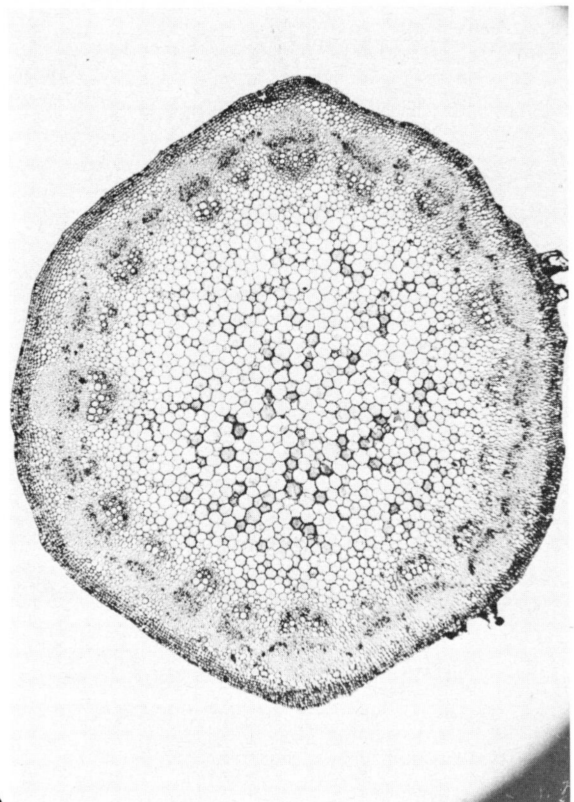

UMBROS|US, -A, -UM Growing in shady places.

UNCINAT|US, -A, -UM Hooked, as the spines of certain cacti.

UNDERSHRUB Another name for a SUBSHRUB.

UNDULAT|US, -A, -UM Wavy, e.g. the margins of the leaves.

UNGUICULAR|IS, -A, -UM Clawed, usually referring to the tapered base or stalk of a petal.

UNICELLULAR Plants of one cell only, e.g. many ALGAE, particularly *Pleurococcus* which forms the green powder on old fences and bark.

UNICORN PLANT *(Martynia louisiana)*. A robust, clammy-glandular hairy annual to 900mm in the *Martyniaceae* from southern USA. It has broadly OVATE-CORDATE, wavy margined leaves and short, terminal RACEMES of 50mm long, pale-purple bell-shaped flowers having white tubes variegated green, yellow and violet. The 75–150mm long fruits have long tapering beaks to half their lengths. When dry, they split open and the beak becomes two, curved horns.

UNIFLOR|US, -A, -UM One flower to each stem.

UNIFOLI|US, -A, -UM One-leaved, or apparently so.

UNILOCULAR One-celled, or an ovary of one carpel.

UNISEXUAL Having single-sexed flowers, though both male and female blooms may be borne on the same plant.

UPAS TREE *(Antiaris toxicaria)*. An evergreen tree in the *Moraceae* from Malaysia. It has oblong-obovate leaves and spherical heads of tiny, greenish flowers. The latex from the trunk provides a virulent arrow-poison. Owing to garbled and false information, received by the botanist Rumphius in the late 17th century, and his subsequent writings, this tree acquired an undeserved reputation. It was said to give off an effluvium lethal to all life, the land for miles around it being a virtual desert.

URBIC|US, -A, -UM Growing near towns.

URCEOLATE [URCEOLAT|US, -A, -UM] Pitcher or urn-shaped, i.e. globose flowers constricted at the mouth.

UREDOSPORE
see RUST FUNGI

URENS Stinging, e.g. of stinging nettle.

URNIGER|US, -A, -UM Urn-shaped, e.g. seed capsules.

USITATISSIM|US, -A, -UM Most useful.

UTIL|IS, -E Useful.

UTRICLE [UTRICULAT|US, -A, -UM] A small bladder; or bladder-like; e.g. the leafy traps of bladderwort or the inflated PERIGYNIA around the fruits of sedge.

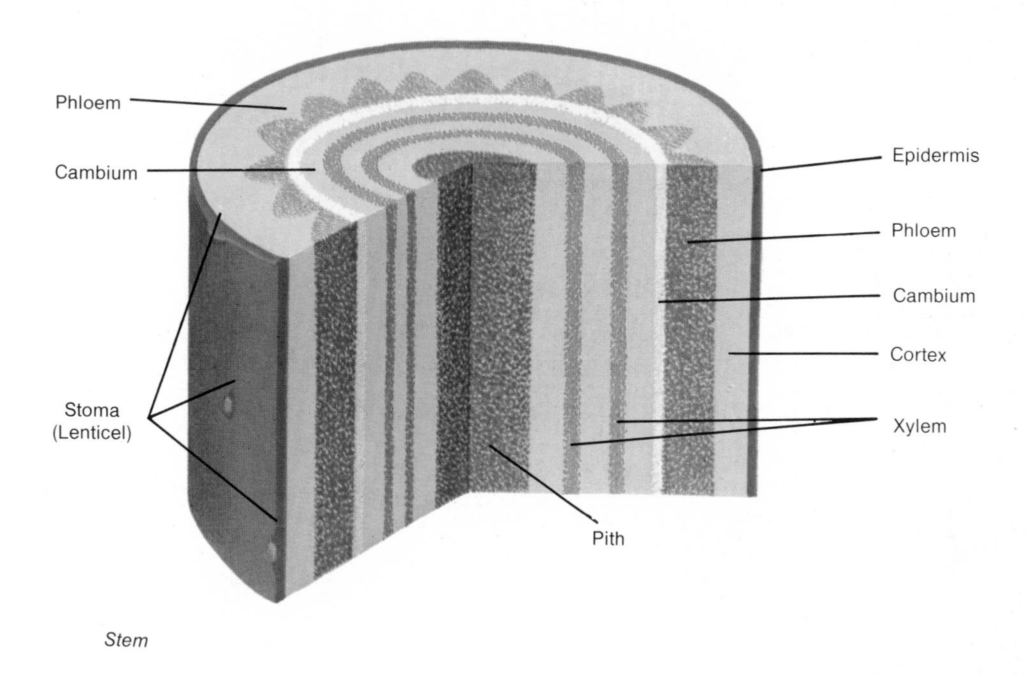

Stem

▲

UVA-URSI Literally 'bears' grapes', referring to fruits, as those of bearberry *(Arctostaphylos uva-ursi)*.

VACUOLE
see CELLS

VAGANS Wandering, of wide distribution.

VALERIAN Several species of *Valeriana,* perennial plants in the *Valerianaceae* from Europe and Asia. Common valerian grows to 1.2m with stalked, PINNATE leaves 100–200mm long and dense, terminal CYMES of 6mm long, tubular 5-lobed, pale pink flowers. *See also* RED VALERIAN.

VALVES The segments into which a fruit splits on ripening.

VANILLA *(Vanilla planifolia* syn. *V. fragrans)*. An evergreen climber in the *Orchidaceae* from central America, but cultivated in other tropical countries. It has fleshy, 125–175mm long oblong-ovate leaves and AXILLARY spikes of 50mm wide, green flowers with 5 narrow tepals and rolled LABELLUMS. The pods are semi-cylindrical, about 150mm long and need a slow drying process to prepare them for the market, where they are sold for the flavouring vanilla. The pods turn black during the drying process.

VARIABIL|IS, -E Variable in one or more characters.

VARIATION All plants are subject to variation, sometimes small, sometimes great. Generally this is due to MUTATION, soil, climate or disease, e.g. poor soil can cause stunting, rich soil excessively large plants or leaves. Strong winds can produce stunting or lop-sided growth. Diseases, particularly viruses, can cause severe crippling, or more subtle changes such as leaves with fewer lobes as with blackcurrant (reversion virus) or the curious striping or streaking of 'broken' tulips.

● **VARIEGATION** The white or yellow streaks, spots or blotches on leaves due to lack of CHLOROPHYLL in one or more layers of the leaf tissue. This condition has three major causes: mutation, virus, or a mineral deficiency that upsets the formation of chlorophyll. Variegated or golden privet is of mutation origin, spotted (Japanese) laurel is due to virus and chlorosis, while a patchy yellowing of the leaf tissue on normally green plants is often due to lack of manganese or magnesium.

VARNISH A mixture of natural resin (now also synthetic in origin) and a drying oil, usually linseed (FLAX) or spirit. Among plants which provide the resins are pines and other conifers, CANDLE NUT, and SUMAC.

▲ **VASCULAR BUNDLE [VASCULAR TISSUE]** In the young stem, groups of water conducting tissue linking the roots with the leaves are arranged in a circle just inside the epidermis. Each bundle consists of XYLEM vessels to the inside and PHLOEM to the outside, the two separated as collateral bundles. In some climbing plants, e.g. marrow and white bryony, there is a second inner group of phloem and the bundle is then said to be bi-collateral. Both xylem and phloem cells are thickened with LIGNIN and form also strengthening tissue. Phloem consist mainly of tubular cells (sieve tubes) end to end; the cross walls between them

▲

(sieve plates) being perforated to allow a freer flow of waters and sugars down from the leaves. Xylem consist of tracheids and vessels, non-living cells also arranged end to end, the vessels without any cross walls to allow a free flow of water and minerals up from the root. After secondary thickening, the xylem and phloem are arranged in continuous rings within the stem.

VAUCHERIA Thread-like, much-branched green algae mainly inhabitants of fresh water, but some marine. The threads are like continuous hollow tubes, technically known as COENOCYTES.

VEGETABLE HORSE HAIR Another name for SPANISH MOSS.

VEGETABLE MARROW
see CUCURBITS

VEGETABLE OYSTER
see SALSIFY

VEGETABLE SHEEP *(Raoulia eximia)*. A dense evergreen shrub in the *Compositae* from New Zealand. It has densely, white-woolly rosettes of small OBOVATE leaves crowded together in irregular hummocks which at a distance resemble a resting sheep. The tiny, sessile flower heads are reddish.

VEINS Strands of vascular tissue that form a network in leaves, bracts and petals. They not only convey water to the cells, but give strength and rigidity. Veins may be contained entirely within the leaf tissue or raised above the surface, usually on the lower side. In some cases, notably large-leaved water plants such as *Victoria* and giant Chinese water lily *(Euryale ferox)*, the veins are remarkably deep and girder-like and arranged with such mathematical precision that they have given inspiration to architects. *See* VENATION.

VELUTIN|US, -A, -UM Velvety; having a dense layer of fine, short, soft erect hairs.

● **VENATION** The arrangement or pattern of veins in a leaf. In simple leaves there is a central main vein with smaller, lateral branches. In PALMATE or DIGITATE leaves there is a main vein to each lobe or leaflet. In each case the lateral veins branch and branch again, forming an intricate network. Such leaves are described as net-veined and are typical of the DICOTYLEDONS. In many MONOCOTYLEDONS, a number of straight veins of equal size run the length of each leaf, smaller laterals linking them together. These are called parallel-veined leaves, e.g. orchids, lily of the valley, Solomon's seal. *See* VEINS.

VENENOS|US, -A, -UM Very poisonous.

VENOS|US, -A, -UM Having many or conspicuous veins.

VENTRICOS|US, -A, -UM Swollen or inflated, especially on one side, e.g. the tubular flowers of figwort and foxglove.

VENUS'S COMB
see SHEPHERD'S NEEDLE

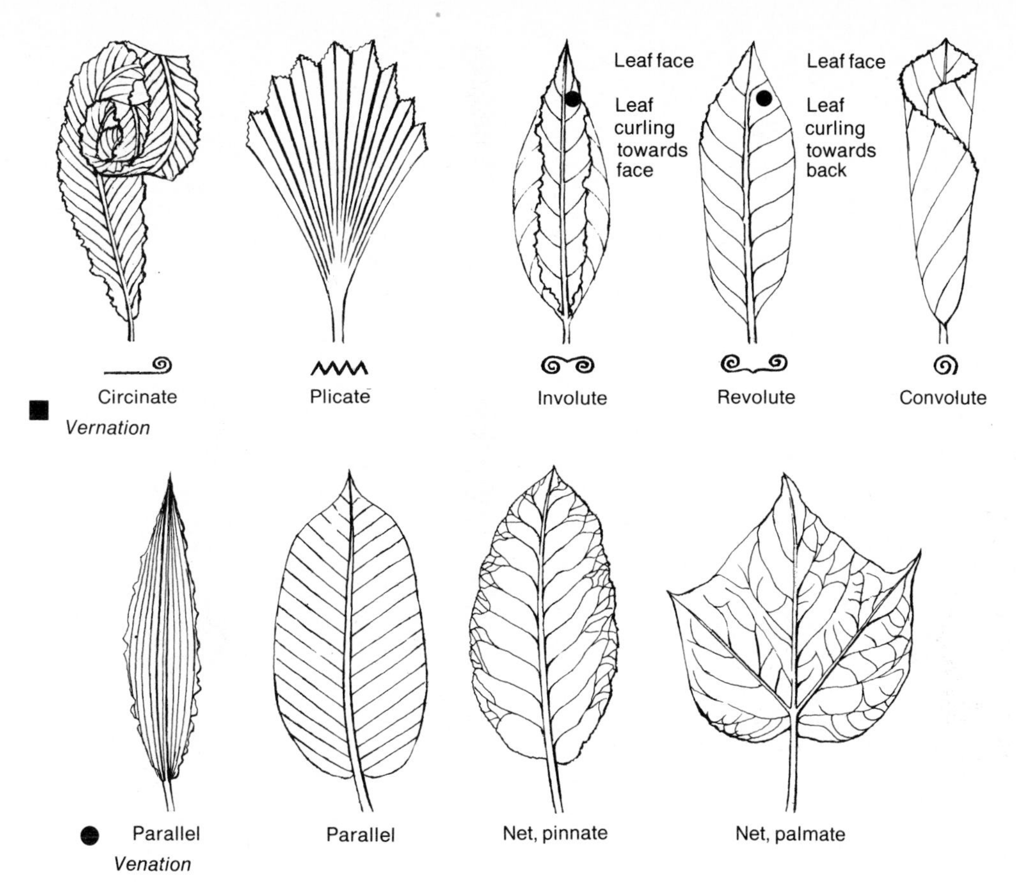

▲ **VENUS'S FLY TRAP** *(Dionaea muscipula)*. Prpbably the most remarkable of all the CARNIVOROUS PLANTS; a perennial in the *Droseraceae* from damp and mossy places in the Pine Barrens of the Carolinas, USA. It forms small erect rosettes of leaves with winged stalks and oblong blades, the margins of which bear inward-pointing, spine-like teeth. Each blade is poised like an open book, on each half of which are three, trigger hairs. When an insect alights on the leaf and touches these hairs, it snaps shut like a man-trap. Capture of prey stimulates small, red glands to secrete a digestive fluid. UMBEL-like clusters of 19mm wide, 5-petalled flowers are carried on stems to 150mm tall.

VENUS'S LOOKING-GLASS *(Legousia hybrida* syn. *Specularia hybrida)*. An erect, hispidly hairy annual to 300mm in the *Campanulaceae* from Europe, N. Africa and Asia. It has 9–25mm long oblong, sessile, undulate leaves and slender, tubular, red-purple flowers having 5 lobes.

VERBENA A genus of 250 annual, perennial and shrubby plants mostly from N. and S. America, a few in Europe and Asia. They have LINEAR to OVATE leaves in pairs, or rarely in WHORLS, and small tubular, 5-lobed flowers, sometimes highly coloured and fragrant. Several species and hybrids are grown as ornamentals notably the CULTIVARS of *V. X hybrida* (florists' or common garden verbena). Derived from several N. American species it has CORYMBS of blue, purple, crimson, pink and scarlet flowers on decumbent stems to 300mm or more long. Vervain *(V. officinalis)* is an erect, 300–600mm perennial from Europe, Asia and N. Africa. It has OBLANCEOLATE leaves and branched spikes of 3mm wide, pale lilac flowers. *See also* LEMON-SCENTED VERBENA.

VERECUND|US, -A, -UM Of modest appearance.

VERNAL|IS, -E Appearing in the spring.

■ **VERNATION** The way young leaves are folded in the bud. They may be circinate (rolled from top to bottom as in ferns, plicate (pleated longitudinally as in palms), revolute (rolled from the sides to the middle of the lower surface), involute (rolled from the sides to the middle of the upper surface), convolute (rolled from the sides, but one side of the leaf wrapped around the other), or conduplicate (folded forwards in a U-shape, each leaf clasping the next youngest and all those within).

VERNICOS|US, -A, -UM Varnished.

VERN|US, -A, -UM Of the spring.

VERRUCOS|US, -A, -UM Covered with wart-like outgrowths.

VERSATILE Used of freely pivoting anthers which are joined to the filaments only by their middles.

VERTICIL A WHORL or ring of flowers or leaves.

VERTICILLASTER A false WHORL, typical of the head-nettle family *(Labiatae)*, formed by two, almost SESSILE, opposite CYMES from the upper leaf or bract AXILS.

VERVAIN
see VERBENA

● 

■ 

VESC|US, -A, -UM Small, thin or weak, or alternatively edible or to eat.

VESICARI|US, -A, -UM Inflated or bladder-like.

VESSEL
see VASCULAR BUNDLE

VESTAL|IS, -E White

VESTIT|US, -A, -UM Clothed or covered, e.g. with hairs.

● **VETCH [TARE]** Several species of *Vicia*, a genus of 150 annuals and perennials in the *Leguminosae* from the northern hemisphere and S. America, some naturalised elsewhere. They have PINNATE leaves with terminal tendrils and either AXILLARY clusters of a few, almost stalkless flowers or long-stalked RACEMES. Individual blossoms are pea-shaped followed by explosive legumes (pods): common vetch *(Vicia sativa)* is a 30–120cm annual with 8–16 LINEAR to OBOVATE leaflets and AXILLARY clusters of 1–4, purple flowers; tufted vetch *(V. cracca)* is a 0.4–1.8m perennial with 12–24, linear leaflets and stalked racemes of small, purple-blue flowers. *See also* BROAD BEAN; MILK VETCH.

VEXILLUM The large upper or standard petal of a pea-flower.

VIABILITY The period during which a seed will remain capable of germination and grow into a plant. This may be a few days in the case of a poplar or up to 100 years in some legumes. *See also* DORMANT.

▲ **VICTORIA** A genus of 2–3 giant water lilies in the *Euryalaceae* from S. America. Victoria water lily *(V. amazonica* syn. *V. regina)* is a prickly plant growing in 1.2–1.8m of water with circular leaves up to 1.8m wide, each with a deep, raised margin. The large prickly buds open to 300mm wide, white to cream, many-petalled blooms which age to purple.

VICTORIAL|IS, -E Victorious; used for example for *Allium victorialis* (a wild onion), the bulbs of which were worn by Bohemian miners to ward off evil spirits.

VILLOSE [VILLOS|US, -A, -UM Shaggy, covered with soft, long hairs.

VINE A general term, especially in the USA for any climbing plant. *See also* GRAPE VINE.

■ **VIOLA** A genus of 500 species of diverse form found throughout the world. They are usually tufted or clump-forming, but may be shrubby or semi-climbing. In the Andes, many species have a neat rosette form and closely resemble houseleeks. The leaves are often OVATE-CORDATE, but can be OBLONG to SPATHULATE and variously dissected, from PINNATE to PEDATE. However varied the plants, the flowers have a strong resemblance to each other. They are ZYGOMORPHIC, with a pair of more or less erect petals, 2 spreading lateral ones and a broad, lower, spurred petal which curves forward. The garden violas are of hybrid origin, largely derived from the horned violet *(V. cornuta)*. This has ovate, CRENATE leaves with large, deeply cut STIPULES and 32mm long, slender spurred, violet-mauve flowers.

▲

VIOLACE|US, -A, -UM The colour of violets.

VIOLET Several species of *Viola*. The best known is sweet violet *(V. odorata)* with long-stalked, ovate-cordate, hairy leaves and sweetly fragrant 19mm long violet-purple flowers. Common dog violet *(V. riviniana)* forms mats of slender stems and bears smaller, blue-purple flowers.

VIPER'S BUGLOSS Two species of *Echium*, erect, HISPID hairy biennials in the *Boraginaceae* from Europe to Turkey, naturalised elsewhere: viper's bugloss itself *(E. vulgare)* grows 300–900mm with lanceolate to oblong leaves. The stems ones SESSILE, and narrowly pyramidal PANICLES, the branches of which are CYMES. The funnel-shaped, 16mm long flowers are pinkish in bud, opening bright blue; purple viper's bugloss *(E. lycopsis)* is similar but the flowers are red, becoming purple-blue, borne in longer, stalked CYMES.

VIPER'S GRASS *(Scorzonera hispanica)* A perennial plant in the *Compositae* from S. Europe. It has carrot-shaped black-brown roots with white flesh and a rosette of LANCEOLATE leaves. The yellow, dandelion-like flower heads are borne on branched stems to 900mm. The plant is grown for its edible roots.

VIRENS Green.

VIRGAT|US, -A, -UM Twiggy.

VIRGINALIS, -E [VIRGINE|US, -A, -UM] White, virginal.

VIRGINIA CREEPER *(Parthenocissus quinquefolia)*. A deciduous climber to 20m in the *Vitidaceae* from east N. America. It has PALMATE leaves with 5 OBOVATE, slender-pointed leaflets which turn bright red in autumn. The tendrils bear small suckers which cling readily to trees and walls. The tiny, 5-petalled, green flowers give way to small blue-black grapes. *P. tricuspidaria* syn. *Ampelopsis veitchii*, from China and Japan, has OVATE, 3-lobed leaves and is also sometimes called Virginian creeper.

VIRGINIAN STOCK *(Malcolmia maritima)*. A slender maritime annual in the *Compositae* from (despite its vernacular name) S. Europe. It has 100–300mm stems, elliptic to lanceolate leaves and RACEMES of rose-purple, 4-petalled flowers. Lilac, pink, red and white forms are grown in gardens.

VIRGIN'S BOWER Several kinds of *Clematis*, including *C. viticella* and its hybrid *C. X jackmannii*.

VIRID|US, -A, -UM Green.

VIRUS Extremely minute, disease-causing organism which will pass through a filter designed to hold bacteria. They are parasitic and once inside a host plant can quickly duplicate themselves. Their effect on plants is various, from mild mottling or streaking of leaves or petals to severe deformation and eventual death. Many viruses are spread by sucking insects, e.g. greenfly or aphids, and some can be spread merely by handling the leaves.

VISCOS|US, -A, -UM Sticky from the exudations of glands.

●

▲

VITIFOLI|US, -A, -UM Having leaves like the grape vine.

VITTAT|US, -A, -UM Striped longitudinally.

VIVIPARY [VIVIPAR|US, -A, -UM] Producing young plants or bulbils in place of or with flowers and seeds or spores, e.g. spider plant.

VOLUBIL|IS, -E With twining stems.

VULGAR|IS, -E [VULGAT|US, -A, -UM] Common.

WAKE ROBIN An alternative name used for both WOOD LILY *(Trillium grandiflorum)* and LORDS AND LADIES.

WALLFLOWER *(Cheiranthus cheiri)*. A sub-shrubby perennial in the *Cruciferae* from the E. Mediterranean, but much cultivated and naturalised elsewhere. It grows 300–600mm or more tall, with oblong-lanceolate leaves and terminal RACEMES of orange-yellow, 4-petalled fragrant flowers. There are several CULTIVARS in shades of yellow, ivory, purple, red and mahogany. Siberian wallflower *(Erysimum allionii)* is similar, but more compact and with brilliant orange flowers.

WALL PEPPER
see STONECROP *(Sedum acre)*

WALL RUE *(Asplenium ruta-muraria)*. A small, tufted, evergreen fern in the *Aspleniaceae* from the north temperate zone. Has bipinnate leaves 32–125mm high, each of the 6–10 PINNAE divided into 3–5 pinnules. The tiny SORI are LINEAR, borne on the backs of the pinnules.

● **WALNUT** Several species of *Juglans*, a genus of 15 deciduous trees in the *Juglandaceae* from the north temperate zone, with pinnate leaves and MONOECIOUS flowers, the males in catkins, the females solitary or in small clusters. Common walnut *(J. regia)* grows to 27m or more with 7–9, obovate to elliptic leaflets 56–125mm long. The green, ovoid fruits are plum-like to 50mm long, the 'stone' being the familiar walnut. There are several cultivars.

WANDFLOWER Several species of *Dierama*, notably *D. pulcherrimum*, an evergreen perennial in the *Iridaceae* from S. Africa. It has dense tufts of slender, LINEAR leaves to 600mm long and gracefully arching, wiry stems to 1.8m bearing pendulous PANICLES or funnel-shaped rose-purple to red flowers with 6 tepals.

WARD KINGDON F. *See* KINGDON WARD, F.

▲ **WART DISEASE** *(Synchytrium endobioticum)*. An *Archimycetes* fungus parasitic on potatoes, forming warty or cauliflower-like outgrowths on the tubers. Once the disease is present its spores contaminate the soil for 10–12 years. It is a notifiable disease and must be reported to the appropriate authorities. There are many CULTIVARS of potato immune to the fungus, including the well-known 'Majestic', 'Golden Wonder' and 'Arran Pilot'.

WATER CHESTNUT [WATER CALTROPS] *(Trapa natans)*. A floating aquatic in the *Trapaceae* from Europe to the Middle East. It forms rosettes of RHOMBOID leaves, the long stalks of which have ellipsoid air bladders. The small, white, 4-petalled flowers give way to roughly triangular woody fruits with 2–4 spine-tipped horns.

WATERCRESS *(Rorippa nasturtium- aquaticum* syn. *Nasturtium aquaticum)*. An aquatic perennial in the *Cruciferae* from Europe, Asia and N. Africa, naturalised elsewhere and sometimes a serious river weed as in New Zealand. It has LYRATE-PINNATIFID leaves on somewhat fleshy stems to 600mm long. The small white, 4-petalled flowers are carried in erect RACEMES.

WATER DROPWORT Several species of *Oenanthe*, aquatic or wet ground perennials in the *Umbelliferae* from Europe and Asia. They are erect plants with PINNATE or bipinnate, sometimes tripinnate leaves, the lower ones occasionally submerged. The tiny white flowers have 5, notched petals and are carried in UMBELS: common water dropwort *(O. fistulosa)* has 300–600mm hollow stems with long-stalked, pinnate leaves, the upper ones having LINEAR leaflets, and small, dense topped, flat umbels which become spherical in seed; fine-leaved water dropwort *(O. aquatica)* is a robust, well branched STOLONIFEROUS plant to 1.5m with tripinnate leaves having LANCEOLATE to OVATE leaflets.

WATER FERN Two plants bear this name. (1) *(Azolla filiculoides)*. A small free-floating aquatic plant in the *Azollaceae* from the Americas, naturalised in Europe. Individual plants are moss-like, densely branched, 13–50mm across with a few trailing roots. They quickly form extensive colonies covering large areas of water. The ovate leaves are minute and overlapping, turning red in autumn. The even smaller SORI are borne in pairs. (2) *(Ceratopteris)*. Tufted, submerged and floating ferns in the *Parkeriaceae* from the tropics and subtropics. They have edible bi- or tripinnate fronds, in some cases with inflated PINNAE, and solitary sori. *C. thalictroides* is often used in aquaria and has bi- or tripinnate fronds with LANCEOLATE to LINEAR pinnae.

WATER HAWTHORN
see CAPE PONDWEED

WATER HYACINTH *(Eichhornia speciosa* syn. *E. crassipes)*. A free floating aquatic in the *Pontederiaceae* from S. America and naturalised elsewhere. It forms rosettes of fleshy, orbicular leaves attached to spongy, inflated leaf stalks. The purple-tinted, feathery roots trail in the water, sometimes rooting in the mud. The widely funnel-shaped, 38mm violet-purple flowers are borne in short, dense RACEMES. Water hyacinth can grow and multiply at great speed and frequently becomes a major pest, clogging rivers, lakes and reservoirs.

WATER LEMON
see PASSION FLOWER

WATER LETTUCE *(Pistia stratioites)*. A free-floating aquatic in the *Araceae* from the tropics and sub-tropics. It forms neat rosettes of pale green, OBOVATE, blunt-ended leaves which resemble half grown lettuces, with feathery, trailing roots. Small greenish SPATHES each bear one female and a WHORL of male flowers, all petal-less.

WATER LILY (1) *(Nymphaea)*. A genus of 50 RHIZOMATOUS or tuberous-rooted aquatics in the *Nymphaceae* from temperate and tropical regions. They have orbicular to broadly ovate, cordate-based floating leaves and bowl-shaped flowers, usually of many petals which either float or

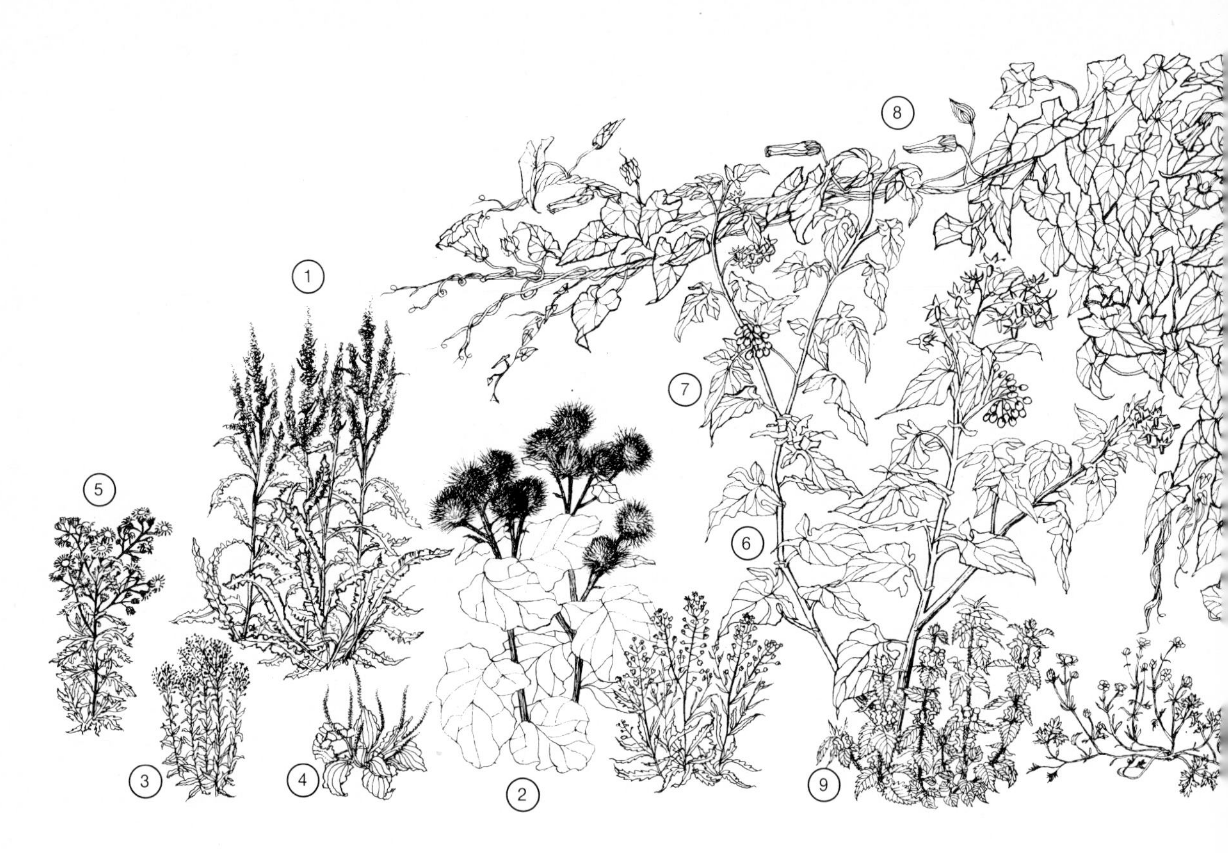

stand just above the water: white water lily *(N. alba)* from Europe has 100–200mm white, rarely pink or red, floating flowers and is a primary parent of the hybrid cultivars grown in temperate countries; the true Egyptian lotus *(N. lotus)* has large PELTATE leaves and 125–200mm white or pink-tinted flowers. Cape water lily *(N. capensis)* from S. Africa and Malagasy has blue flowers which stand just above the water.

◀ ○ (2) *(Nuphar)*. A closely related, rhizomatous genus of 25 species from the north temperate zone. They have globular flowers of 5, broad, petal-like PERIANTH segments and numerous STAMINODES. The ovate to oblong-cordate leaves are sometimes submerged: yellow water lily or brandy bottle *(N. lutea)* has 38–63mm wide yellow flowers which stand above the water and bottle-like fruits.

WATER MILFOIL Several species of *Myriophyllum,* a genus of 45 perennials in the *Haloragaceae*, of cosmopolitan distribution; mainly aquatic but some bog plants (e.g. 'soft herb' in New Zealand). They have WHORLS of PINNATE leaves, the leaflets often LINEAR to thread-like, and tiny, greenish or reddish flowers either petal-less or with 2–4 petals. Several of the tropical species are popular aquarium plants especially *M. proserpinacoides* with slender stems to 600mm and whorls of 5 finely dissected leaves. Common or spiked water milfoil *(M. spicatum)* from the north temperate zone grows 0.6–1.8m or more in deeper water, with leaves in whorls of 4 and tiny, dull red flowers.

WATER PENNYWORT
see MARSH PENNYWORT

WATER PEPPER *(Polygonum hydropiper)*. An erect, 300–600mm annual in the *Polygonaceae* from the north temperate zone, often in ponds and ditches. It has LANCEOLATE, wavy leaves to 100mm long and slender, nodding RACEMES of tiny, greenish flowers having a 6-lobed PERIANTH.

WATER PLANTAIN *(Alisma plantago-aquatica)*. An erect aquatic or wet ground perennial in the *Alismataceae* from the north temperate zone and Australia. Has rosettes of long-stalked, OVATE leaves and stems to 105cm carrying many WHORLS of 3-petalled, pale lilac flowers about 13mm wide. Narrow leaved water plantain *(A. lanceolatum)* is similar, but with lanceolate leaves and pink flowers.

WATER PORES
see CHALK GLANDS

WATER SOLDIER *(Stratiotes aloides)*. A submerged aquatic in the *Hydrocharitaceae* from Europe and N. W. Asia. It has agave-like rosettes of narrow, tapered, 150–500mm long spiny, SERRATE leaves and 25–38mm wide 3-petalled, white flowers just above the water. The plant is normally submerged, but rises to the surface at flowering time. Only female plants are found in Great Britain.

WATER STARWORT Several species of *Callitriche*, annual or perennial, generally aquatic plants in the *Callitrichaceae,* of cosmopolitan distribution. The form of the plants depends much on the depth of water and the species are not easy to identify. In general they are slender plants with pairs of linear or narrowly SPATHULATE leaves sometimes bunched or in rosettes at the stem tips when they

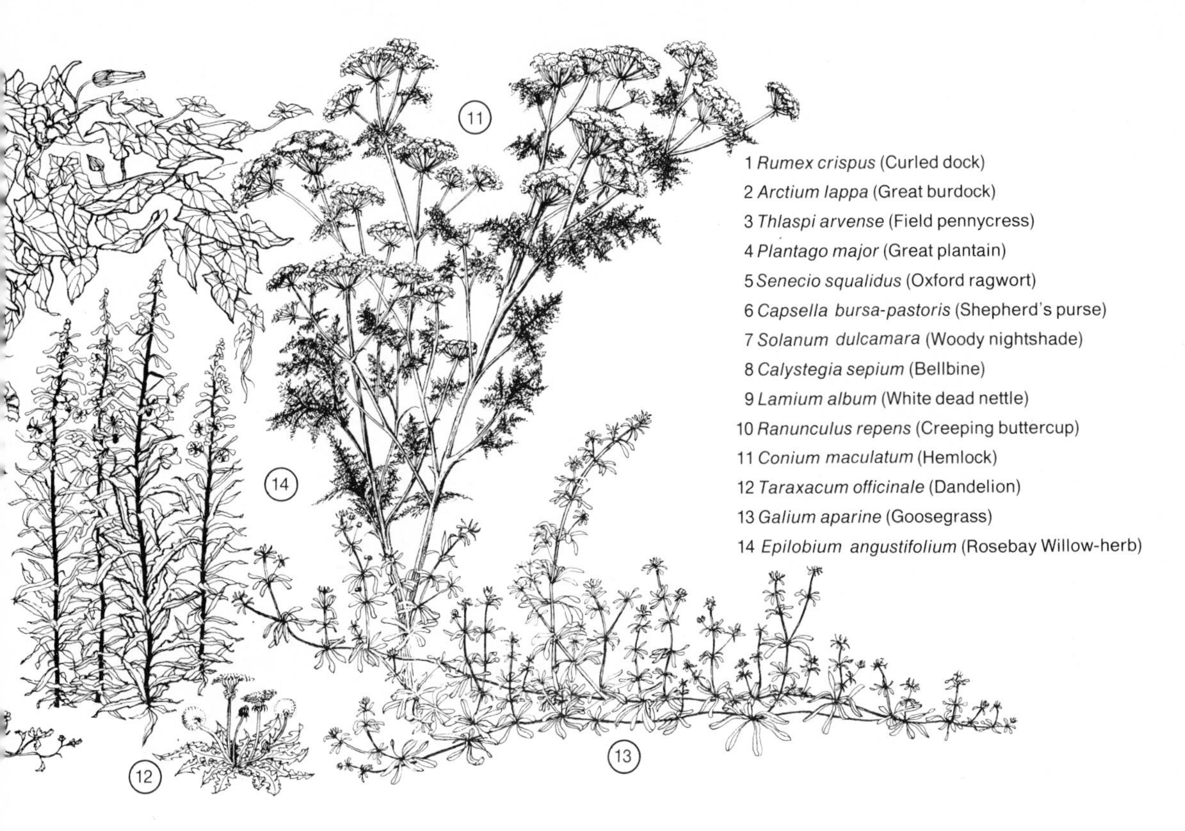

reach the water surface. The minute, green, AXILLARY flowers are MONOECIOUS, the males with one stamen only. Common starwort *(C. stagnalis)* forms loose mats on wet mud or grows to 600mm in water, the lower leaves are spathulate, the upper ones (which often form floating rosettes) are broadly elliptic.

WATER THYME
see CANADIAN WATERWEED

WATER VIOLET *(Hottonia palustris)*. An evergreen, aquatic perennial in the *Primulaceae* from Europe and Asia. It floats just beneath the surface with WHORLS of PINNATE or bipinnate leaves 50–100mm long, the leaflets LINEAR. The erect, leafless aerial stems can exceed 300mm with several whorls 19mm wide, lilac, yellow-throated, 5-lobed flowers.

WATERWEED A general term for many submerged aquatics but often applied to Canadian pondweed.

WATTLE A vernacular name for many of the Australian *Acacia* species.

WAX FLOWER *(Hoya carnosa)*. An evergreen climber to 6m or more in the *Asclepiadaceae* from Australia. It has firm, fleshy, elliptic-oblong, 50–100mm leaves and pendent, axillary UMBEL-like heads of fragrant, waxy-looking flowers. Individual blooms are like white, fleshy stars with 5, blunt-tipped rays and a red centre.

WAX GOURD
see CUCURBITS

WAX TREE
see SUMAC *(Rhus succedana)*.

WAYFARING TREE *(Viburnum lantana)*. A deciduous shrub to 3.6m or more in the *Caprifoliaceae* from Europe, invariably on chalky or limy soil. It has 50–100mm long OVATE to OBOVATE leaves, somewhat rugose above and hairy beneath. The small, funnel-shaped, cream-white flowers are densely arranged in umbel-like compound CYMES. The oval, flattened fruits are DRUPES changing from green to red and then to black-purple.

WEATHER PLANT
see CRAB'S EYE VINE

● **WEEDS** As a general definition, a weed is a plant which is growing where it is not wanted. In a stricter sense it is a plant of no value, competing for water, food and light with purposely cultivated crops or ornamental plants. Often it has creeping RHIZOMES or roots, such as couch grass, ground elder or bindweed which are very difficult to remove. Alternatively it may be an abundant seeder such as chickweed, groundsel or sow-thistle.

WELD [DYER'S ROCKET] *(Reseda luteola)*. An erect biennial to 1.5m in the *Resedaceae* from Europe, W. Asia and N. Africa, and introduced into N. America, it has a rosette of undulate, narrowly OBLANCEOLATE leaves and a ribbed often unbranched stem bearing a long tapering RACEME of greenish-yellow flowers with 4 petals 3 of which are deeply lobed. Formerly much grown for its yellow dye.

WELSH ONION *(Allium fistulosum)*. A clump-forming, bulbous plant in the *Alliaceae* from Siberia but much grown elsewhere. It has narrow, oblong bulbs and tubular leaves to 300mm or more, rounded in cross-section. The greenish-white, 6-tepalled flowers are borne in globose UMBELS. The name comes from the German *welsch,* meaning 'foreign'. Very similar to and often confused with it is the ever-ready or everlasting onion *(A. cepa perutile)*. This is easily distinguished by its leaves, semi-circular in cross-section, and by the rareness with which it flowers.

WELSH POPPY
see MECONOPSIS

● **WELWITSCHIA** *(W. bainesii,* syn. *W. mirabilis)*. A curious woody-based, perennial *Gymnosperm* in the *Welwitschiaceae* from S. W. Africa. It comes from very arid regions and when mature, forms a stout trunk almost buried in the sand. The top of this trunk is bilobed and bears 2 leathery, strap-like leaves which last for the life span of the plant (up to 100 years). They grow from the base only, the tip gradually wearing away. Winds often split the leaves into thong-like segments. The MONOECIOUS flowers ▲ are in red, cone-like spikes, followed by winged seeds.

WEYMOUTH PINE
see PINE

WHEAT Several species of *Triticum*, important annual grain-bearing grasses in the *Gramineae* probably from S. W. Asia, but long cultivated and not known in the wild. They are erect, tufted plants to 900mm with broad, flat, LINEAR leaves and dense, cylindrical spikes (ears) of 2 to many flowered spikelets with or without AWNS. The familiar ovoid seeds (grain) are deeply grooved on one surface: bread wheat *(T. aestivum* syn. *T. vulgare)* is the most widely grown and has hundreds of CULTIVARS. Some are sown in the autumn and harvested the following summer (winter wheats); others are sown in spring and harvested in late summer (spring wheat). The grain they yield is the source of high quality flour for bread-making, biscuits and pastry. Like many of the food plants long used by man, bread wheat is probably of complex hybrid origin: durum wheat *(T. durum)* is similar in appearance but usually awned and with long, hard grains that contain much gluten, a substance in flour which becomes sticky and elastic when wetted. It is used for making all kinds of pasta, particularly macaroni and spaghetti; rivet, cone or English wheat *(T. turgidum)* is a tall, vigorous awned species, once much grown in S. England, but producing poor quality bread flour now superseded by *T. aestivum.* It is still grown in some parts of Europe as a stock feed; the same comment holds for emmer wheat *(T. dicoccum)*, a small eared, narrow-grained species important in early times and the main ancestor of modern wheats.

WHIN An alternative name for GORSE.

WHITEBEAM *(Sorbus aria)*. A deciduous tree to 15m, sometimes more, in the *Rosaceae* from Europe. It has 50–113mm long OVATE, prominently veined leaves with double CRENATE-SERRATE margins and strikingly white-hairy undersides. The small, 5-petalled cream flowers are borne in broad compound CORYMBS and followed by ovoid scarlet fruits to 16mm long.

▲

▲

■

WHITE BRYONY *(Bryonia dioica)*. A herbaceous, perennial climber in the *Cucurbitaceae* from Europe, W. Asia and N. Africa. It has a thick, tuberous rootstock and stems to 6m or more, with PALMATE, 5-lobed leaves and twining tendrils. The pale green flowers are DIOECIOUS with 5 petals, the males in short CYMES the females in small, UMBEL-like clusters. The fruit is a red berry to 8mm wide.

WHITE FORK MOSS *(Leucobryum glaucum)*. A striking, silvery-green, cushion-forming moss of temperate woodland and wet moorland. Large specimens may be about 50mm high and 300mm across, formed of crowded, erect stems bearing tiny, LINEAR leaves.

WHITLOW GRASS Species of *Draba* and *Erophila*, genera of annual and perennial plants in the *Cruciferae* from the northern hemisphere and S. America. They are small rosette- or cushion-forming plants with short RACEMES of white or yellow 4-petalled flowers: common whitlow grass *(E. verna)* is a variable winter annual with 13mm long SPATHULATE leaves in rosettes and tiny white flowers on 25–75mm stems; yellow whitlow grass *(Draba aizoides)* forms small hummocks of rosettes with rigid linear leaves and 50–100mm stems bearing bright yellow, 8mm wide blossoms.

WHORL Leaves, bracts or flowers arranged in a ring of three or more from the same point on a stem.

WHORTLEBERRY
see BILBERRY

■ **WILLOW** *(Salix)*. A genus of 500 species of deciduous trees, shrubs and sub-shrubs in the *Salicaceae* mainly from the north temperate zone, a few in S. America. They have simple, LINEAR to OVATE leaves – many LANCEOLATE – and DIOECIOUS flowers in stiff catkins, either with or before the young leaves. The small, individual flowers are petal-less, borne in the AXILS of silky-haired bracts. The tiny seeds have a plume of hairs for wind distribution. Young stems of some willows are used for basket-making (osiers) and the wood of others for cricket bats. Several of the water-loving trees have been planted to prevent erosion of river banks, and many species and hybrids are grown as ornamentals: bay willow *(S. pentandra)* is a shrub or tree to 7.5m with varnished brown twigs and 50–100mm long, dark glossy green ovate to lanceolate leaves; crack willow *(S. fragilis)* is a tree to 15m with olive twigs easily broken at their junction with the stems and 63–150mm long lanceolate, long-pointed leaves; dwarf or least willow *(S. herbacea)* forms mats of underground stems and erect shoots to 50mm with broadly ovate or obovate leaves 6–19mm long; goat willow or great sallow *(S. caprea)* is a large shrub or small tree to 9.1m with stout twigs, 50–100mm long oval-oblong to obovate leaves and ovoid, silky-hairy catkins known as pussy willow; weeping willow covers several different willows with pendulous branches, the commonest in Britain being the white willow hybrid, *S. X chrysocoma* with yellow twigs and lanceolate leaves; white willow *(S. alba)* grows to 24.4m and has 50–100mm lanceolate leaves covered with silky white hairs; cricket-bat willow *(S. alba coerulea)* has bluish green leaves, somewhat glaucous beneath; common osier *(S. viminalis)* is a shrub or tree to 9.1m, often pollarded to produce long vigorous stems for basket-making.

● **WILLOW-HERB** *(Epilobium)*. A genus of 215, mainly perennial plants in the *Onagraceae* from the temperate regions of the world. They vary from prostrate to erect, usually with simple leaves in pairs and HYPOGYNOUS flowers, the 4 sepals and 4 petals on the top of slender, semi-cylindrical ovaries. The minute seeds have long hairs for wind dispersal: rose-bay willow-herb or fireweed *(E. angustifolium* syn. *Chamaenerion angustifolium)* is distinct in having the lanceolate leaves spirally arranged, and an extensive root system which produces ADVENTITIOUS stems. The 19–25mm wide rose-purple flowers are borne in spire-like RACEMES; great hairy willow-herb *(E. hirsutum)* has white, fleshy underground STOLONS, erect, robust stems to 1.5m and 63–125mm long, oblong-lanceolate leaves in pairs. The purple-red flowers are 19mm wide.

▲ **WILSON, ERNEST HENRY (1876–1930)** Born in Gloucestershire, he worked in a nursery in Solihull upon leaving school and moved to the Birmingham Botanic Gardens in 1892. From there in 1897 he moved to Kew and shortly afterwards was taken on by the famous horticultural firm of James Veitch and Sons as a collector. After a short period of training at their nursery, he was sent to China, travelling via the USA where he stayed with Sargent at the Arnold Aboretum in Boston. In the summer of 1899 he continued on his way to China and spent several months with Augustine Henry who had botanised widely in that country. On his advice he journeyed in 1901 to Hupeh and Szechuan provinces, bringing back enough pressed plants and seeds to encourage Veitch to send him again in 1902.

The chief object of his first trip had been the Dove tree *(Davidia involucrata)*; this time he was asked to concentrate on *Meconopsis integrifolia* the yellow poppy. Once again he was successful and most specimens of both in cultivation are descended from seeds he introduced.

He visited China again 1907–10, now working for the Arnold Aboretum and Sargent. During the third of these visits he broke his leg in a remote part of the country, which left him with a permanent limp. This did not however deter him and he made further visits to China, Japan, S. Africa, India and Australasia. In 1927, upon the death of Professor Sargent, he was appointed Keeper of the Arnold Arboretum, but together with his wife was tragically killed in a car crash. He introduced many first rate plants into cultivation, among them the regal lily *(Lillium regale)*, *Magnolia wilsonii*, *Hydrangea sargentiana* and *Cerastostigma willmottiae*.

Chief books: *A Naturalist in Western China* (in USA, *China, Mother of Gardens*) (1913), *The Lilies of Eastern Asia* (1925).

WILTING The cells of the plant's young tissues are kept firm entirely by the pressure of water within them, so that if the water supply fails through dryness at the roots or a greater demand by the leaves than the roots can supply, the shoot tips and leaves sag or wilt. Disease may also cause wilting indirectly by clogging or rotting the vascular tissue through which the water is passed.

WIND FLOWER
see ANEMONE

WINEBERRY [JAPANESE WINEBERRY] *(Rubus phoenicolasius)*. A close ally of blackberry from China and Japan, with reddish stems, long, slender, reddish sepals and tiny, pink petals and 19mm long, conical sweet, red fruits.

WINTER BUDS The small, tightly furled shoots, protected by hard, scaly bracts which are produced by deciduous trees and shrubs in the autumn to survive the winter. They quickly break into growth when the warmer, longer days of spring arrive.

WINTER CHERRY
see BLADDER CHERRY

WINTER CRESS [YELLOW ROCKET] Several species of *Barbarea*, biennials in the *Cruciferae* from Europe and Asia, some widely naturalised elsewhere. They have stalked, LYRATE-PINNATIFID basal leaves and lobed or pinnatifid SESSILE stem ones. The tiny, 4-petalled yellow flowers are carried in erect RACEMES. The leaves are edible and can be eaten in salads as a substitute for watercress, Land cress, American winter cress *(B. verna)*, grows to 600mm with pinnatifid stem leaves and flowers to 9mm.

■ **WINTERGREEN** Species of *Pyrola*, *Moneses* and *Orthilia*, closely related genera in the *Pyrolaceae* from the north temperate and Arctic zones. They are small, evergreen perennials, generally tufted, with leathery long-stalked broadly ovate to orbicular leaves and racemes of bell-shaped, 5-petalled flowers. (*Moneses* has solitary, saucer shaped white blossoms). Common wintergreen *(P. minor)* has ovate 25–38mm leaves and 100–300mm stems bearing dense racemes of pinkish flowers.

WINTER HELIOTROPE *(Petasites fragrans)*. Very much like BUTTERBUR, but more or less evergreen with leaves to 200mm wide and very fragrant lilac flowers in winter.

WINTER'S BARK *(Drimys winteri)*. An evergreen tree to 7.5m in the *Winteraceae* from S. America. It has oblong, elliptic 125–250mm long leaves, glaucous beneath, and white, cherry-like flowers in pendulous UMBELS. The medicinal bark is used for indigestion, diarrhoea, and as a stimulant in its native S. America.

WISTERIA A genus of 10 deciduous climbers in the *Leguminosae* from E. Asia and east N. America. They have PINNATE leaves with 9–19 leaflets and pendulous RACEMES of lilac, violet or white fragrant pea-flowers. *W. floribunda* from Japan grows to 9m with blue-violet, pink or white flowers in 45–105cm long racemes; to 180cm in Japan.

WITCHES' BROOM Dense, twiggy, broom-like masses in many kinds of trees, particularly birches, elms, pines and other conifers. They are variously caused by mites, fungi or mutation. Plants raised from the mutational brooms are the origin of many pigmy conifers.

WITCH HAZEL *(Hamamelis virginiana)*. A deciduous shrub to 4.5m in the *Hamamelidaceae* from east N. America. It has 75–150mm long OBOVATE, hazel-like leaves and small clusters of fragrant flowers with 4, slender, strap-shaped yellow petals which open as the leaves fall. The ovoid seed capsules are explosive.

○ **WOAD** *(Isatis tinctoria)*. A biennial or short-lived perennial in the *Cruciferae* from Europe, naturalised elsewhere. It has a rosette of stalked, LANCEOLATE leaves and large, terminal PANICLES of tiny, 4-petalled, yellow flowers. The black-brown fruits are one-seeded winged SILIQUAS. It is famed for its blue dye, made from the leaves.

WOLFSBANE
see ACONITE

WOODBINE
see HONEYSUCKLE

WOODLAND In the strictest sense, areas which are naturally covered with trees, usually in temperate climates of light to moderate rainfall. These may be dominated by a few tree species, e.g. oak or pine, or they may be a mixture of many species. Generally, the vegetation beneath the canopy of tree branches in relatively light. There may be smaller shade-tolerating trees forming an understory, for example yew in Great Britain, or shrubs such as hawthorn, hazel, spurge laurel and butchers' broom forming a shrub layer. When there is enough light, smaller perennial plants such as primroses, anemones, bluebells, dog's mercury, etc, form the herb layer. There may be a moss layer beneath the herbs. Where woodland has had the largest timber felled or where the shrubs are regularly COPPICED, the shrubs and herbs grow more luxuriantly and plants like blackberry form dense tangles in the increased light.

WOOD LILY *(Trillium)*. A genus of 30 woodland perennials in the *Trilliaceae* mainly from N. America, a few in Asia. It is an easily recognised genus, all the species having their leaves, sepals, petals and stamens in threes and a trilocular ovary. The flowers are always solitary, borne above a WHORL of three leaves. Common wood lily or wake robin *(T. grandiflorum)* has 300–450mm tall stems with 3 SESSILE, OVATE leaves and 50mm long, white flowers which age pinkish. There are double-flowered forms and curious mutants with the leaves and petals in multiples of 4 and 6.

● **WOODRUFF** *(Galium odoratum* syn. *Asperula odorata)*. A hayscented, woodland perennial in the *Rubiaceae* from Europe, Asia and N. Africa. It forms dense colonies from creeping RHIZOMES, the erect, slender stems bearing WHORLS of 25mm long LANCEOLATE leaves. Tubular, fragrant white flowers with 4 arching lobes are borne in terminal CYMES.

WOODRUSH *(Luzula)*. A genus of 80 grass-like perennial plants in the *Juncaceae* of cosmopolitan distribution. They are generally tufted plants with LINEAR tapering leaves and dense to loose PANICLE-like cymes of tiny flowers composed of 6, petal-like green or brown PERIANTH segments: field woodrush *(L. campestris)* is often common in grassland, with arching, long-haired leaves to 100mm or more and stems up to 150mm bearing dense cymes of chestnut-brown flowers.

WOOD SAGE *(Teucrium scorodonia)*. A perennial plant with somewhat woody rhizomes in the *Labiatae* from W. Europe. It has pairs of 32–63mm long OVATE-CORDATE, finely RUGOSE leaves on stems to 600mm and one-sided RACEMES of yellow-green flowers. Each blossom is tubular, with one large lobed lower lip, the terminal lobe concave.

▲ **WOOD SORREL** *(Oxalis acetosella)*. A slenderly-RHIZOMATOUS perennial in the *Oxalidaceae* from Europe and Asia. It forms patches of TRIFOLIATE leaves with OBCORDATE leaflets and 5-petalled, funnel-shaped lilac-veined, white flowers to 16mm long.

WOODY NIGHTSHADE
see BITTERSWEET

WOODY PERENNIAL A herbaceous plant with a woody crown and rootstock, e.g. purple loosestrife.

WORMWOOD
see ARTEMISIA

■ **WOUNDWORT** Several species of *Stachys*, annuals and perennials in the *Labiatae* of cosmopolitan distribution. They have pairs of LANCEOLATE to OVATE leaves and WHORLED spikes of tubular, 2-lipped flowers: hedge woundwort *(S. sylvatica)* is a roughly hairy rhizomatous plant to 105cm with ovate-cordate stalked leaves and 13mm long red-purple flowers.

WRACK A name covering several different sorts of brown seaweed in the *Fucaceae* from the zone between the tides in temperate waters. They have repeatedly forking strap-shaped or ribbon-like fronds and terminal ovoid swellings known as receptacles and which bear the sex cells: toothed or serrated wrack *(Fucus serratus)* has fronds to 0.6, rarely to 1.5m long with sharply toothed margins; bladder wrack *(F. vesiculosus)* is similar, but lacks the crenations and has rounded air bladders, often in pairs, in addition to the terminal receptacles; knotted wrack *(Ascophyllum)* has very narrow fronds with air bladders at intervals and stalked, ovoid receptacles borne singly or in small bunches.

XANTHOPHYLL An insoluble yellow pigment found in yellow flowers and autumn leaves.

XEROPHYTES [XEROPHYTIC] Plants adapted to live in arid climates. Some survive by forming water storage tissue, e.g. succulents; others by reducing the surface area through which water is lost, e.g. the rolled leaves of some grasses. Other ways of cutting down water loss are a thick, waxy cuticle, sunken STOMATA, or a covering of hairs.

XYLEM
see VASCULAR BUNDLE

YAM *(Dioscorea)*. A genus of 600 tuberous rooted twining climbers in the *Dioscoriaceae* from tropical to warm temperate zones. Many have broadly OVATE-CORDATE leaves, others are narrower and lobed. The small green or white flowers have 6 PERIANTH segments and are borne in AXILLARY RACEMES. Yams are important food plants in W. Africa, S. E. Asia, the West Indies and many Pacific islands. The food value of the tubers is mainly starch and they are peeled, boiled and mashed, roasted or fried: Greater yam *(D. alata)* has ovate-cordate leaves to 300mm long and large, rounded to serpentine tubers weighing up to 9kg or more. It is mostly grown in S. E. Asia and the Pacific; white Guinea yam *(D. rotundata)* with cylindrical tubers to 4.5kg, is grown in W. Africa and the W. Indies.

YARD LONG BEAN
see COW PEA

○ **YARROW** *(Achillea millefolium)*. An erect perennial with creeping underground STOLONS in the *Compositae* from Europe, Asia and naturalised in N. America, Australia and New Zealand. It has LANCEOLATE, densely bi- or tri-pinnatisect leaves and 300–600mm stems bearing flattened, crowded CORYMBS of small, white or pink daisy-like flower heads.

YEAST Several kinds of fungi belonging to the *Ascomycetes* and related to penicillin. The most important yeasts are members of the genus *Saccaromyces*, unicellular species with spherical or ovoid cells. They multiply at great speed under ideal conditions by a process known as budding. A small knob appears which grows into a new cell and breaks away from the parent, though in some species the cells form chains. During their life processes, yeasts convert sugar into alcohol and carbon dioxide. Both these by-products have their uses, the alcohol in brewing and the gas in bread making.

YELLOW ARCHANGEL *(Lamiastrum galeobdolon)*. A vigorous, stoloniferous perennial in the *Labiatae* from Europe to Iran. It has erect, 300–600mm angular stems, hairy ovate leaves and WHORLED spikes of 19mm long, yellow, 2-lipped flowers, the upper lip larger and hooded.

YELLOW BIRCH
see BIRCH

YELLOW BIRD'S-NEST [PINE SAP] *(Monotropa hypopitys)*. An erect, SAPROPHYTIC plant in the *Monotropaceae* from the north temperate zone. The whole plant is yellowish or ivory white, the 75–300mm stems bearing ovate-oblong, SESSILE scale leaves. The 9–13mm long bell-shaped flowers have 4 or 5 petals and are borne in short, nodding RACEMES. Corpse plant or Indian pipe *(M. uniflora)* from N. America is similar but with solitary flowers.

YELLOW CEDAR [YELLOW CYPRESS]
see CYPRESS

YELLOW CRESS
see WINTER CRESS

YELLOW FLAG
see IRIS

● **YELLOW RATTLE** Several species of *Rhinanthus*, erect, semi-parasites in the *Scrophulariaceae* from Europe and Asia, (others in N. America). They have pairs of LANCEOLATE, CRENATE leaves and short, leafy-bracted racemes of tubular, 2-lipped, yellow flowers, the upper lip larger and hooded with two teeth. The CALYCES are inflated and rattle when dry. Common yellow rattle *(R. minor)* is a variable plant to 375mm with 13mm long flowers having purple teeth.

YELLOW ROCKET
see WINTER CRESS

YELLOW SCALES *(Xanthoria parietina)*. A common yellow lichen on rocks, walls, roofs and trees from temperate regions. It forms scale-like patches of flattened, lobed branches often bearing small, saucer-shaped APOTHECIA. It thrives best where the air is laden with dust or spray bearing mineral salts and is commonest around farm buildings and by the sea.

▲ **YEW** *(Taxus)*. A genus of 10 DIOECIOUS, evergreen coniferous trees in the *Taxaceae* from north temperate and mountainous regions. They have dense, LINEAR leaves, solitary female flowers and groups of male flowers in small, globular cones which shed vast clouds of pollen on windy days in early spring. The ovoid red, berry-like fruits

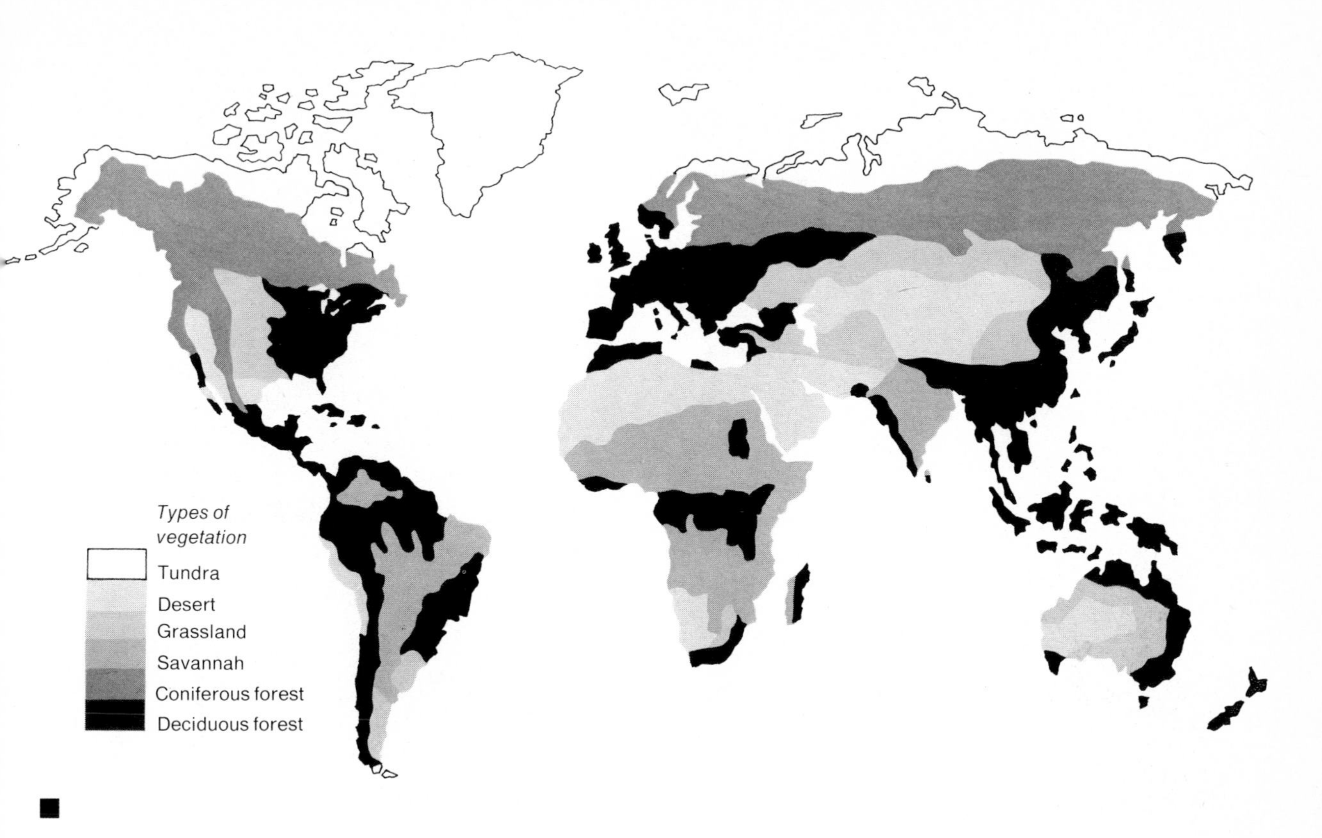

are formed of a single cup-shaped, sticky, fleshy ARIL with a single seed in the middle. Common yew *(Taxus baccata)*, grows to 18.2m or more, often with massive, irregular fluted, reddish trunks and dark green leaves to 25mm long.

YLANG-YLANG
see MACASSAR OIL

YORKSHIRE FOG
see SOFT GRASS

YOUTH AND AGE
see ZINNIA

ZINNIA A genus of 20 annuals, perennials and sub-shrubs in the *Compositae* from southern USA and Chile, with LANCEOLATE to OVATE leaves in pairs or sometimes WHORLS, and terminal daisy-like heads of disk and ray florets. Some of the annual species are popular garden plants, notably *Z. elegans* (youth and age) from Mexico. It grows to 900mm with sessile, ovate, cordate leaves and 50–125mm wide flower heads in shades of pink, red, purple, buff, white and green.

■ **ZONES OF VEGETATION** Although much depends upon local conditions, the earth can be roughly divided into temperature zones each characterised by a particular group of plants. For convenience, this zonation can be equated with lines of latitude. The zone from 0°–10° latitude is characterised by the wet, humid jungles and wet tropical woodlands; from 10°–20°, by savannas (tropical grasslands); from 20°–30°, deserts and sub-tropical woodlands; from 30°–40°, steppes and evergreen woodland; from 40°–55°, deciduous woodland; from 55°–65°, coniferous woodland; and from 65°–75°, by tundra, mainly with mosses and lichens. This zonation can be directly linked to altitudinal zones on mountains: for example, high mountains (above 4,800m) in the tropics can exhibit all or several of the above plant groups. Distinguish from FLORISTIC REGION.

ZOOSPORANGIA Single specialised cells found in certain fungi and filamentous algae which produce zoospores.

ZOOSPORES The contents of a zoosporangium which contracts into an oval spore and grows one to several *flagellae.* The zoosporangium then splits or opens and a zoospore swims away. Later it settles down, forms a new cell wall and grows into a new plant.

ZYGOMORPHIC Applied to flowers with irregularly arranged petals, such as those of pea, pansy and snapdragon with petals of different shapes and sizes. Such a flower can be cut into mirror halves in one vertical plane only. *See* ACTINOMORPHIC.

ZYGOSPORE The fusion of undifferentiated male and female cells as in certain primitive algae and fungi. The result is often a thick walled spore which passes through a dormant phase before germinating.

ZYGOTE The result of a fusion between male and female GAMETES; a newly-fertilised female cell or egg which then develops into the embryo plant.

Latin Name Index

A

B

C

D

E

Eriodendron kapok
Eriophorum cotton grass
Erodium storksbill
Erophila whitlow grass
Eryngium sea holly
Erysimum wallflower
Erysiphe mildew
Erythraea centaury
Erythrina coral tree
Erythronium dog's tooth violet
Erythroxylon cocaine
Eschscholzia Californian poppy
Espeletia paramo
Eucalyptus gum
Eugenia cloves, rose apple
Euonymus spindle
Eupatorium hemp agrimony
Euphorbia spurge
Euphrasia eyebright
Exocarpus sandalwood

F

Fagopyrum buckwheat
Fagus beech
Fatsia Japanese aralia
Ferula asafoetida, giant fennel
Festuca fescue
Ficus banyan, bo tree, fig
Filago cudweed
Filices ferns
Filipendula meadowsweet
Fistulina beef steak fungus
Foeniculum fennel
Forsythia spring bells
Fortunella kumquat
Fouquieria ocotilla
Fragaria strawberry
Frangula buckthorn
Frankenia sea heath
Fraxinus ash
Fritillaria fritillary
Fucus wrack
Fumaria fumitory
Funkia hosta

G

Gaillardia blanket flower
Galanthus snowdrop
Galega goats' rue
Galeobdolon yellow archangel
Galeopsis hemp nettle
Galinsoga gallant soldier
Galium Cleavers, crosswort, bedstraw, woodruff
Galtonia summer hyacinth
Garcinia mangosteen
Gaultheria partridge berry, shallon
Gaylussacia huckleberry
Geastrum earth star
Genista Dyer's greenweed, petty whin
Gentiana gentian
Geranium cranesbill, Pelargonium
Geum avens
Ginkgo maidenhair tree
Glaucium horned poppy
Glaux sea milkwort
Glechoma ground ivy
Gleditschia (*Gleditsia*) honey locust
Glyceria flote grass
Glycine soy bean
Glycyrrhiza liquorice
Gnaphalium cudweed
Gossypium cotton
Grevillea silky oak
Grossularia gooseberry
Guaicum lignum vitae
Gunnera prickly rhubarb
Gymnadenia fragrant orchid
Gymnocarpium oak fern
Gymnocladus Kentucky coffee
Gynerium pampas
Gypsophila chalk plant

H

Haemanthus blood lily
Halimione sea purslane
Hamamelis witch hazel
Harpagophytum grapple plant
Hedera ivy
Helianthemum rock rose
Helianthus artichoke, sunflower
Helichrysum curry plant, everlasting flower
Helictotrichon oat grass
Heliotropium heliotrope
Helleborus Christmas rose, hellebore
Helxine mind your own business
Hemidesmus sarsaparilla
Hemitelia tree fern
Hemerocallis day lily
Heracleum hogweed
Heritiera nettle tree
Hesperis dame's violet
Heuchera alum root
Hevea rubber
Hibiscus roselle
Hieracium hawkweed
Hierochloe holy grass
Himantoglossum lizard orchid
Hippocastanum horse chestnut
Hippophae sea buckthorn
Hippuris marestail
Hoheria lace bark
Holcus soft grass
Honkenya sea sandwort
Hordeum barley, barley grass
Hottonia water violet
Houstonia bluets
Hoya wax flower
Humulus hop
Hyacinthus hyacinth
Hydrocharis frog bit
Hydrocotyle marsh pennywort
Hymenophyllum filmy fern
Hyoscyamus henbane
Hypericum St. John's wort
Hyphaene palm
Hypholoma sulphur tufts
Hypochaeris cats ear
Hyssopus hyssop

I

Iberis candytuft
Ilex holly, maté
Impatiens balsam
Indigofera indigo
Inula elecampane, ploughman's spikenard, samphire
Ipomoea jalap, morning glory, sweet potato
Isatis woad
Isoetes quillwort
Ixia African corn lily

J

Jambosa rose apple
Jasione sheepsbit
Jasminum jasmine
Jovibarba houseleek
Juglans walnut
Juncus rush
Juniperus juniper

K

Kalmia calico bush
Kentia palm
Kentranthus red valerian
Kerria Jew's mallow
Kigelia sausage tree
Knautia scabious
Kniphofia red hot poker
Kochia burning bush
Koeleria hair grass
Kolkwitzia beauty bush

L

Lachenalia cape cowslip
Lactuca lettuce, prickly lettuce
Lagenaria cucurbits
Lagurus harestail grass
Lamiastrum yellow archangel
Laminaria oar-weed
Lamium dead nettle
Lapsana nipplewort
Larix larch
Lathraea toothwort
Lathyrus everlasting pea, sea pea, sweet pea
Laurus bay laurel
Lavandula lavender
Lavatera mallow
Lawsonia henna
Lecanora black shields
Ledum Labrador tea
Legousia Venus's looking glass
Lemna duckweed
Lens lentil
Leontodon hawkbit
Leontopodium edelweiss
Leonurus motherwort
Lepidium dittander, garden cress, pepperwort
Lepiota parasol mushroom
Lepista blewits
Leucadendron silver tree
Leucobryum white fork moss
Leucojum snowflake
Lewisia bitterroot
Liatris snakeroot
Libocedrus incense cedar
Ligusticum lovage
Ligustrum privet
Lilium lily
Limonium sea lavender
Linaria toadflax
Linnaea twin flower
Linum flax
Lippia lemon scented verbena
Liriodendron tulip tree
Listera twayblade
Lithops living stones
Littorella shore weed
Lobularia madwort
Loiseleuria trialing azalea
Lolium rye grass
Lonicera fly honeysuckle, honeysuckle
Lophophora cactus, mescal button
Lotus bird's foot trefoil
Luffa loofah
Lunaria honesty
Lupinus lupin
Luzula woodrush
Lychnis catchfly, ragged robin
Lycium Duke of Argyll's tea tree
Lycoperdon puff ball
Lycopersicum tomato
Lycopodium club moss
Lycopsis bugloss
Lycopus gipsywort
Lygodium climbing fern
Lysichiton skunk cabbage
Lysimachia loosestrife, moneywort
Lythrum loosestrife

M

Macadamia Queensland nut
Maclura osage orange
Mahonia Oregon grape
Malcolmia Virginian stock
Malus apple, crab apple
Malva mallow
Mammea mammee apple
Mandragora mandrake
Mangifera mango
Manihot arrowroot, cassava
Maranta arrowroot
Marasmius champignon
Marchantia liverwort
Marrubium horehound
Martynia unicorn plant
Matricaria feverfew, mayweed, pineapple weed
Matteuccia ostrich fern
Matthiola stock
Melandrium campion
Medicago lucerne, medick
Melampyrum cow wheat
Melica melick grass
Melilotus melilot
Melissa balm
Melittis balm
Mentha mint
Menyanthes bog bean
Mercuralis mercury
Mertensia oyster plant
Mesembryanthemum Hottentot fig, ice plant
Mespilus medlar
Metasequoia dawn redwood
Metroxylon fern
Mimulus monkey flower
Mirabilis jalap
Mitchellia partridge berry
Mnium thread moss
Molinia purple moor grass
Momordica balsam apple, balsam pear
Monarda bergamot
Moneses wintergreen
Monotropa yellow bird's nest
Montia blinks, Claytonia
Morchella morel
Morus mulberry
Mucor mould
Murraya curry leaf
Musa banana, hemp
Muscari grape hyacinth
Myosotis forget-me-not
Myosurus mousetail
Myrica bog myrtle
Myriophyllum water milfoil
Myristica nutmeg
Myrrhis sweet Cicely
Myrtus myrtle

N

Narduus mat grass
Narthecium asphodel
Nasturtium water cress
Nectandra greenheart
Nectria coral spot
Nelumbium lotus
Nelumbo lotus
Nemophila baby blue eyes
Neottia bird's nest orchid
Nepenthes pitcher plant
Nephelium lychee, rambutan
Nepeta catmint
Nerine Guernsey lily
Nerium oleander
Nicandra shoo fly
Nicotiana tobacco
Nigella love in a mist
Nitrobacter nitrifying bacteria
Nitrosomonas nitrifying bacteria
Nothofagus southern beech
Nuphar water lily, spatterdock
Nymphaea water lily
Nyssa ironwood

O

Ochroma balsa
Ocimum basil
Odontites bartsia
Oenanthe water dropwort
Oenothera evening primrose
Olea olive
Olearia daisy bush
Onobrychis sainfoin
Onoclea sensitive fern
Ononis rest harrow
Onopordon Scotch thistle
Ophioglossum adder's tongue
Ophrys bee orchid, fly orchid, mirror orchid, spider orchid
Opuntia cactus
Orchis Early purple orchid, Lady orchid, Military orchid, monkey orchid
Origanum marjoram
Ornithogalum star of Bethlehem
Ornithopus bird's foot
Orobanche broomrape
Orontium golden club
Orthilia wintergreen
Oryza rice
Osmunda royal fern
Ostrya ironwood
Oxalis wood sorrel

P

Paeonia peony
Paliurus Christ's thorn
Panax ginseng
Pancratium sea daffodil
Pandanus screw pine
Panicum millet
Papaver poppy
Parietaria pellitory
Paris herb Paris
Parnassia grass of Parnassus
Parthenocissus Virginian creeper
Passiflora passion flower
Pastinaca parsnip
Pedicularis lousewort
Peltigera lichen
Penicillium mould
Pennisetum millet
Pentaglottis alkanet
Pereskia Barbados gooseberry
Peronospora mildew
Persea avocado
Petalostemon prairie clover
Petasites butterbur, winter heliotrope
Petroselinum parsley
Peucedanum dill, milk parsley
Phalaris canary grass, reed grass
Phallus stinkhorn
Pharbitis morning glory
Phaseolus butter bean, mung bean, runner bean, snail flower
Philadelphus mock orange
Phleum cat's tail grass
Phlomis Jerusalem sage
Phoenix palm
Phormium New Zealand flax
Phragmites reed
Phygelius cape figwort
Phyllitis hart's tongue
Phyllocactus cactus
Physalis bladder cherry, cape gooseberry
Phytelephas palm
Phyteuma rampion
Phytolacca pokeweed
Phytophthora blight
Picea spruce
Picrasma quassia

Pilea artillery plant
Pilularia pillwort
Pimenta all-spice
Pimpinella burnet saxifrage
Pinguicula butterwort
Pinus pine
Piper pepper
Piptoporus razor strop fungus
Pistacia mastic tree
Pistia water lettuce
Pisum pea
Pithecolobium rain tree
Plantago plantain
Platanus plane
Platycerium stag's horn fern
Platycodon balloon flower
Pleurotus oyster mushroom
Plumbago leadwort
Plumeria frangipani
Poa meadow grass
Podocarpus totara
Podophyllum may apple
Pogostemon patchouli
Poinciana flamboyant tree
Polemonium Jacob's ladder
Polygala milkwort
Polygonatum Solomon's seal
Polygonum black bindweed, knotgrass, knotweed, water pepper
Polypodium polypody
Polypogon beard grass
Polyporus dryad's saddle
Polystichum holly fern, shield fern
Polytrichum hair moss
Pontederia pickerel weed
Populus poplar
Porphyra laver
Potamogeton pondweed
Portulacca purslane
Posidonia poseidon weed
Potentilla creeping cinquefoil, marsh cinquefoil, silverweed, shrubby cinquefoil, tormentil
Poterium salad burnet
Proboscidea unicorn plant
Prunella self heal
Prunus almond, apricot, bird cherry, blackthorn, cherry, cherry laurel, peach, plum
Psamma marram grass
Pseudotsuga Douglas fir
Psidium guava
Psoralea bitumen plant
Pteridium bracken
Pteridophyta ferns
Puccinia rust fungi
Pueraria kudzu
Pulicaria fleabane
Pulmonaria lungwort
Pulsatilla pasque flower
Punica pomegranate
Pyracantha firethorn
Pyrola wintergreen
Pyrus pear

Q

Quercus oak

R

Radiola all seed
Ramalina sea ivory
Ranunculus buttercup, celandine, crowfoot, fair maids of France, spearwort
Raoulia vegetable sheep
Raphanus radish
Raphis palm
Ravenala travellers' tree
Reseda mignonette, weld
Rhamnus buckthorn, cascara sagrada
Rheum rhubarb
Rhinanthus yellow rattle
Rhizophora mangrove
Rhododendron alpenrose, American laurel, rhododendron
Rhus lacquer tree, sumach
Ribes currant, gooseberry
Ricinus castor oil plant
Robinia false acacia
Roccella litmus
Romneya Californian poppy
Romulea sand crocus
Rorippa watercress
Rosa rose
Rosmarinus rosemary
Roystonea palm
Rubia madder
Rubus blackberry, cloudberry, dewberry, loganberry, raspberry, wineberry
Rumex dock, sorrel
Ruscus butcher's broom
Russula sickener
Ruta rue

S

Sagina pearlwort
Sagittaria arrowhead
Saintpaulia African violet
Salicornia glasswort
Salix willow
Salsola saltwort
Salvia clary, sage
Sambucus elder
Samolus brookweed
Sanguinaria bloodroot
Sanguisorba salad burnet
Sanicula sanicle
Sansevieria mother in law's tongue
Santalum sandalwood
Santolina cotton lavender
Sapindus soapberry
Saponaria soapwort
Sargassum seaweed
Sarothamnus broom
Sarracenia pitcher plant
Sassa bamboo
Satureja savory
Saururus sausage tree
Saxifraga saxifrage
Scabiosa scabious
Scandix shepherd's needle
Schinus mastic tree
Schizanthus poor man's orchid
Schizostylis Kaffir lily
Schoenus bog rush
Sciadopitys umbrella fir
Scilla squill
Scirpus bulrush, deer grass
Scleranthus knawel
Scolopendrium hart's tongue
Scorzonera vipers' grass
Scrophularia figwort
Scutellaria skullcap
Secale rye
Sedum ice plant, stonecrop
Selaginella clubmoss, resurrection plant
Sempervivum house leek
Senecio Cineraria, giant groundsel, groundsel, ragwort
Sequoia redwood
Sequoiadendron big tree
Serratula sawwort
Sesamum sesame
Setaria millet
Setcreasea purple heart
Silene campion, catchfly
Silybum milk thistle
Sinapis charlock, mustard
Sinningia gloxinia
Sisymbrium hedge mustard, rocket
Sisyrinchium blue eyed grass
Smyrnium alexanders
Solanum aubergine, bittersweet, kangaroo apple, nightshade, potato, tomato
Soleirolia mind your own business
Solidago golden rod
Sonchus sow thistle
Sophora kowhai, pagoda tree
Sorbus rowan, service tree, whitebeam
Sorghum millet
Sparganium burreed
Spartina cord grass
Spartium Spanish broom
Specularia Venus's looking glass
Spergula spurrey
Spergularia sand spurrey
Spinacia spinach
Spinifex tumbleweed
Spiranthes lady's tresses
Sprekelia Jacobean lily
Stachys artichoke, woundwort
Stapelia carrion flower
Staphylea bladdernut
Stellaria chickweed, stitchwort
Stereum silver leaf fungus
Stipa esparto
Strelitzia bird of paradise
Streptocarpus cape primrose
Strongylodon jade vine
Strychnos strychnine
Suaeda seablite
Subularia awlwort
Succisa devil's bit scabious
Swietenia mahogany
Symphoricarpos snowberry
Symphytum comfrey
Symplocarpus skunk cabbage
Synchytrium wart disease
Syringa lilac

T

Tacca arrowroot
Tagetes African marigold, French marigold
Tamarix tamarisk
Tamus black bryony
Tanacetum tansy
Taraxacum dandelion
Taxodium swamp cypress
Taxus yew
Tectona teak
Testudinaria Hottentot bread
Tetragonia New Zealand spinach
Teucrium wood sage
Thalictrum meadow rue
Thea tea
Thelypteris beech fern, marsh fern, oak fern
Theobroma cocoa
Thesium bastard toadflax
Thlaspi penny cress
Thuja arbor vitae
Thymus thyme
Tilia bass, lime
Tillandsia Spanish moss
Tofieldia asphodel
Tolmiea pick a back plant
Tortula hair moss
Trachycarpus palm
Tradescantia spiderwort
Tragopogon goatsbeard, salsify
Trapa water chestnut
Tremella jelly fungus
Tribulus caltrops
Tricholoma blewits, St. George's mushroom
Trichomanes Killarney fern
Trichophorum deer grass
Trichosantha snake gourd
Trientalis wintergreem
Trifolium clover, shamrock
Triglochin arrow grass
Trillium wood lily
Tripleurospermum mayweed
Trisetum yellow oat
Triticum wheat
Tritoma Kniphofia
Trollius globe flower
Tropaeolum canary creeper, nasturtium
Tsuga hemlock spruce
Tulipa tulip
Typha reed mace

U

Ulex gorse
Ulmus elm
Ulva sea lettuce
Umbilicus navelwort
Urginea squill
Urtica nettle
Utricularia bladderwort

V

Vaccinium bilberry, blueberry, cowberry, cranberry
Valeriana valerian
Valerianella lamb's lettuce
Vallisneria tape grass
Verbascum mullein
Veronica brooklime, speedwell
Viburnum guelder rose, laurestinus, snowball tree, wayfaring tree
Vicia broad bean, vetch
Vigna cow pea
Vinca periwinkle
Viola pansy, violet
Viscaria catchfly
Viscum mistletoe
Vitis grape vine

W

Wahlenbergia ivy leaved bellflower

X

Xanthium cocklebur
Xanthorea yellow scales
Xanthorrhoea grass tree

Y

Yucca Adam's needle

Z

Zantedeschia arum lily
Zea maize
Zingiber ginger
Zizania Canada rice
Zizyphus jujube
Zostera eel grass